D1037478

The People's Car

The People's Car

A Global History of the Volkswagen Beetle

Bernhard Rieger

HARVARD UNIVERSITY PRESS
Cambridge, Massachusetts & London, England
2013

Copyright © 2013 by the President and Fellows of Harvard College
All rights reserved
Printed in the United States of America

Library of Congress Cataloging-in-Publication Data

Rieger, Bernhard, 1967–
The people's car : a global history of the Volkswagen Beetle / Bernhard Rieger.
p. cm.
Summary: "The People's Car is a transnational cultural history tracing the Beetle from its origins in Nazi Germany to its role in the postwar West German 'economic miracle' to its popularity in midcentury Europe and the U.S., second career in Mexico and Latin America, and revival in the late 1990s"—Provided by publisher.
Includes bibliographical references and index.
ISBN 978-0-674-05091-4 (alk. paper)
1. Volkswagen Beetle automobile—History. I. Title. II. Title: Global history of the Volkswagen Beetle.
TL215.V6R54 2013
629.222'2—dc23 2012029928

To my mother and the memory of my father

Contents

The People's Car

Prologue

"Some Shapes Are Hard to Improve On"

"This model will open up the automobile to millions of new customers on low incomes," Adolf Hitler predicted in February 1938 as he presented the prototype of a small, inexpensive family vehicle amid much fanfare at the Berlin Auto Show. Commissioned by the führer and designed by Ferdinand Porsche, the bug-shaped car unveiled on the eve of World War II was indeed to become the "people's car"—or *Volkswagen*—that fulfilled the dream of private auto ownership for millions. Yet it ultimately did so under circumstances Hitler never anticipated. Since the Third Reich never produced the vehicle, it was only after National Socialism's downfall that Porsche's brainchild turned into the global hit that everyone now knows as the "Beetle." In the postwar period, the Beetle not only played a prominent role in taking Western Europe into the age of mass motorization but also triumphed in the United States, where it became the leading small car. By the late sixties, both those living in suburban affluence and members of the counterculture rebelling against suburbia as the epitome of conformity drove Volkswagens en masse. Between 1938 and 1968, the Beetle—and only the Beetle—exerted a profound appeal among customers across the political spectrum from the extreme Right to the Left. In Latin America, meanwhile, the Volkswagen dominated the roads first in Brazil and subsequently in Mexico as late as the 1990s. When the curtain eventually fell on Mexican production in 2003, more than twenty-one million Beetles had rolled off assembly lines across the world. The Beetle's charm, however, by no means ceased with the end of manufacture. Hundreds of thousands of people come together all over Europe and the United States each year to display, admire, and drive lovingly restored old Volkswagens. Since 1998, admiration for the original VW has also fueled sales of the New Beetle, the first of a growing number of revival cars inspired by automotive nostalgia. As millions across the world bought and drove the old Beetle, it turned into far more than a machine for enhancing individual mobility. Like Coca-Cola, it is a global icon.[1]

On a global journey with many twists and turns, the Beetle became a commercial success that rightfully claims a prominent place among the renowned automobiles of the twentieth century. Ferdinand Porsche's small creation developed into the first automobile to outsell the legendary Model T that Ford had produced between 1908 and 1927. Numerous similarities and connections exist between these two famous automobiles.[2] In mechanical terms, the Model T and the Volkswagen were robust automobiles that many owners found reliable as well as easy to maintain. In their respective heydays, the Model T and the Beetle became objects of profound affection that expressed itself in countless loving nicknames. Their appeal lay partly in their comparatively low purchase price, which brought individual car ownership within the reach of broad sections of society. Methods of standardized mass production put Ford and Volkswagen in a position to cut costs, thereby transforming cars from luxury items into everyday commodities. As the first affordable quality vehicles available to Americans and Germans respectively, both automobiles developed a tremendous popular appeal that established them firmly within the pantheon of national icons in their homelands.

Henry Ford's success in the United States between the 1900s and the 1920s attracted interest and envy on the other side of the Atlantic. When the British, French, and German publics began to discuss plans for a "people's car" in the interwar period, they strove to replicate Ford's accomplishments under European conditions. The German translation of *My Life and Work,* a work that outlines the principles behind Henry Ford's factory in Highland Park, Michigan, turned into a best seller in Weimar Germany. Among its avid readers was no other than Adolf Hitler. A long-standing admirer of Henry Ford, Hitler initiated the drive for a "people's car" in National Socialist Germany. Without the dictator's staunch support and approval, the design Ferdinand Porsche developed between 1934 and 1938 would have been unthinkable. And when Ferdinand Porsche explored plans for an automobile factory suitable for the mass production of his prototype, he traveled to Detroit, where, among other car plants, he visited Ford's factory at River Rouge.[3]

However, a comparison between Volkswagen and Ford also reveals significant differences. Most important, the commercial success of Ford's Model T was primarily an American phenomenon. To be sure, Ford expanded into a global concern during the Model T's prime between the 1900s and 1920s, but compared to the mass market for automobiles in the United States, sales abroad paled because of the marked economic gap between America and the rest of the world after World War I. By contrast, the Beetle developed into an international best seller once it went into production after 1945, transcending its origins in the Third Reich and winning customers beyond West Germany. The Volkswagen enjoyed an exceptionally long production run, attracting drivers between the end of World War II and the turn of the millennium. The Beetle became the first automotive classic to inspire a retro vehicle. While the Model T stands out as a national product that gave rise to a global myth, the Volkswagen developed into a global commodity whose international appeal derived from numerous cultural permutations throughout its long life.

Its origins in the Third Reich, its ascent to icon of the Federal Republic, its appeal in very different environments across the globe, and its persistent market presence gave the Volkswagen Beetle an extraordinarily complex history. The Volkswagen's technical virtues hold important clues for the vehicle's success. Ferdinand Porsche's design from the thirties provided the material foundations for a vehicle that attracted millions of customers in search of a durable, economical car with first-rate driving characteristics. When mass production began after World War II, engineers at Volkswagen's corporate headquarters in Wolfsburg built on the vehicle's main design traits, but persistently modified it to remedy shortcomings and adapt it to changing market conditions. The Beetle thus provides a striking reminder that, long after their initial invention, products retain their attractiveness as a result of unspectacular technical adjustments. At the same time, the management at VW displayed an acute awareness that the best technical design meets with commercial success only if manufactured to a high standard. The vehicle's distinctive technical features, sustained product development, and the production methods championed at Volkswagen jointly played decisive roles in attracting countless customers in search of a good buy.[4]

Nonetheless, the Beetle's aura transcends functionality. In a world full of practical everyday goods, only few commodities inspired the affection surrounding the VW. Its "Beetle" nickname holds an important clue for the love this vehicle has commanded over the decades. One of the VW's most striking characteristics is its shape, composed of a rounded front section, an almost vertical windshield, and a gently sloping roof that falls off in a steep curve at the back. From the outset, observers commented on the vehicle's bug-like, rotund silhouette, which, in a world of angular automobiles, lent it a unique, instantly recognizable appearance. Similar to the Coke bottle, the Volkswagen's body ranks highly among the twentieth century's classic designs. "Some shapes are hard to improve on," declared the tagline of a 1963 VW ad that featured the photograph of an egg adorned with the drawing of a Beetle, thereby asserting that the car's contours amounted to a classic form on a par with a timeless symbol of fecundity. Millions of drivers considered the small German automobile not only a good bargain but a "good egg," drawn to it not least by its distinctive, wholesome, and friendly air.

Social and cultural observers have long been puzzled by cars' intangible qualities. Writing of the sleek and luxurious Citroën DS that made its market debut in the mid-1950s, Roland Barthes famously went so far as to elevate cars to "almost the exact equivalent of the great Gothic cathedrals" of the Middle Ages. He found that contemporary society admired the automobile as "the supreme creation" of its day, treating it "as a purely magical object." While no one has ever compared the Beetle to ecclesiastical architecture, Barthes's reading of vehicles as magical artifacts fits into an analytical tradition of commodity culture that reaches back to the nineteenth century.[5]

In a classic passage in *Das Kapital,* Karl Marx wrote that a commodity—another term for "exchangeable good"—may "appear at first . . . very trivial," but on closer inspection emerges as "a very queer thing." Marx was particularly struck by the fact that prices of many commodities "have absolutely no connection with their physical properties." Rather than possessing an intrinsic worth, they gain their value in complex social processes that, in turn, establish myriad social relations. By directing the focus of attention beyond a commodity's material properties, Marx has encouraged analyses to disentangle the social processes determining why societies hold certain goods in high regard while denigrating others. Despite their apparent triviality, he found many commodities full of "metaphysical subtleties and theological niceties." Baffled by their complexity, Marx took refuge in a metaphor and referred to them as "fetishes." Similar to objects that many societies and individuals credit with supernatural powers, commodities "appear as independent beings endowed with life" once they enter "into relation with one another and the human race," he stated. In other words, rather than regarded as dead objects, commodities need to be understood as possessing and leading lives of their own.[6]

Marx laid his finger upon a crucial phenomenon. By likening commodities to fetishes, he drew attention to the intriguing energies that can lodge in retail goods. As they circulate within and between societies, commodities gain social lives and develop dynamics of their own. Societies create material universes in which commodities can become "active agents" as a result of their "communicative, performative, emotive, and expressive capacities," as one scholar has remarked. Material objects often acquire profound personal and collective significance because they make the "abstract . . . concrete, closer to lived experience." Much in the way that mechanical clocks and watches have long embodied specific concepts of "time," automobiles can be seen as physical manifestations of abstract notions ranging from, among others, "speed" and "freedom" to "unconventionality" and "wealth." Undoubtedly counting among the most standardized commodities mass production brought into existence, numerous VWs nonetheless were revered as deeply personal treasures, which owners lovingly washed and waxed, polished and painted. As much as proprietorship itself, the act of driving provided manifold bonding experiences that tied owners to their cars, leading them to venerate their vehicles as objects that they valued for far more than taking them from point A to point B—if only because teenagers and other romantics had a lot of fun in their automobiles when parked somewhere between A and B.[7]

The Volkswagen's stunning commercial triumph arrived only after an unusually long prehistory. While initial German calls for an affordable car appeared as early as the Weimar Republic, the vehicle the Third Reich had designed to advance mass motorization went into production only after World War II. The car's protracted period of origin not only highlights the longevity of the dreams that stood behind a reasonably priced vehicle. Given its prewar origins and postwar success, the Volkswagen straddles the deep chasms that run through recent German history. Although potential critics and competitors could draw on a rich arsenal to stigmatize this automobile on historical grounds, the Beetle transcended its unsavory origins with remarkable ease in Germany and beyond. During its global journey, the Beetle was suspended in a web of memory woven out of public tales related by managers and politicians as well as private accounts recalling events as diverse as gaining one's driver's license, a family's first holiday, and youthful adventures. Upbeat stories of the Beetle's public and private past, however, could prove effective only if certain aspects of its history remained out of sight. At home and abroad, the interplay of memory and amnesia underpinned the Beetle's progress.

The first Volkswagen's international popularity draws attention to prominent trends in globalization. Irrespective of polemics against the "Coca-Colonization" and "McDonaldization" of the world, the spread of international brands and goods has not simply led to global cultural uniformity. Jeremiahs bemoaning the detrimental effects of globalization overlook that it cannot be equated with a process of relentless Americanization. A large number of products from Europe, Asia, and Africa have begun to circulate all over the world as globalization has gathered pace. Notwithstanding America's undeniable worldwide prominence, a wider internationalization characterized world culture after 1945. Few commodities illustrate this trend better than a German automobile that gained international fame, not least in the United States itself. At the same time, the car's history highlights the social inequities that have long characterized globalization. As Volkswagen turned into a car producer that manufactured Beetles in several geographical regions, including Latin America, as part of an evolving international division of labor, it adopted vastly contrasting approaches to industrial relations and worker remuneration in different locations. In short, the first VW highlights mechanisms in globalization that spread an appealing, colorful commodity culture and social inequalities at the same time.[8]

As the Beetle moved from country to country, its global success was intimately linked to its chameleonlike qualities. Similar to animals whose skin adjusts to various surroundings, global commodities must adapt to local conditions if they are to secure commercial success. Customers often turn their backs on goods from elsewhere that stubbornly stand out as alien rather than blend into the local environment. Nothing less than a corporation's future can be at stake in the games of cultural adaptation that determine whether a commodity with an international pedigree manages to gain a lasting foothold abroad. These processes of adaptation can take many shapes and forms. In the case of world food, for instance, nationally specific dishes often undergo fundamental changes in terms of ingredients and cooking methods if they wish to attract diners in foreign lands.

By contrast, the Beetle crossed borders between consumer cultures without major material alterations. While the car remained essentially the same object in technological and aesthetic respects, it nonetheless gained divergent meanings in countries where it featured as an export commodity. Rather than subject itself to material modifications, the Volkswagen experienced intangible cultural adaptations that made it fit into a wide range of national auto cultures. Many global items owe their popularity to such flexibility and to the capacity to generate new meanings throughout their life span. This car's adaptability bordered on the uncanny when it triggered diametrically opposed associations in different places. With its almost infinite ability to fit in in numerous countries, the Beetle could undergo exceptionally thorough cultural metamorphoses, at times altogether shedding its character as an export item. In some countries, including the United States and Mexico, the Volkswagen has come to be revered as a national icon in its own right rather than a German automobile. By the end of the twentieth century, this car had gained multiple national identities—a remarkable transformation, given its Third Reich roots.[9]

In light of its worldwide proliferation throughout the second half of the twentieth century, the small Beetle stands tall among the versatile artifacts that have attained global star status. Of course, Marx overdrew his case when stating that the value of a commodity had "absolutely no connection" with its physical properties. The Volkswagen's success is inconceivable without its technical virtues. References to the vehicle's engineering traits, however, fail to explain the profound affection that greeted the Beetle in highly diverse environments. While the car's technical quality provides a necessary component in the Beetle's success story, it does not offer an exhaustive explanation for the warmth the vehicle elicited among drivers and passengers in many parts of the world. The relationship between the car's material properties (including its unique shape) and its meanings differed from place to place, from time to time. Examining the tentative and shifting links between the car's solid material core and its softer, more malleable social meanings thus provides a leitmotif to the Volkswagen Beetle's journey from intangible dream to global icon.

1

Before the "People's Car"

"[We] call for fundamental measures facilitating the purchase and maintenance of small cars, so auto ownership comes within the reach of every German," the Small Car Club of Germany demanded in October 1927. Founded six weeks earlier in the Berlin suburb of Oberschöneweide, the association lacked neither ambition nor a sense of mission. Rather than pursue the cause of a small minority, it self-confidently claimed to campaign for affordable automobiles in nothing less than "the interest of civilization's progress" so to "raise the social levels of the German people." Most Germans, on whose supposed behalf the group propagated the "idea of the small car," appear to have paid little attention to this high-minded, idealistic rhetoric, however. Only the lonely opening issue of its magazine *Mein Kleinauto* (My Small Car) survives in the stacks of the Berlin state library. Started with high hopes, the Small Car Club of Germany soon faded from the scene, leaving hardly a trace.[1]

That this association remained an ephemeral episode is symbolic of the state of German automotive affairs in the second half of the 1920s. In the year that witnessed the inauguration of the small car club in Oberschöneweide, automobiles remained a rather rare sight on German roads. To be exact, registrations stood at merely one car for every 196 Germans. When one excluded trucks and buses, this ratio fell even further, to one passenger vehicle for every 242 inhabitants. With these figures, the Weimar Republic trailed Western industrial nations by a wide margin. In France and Great Britain, by comparison, the count (including trucks and buses) stood at 1:44 at the time, more than four times the level of Germany's motorization. A comparison with the United States produces the starkest contrast. In 1927, American statisticians calculated a rate of 1 automobile per 5.3 inhabitants, making the United States the most motorized nation in the world by far. Under these conditions, the German public could be forgiven for not paying too much attention to calls to bring the motor vehicle within reach of the average citizen. Such demands possessed a positively utopian ring that contemporaries most likely met with skepticism or ironic smiles.[2]

For a country whose engineers had played a prominent role in the automobile's early days, it was bound to be a source of irritation that mass motorization remained an unrealistic prospect. Not least, the issue rankled national pride. After all, it had been Wilhelm Maybach, Gottlieb Daimler, and Carl Benz who, in the 1880s, had produced the first horseless carriages that were propelled by the combustion engine Nikolaus August Otto and Eugen Langen had developed in the 1870s.[3] When Carl Benz died in 1929 at the age of eighty-five, a German car journal celebrated him as a "genius" who had given the "civilized world . . . a means to conquer time and space. . . . Through [this feat], he became one of the co-founders of a global industry . . . that now ranks among the highest and most prestigious" branches of manufacturing. As the obituary detailed Benz's contribution to the development of the automobile before World War I, it found few achievements to play up for the time after 1918. After World

War I, the German car industry failed to live up to its pioneering promise.[4]

In the twenties, even the most benevolent commentators would have struggled to portray Germany as a car nation. While the public was well aware that Germany lagged far behind other countries in terms of the automobile's proliferation, the topic never rose to the top of the public agenda. Debates about car matters were pushed into the background by the repeated social, political, and economic crises and conflicts that afflicted Weimar Germany. The country's low car ownership levels attracted limited public note because the issue itself elicited little public controversy. Among analysts, a broad consensus existed regarding the numerous material obstacles that made mass motorization a prospect far beyond Germany's horizon. As a result, calls to remedy this situation, for instance by designing a car for the wider population, surfaced merely intermittently. The idea of a "people's car" never gained sharp public contours in the Weimar Republic but remained a rather fuzzy notion.

German observers assessed their country's potential for mass motorization skeptically not least because they were aware of the profound differences between conditions at home and in the United States, where the automobile had become a means of transportation for the wider population as early as the 1910s. America's global supremacy in automotive affairs was beyond doubt in the interwar years. In 1927, Americans owned no fewer than 80 percent of the world's cars, while Detroit's automakers set standards in manufacturing, design, and marketing methods worldwide. German visitors to the United States in the twenties found that, in car matters, a gap as wide and deep as the Atlantic Ocean separated America and the Weimar Republic. It wasn't simply the sheer number of cars that stunned German observers. Commentators also drew attention to the new forms of industrial production, as well as the social consequences that accompanied mass motorization. Long before anyone took concrete measures to make a "people's car" in Germany, the idea, for all its fuzziness, possessed a transnational pedigree. For this reason, our story begins in the United States before turning to Germany, a country that was a very unlikely candidate to give birth to a "people's car."[5]

America's leadership in the automobile sector highlighted a wider shift of the global economic center from Western Europe to the United States in the first half of the twentieth century. Since the late nineteenth century, the American economy had consistently posted higher growth rates than European nations; production figures of key industrial materials indicated the extent to which the United States outpaced Europe. Between 1900 and 1928, the annual output of America's coal mines more than doubled, from 193,208 to 455,678 metric tons, while the production of raw steel shot up from 10,217 to 51,527 metric tons. Britain and Germany, Western Europe's largest and most dynamic national economies, found it impossible to keep up with this expansion. By 1928, annual coal output stood at 241,283 and 150,876 metric tons in the United Kingdom and the Weimar Republic respectively, while British and German steel companies, which produced 8,637 and 14,517 metric tons in 1928 respectively, were positively dwarfed by American concerns.[6]

Above all, the United States' industrial ascendancy derived from domestic factors. In addition to the availability of vast supplies of raw materials in the form of fossil deposits, as well as iron ore, the influx of millions of immigrants swelling the ranks of the workforce before the end of World War I fueled America's economic expansion. Despite calls by the Progressives for the enforcement of antitrust laws, the American legal frameworks favored the formation of large-scale companies that achieved economies of scale and commanded the financial resources to sustain high investment levels. Industrial growth also profited from rapidly expanding domestic demand as the United States turned into the world's largest internal market, in which an unprecedented range of commodities came within the reach of broad sections of society.[7] While a short recession from 1920 to 1921 accompanied America's adjustment to a peacetime economy after World War I, this temporary contraction soon gave way to the boom of the Roaring Twenties. The economic gap that opened up between the United States and Western Europe in the decades surrounding the First World War contributed not only to contrasting car ownership levels on either side of the Atlantic; it also helped turn the motor industry into one of America's most dynamic business sectors. Citing official statistics, a German trade unionist visiting the United States reported that in 1924, "as an outcome of developments in the last decade," one in ten American jobs was directly or indirectly tied to the car sector.[8]

Of course, Ford Motor Company was the first auto giant to burst onto the American business scene, as the famous Model T became an unprecedented best seller. Launched in October 1908, the Model T realized Henry Ford's ambition to build a "universal car" that, according to his authorized biography, *My Life and Work*, "would meet the wants of the multitudes."[9] Above all, the Model T distinguished itself through a rugged dependability that allowed it to function in a transport environment not yet tailored to the requirements of the automobile. "Model T" quickly became a synonym for sturdiness, steadiness, and reliability, gaining its reputation as a car that, as Tom McCarthy has written, was "ready to take a beating and come right back for more," owing to the simple engineering that characterized its design. Powered by a comparatively strong, trustworthy four-cylinder, twenty-horsepower engine, the Model T possessed ample ground clearance to maneuver in rough terrain, including dirt roads that frequently turned into bottomless mud tracks in the fall and spring. The car also profited from the extensive use of heat-treated vanadium steel in its chassis components. This material allowed for the design of a 100-inch wheelbase that resulted in a light and tough automobile sufficiently big to transport four people as well as heavy loads. While drivers welcomed that Ford had taken the as-yet rare step of placing the steering wheel on the left to make it easier to spot oncoming traffic, they considered the vehicle's low running costs its vital virtue. These moderate operating expenses derived from a gas mileage of twenty-five miles per gallon, infrequent breakdowns, and, in the case that something did go wrong, low repair bills.[10]

"Real simplicity," *My Life and Work* proclaimed, "means that which gives the very best service and is the most convenient in use." Since automobile technology was still in its infancy, necessity was the mother of simplicity, because technical solutions with complicated components only increased the likelihood of breakdowns. The Model T's character as a straightforward technological object put drivers in a position to perform many repairs themselves. In keeping with this ethos of basic functionality, the Model T was undoubtedly an eminently practical automobile, but a comfortable one it wasn't. While the company decided to add electric headlights in 1915, it never made a heating system part of the standard equipment in the vehicle. Although a components manufacturer developed a serviceable button-operated electric starter as early as 1912, Ford decided for cost reasons to fit the device only in 1926. Before then, drivers either had to have an electric starter installed as an extra by a service station, or start their cars by forcefully turning a hand crank, an operation that was not just a tedious chore but one fraught with physical danger. At times, the engine backfired while being started, sending the hand-held crank back in a violent jolt that could easily lead to broken wrists and thumbs. To prevent such injuries, drivers were advised not to grip the crank firmly but push it with an open palm.[11]

Americans clearly considered these deficiencies as minor. Between 1908 and 1927, Ford made over fifteen million Model Ts, a global sales record that was only to be broken by the Volkswagen Beetle in 1972. The Model T was produced and sold in unprecedented numbers from the outset. Annual output already reached almost 35,000 by 1911 and surged to 533,706 in 1919. By the time the brief postwar recession interrupted the first great car boom in American history in 1921, two-thirds of all cars on America's roads were Model Ts. While PR stunts like long-distance races demonstrated the car's reliability, Ford profited even more from a word-of-mouth campaign in which satisfied drivers recommended the vehicle to friends and acquaintances interested in acquiring a car. At a time when most cars were notorious for their erratic performance and the public was bound to regard a manufacturer's full-mouthed promises with skepticism, informal and independent endorsements from ordinary citizens were highly effective in establishing the Model T's credentials as a trustworthy product. In 1911, a survey of two thousand drivers revealed that 85 percent of them had bought their Ford based on another owner's personal recommendation.[12]

Beyond the car's technical virtues, the fact that per capita real income increased by 85 percent in the United States between 1890 and 1925 provided a crucial precondition for the Model T's triumph. As more and more Americans in the early twentieth century came to command the disposable means to acquire and maintain a low-priced motor vehicle, Ford was first in targeting a massive market at the lower end of the automotive spectrum. While other auto manufacturers pursued luxury buyers with vehicles costing in excess of $1,000, Ford aimed his product at customers well below this threshold. Introduced at a price between $825 and $850 in 1908, the car retailed at less than $450 in 1920, a dramatic drop opening up new sales territory. To consolidate the company's customer base, Ford Motor Company also extended its nationwide network of dealerships, which held spare parts and offered competent repair services.[13]

Rising paychecks and falling vehicle prices undoubtedly help explain why unprecedented numbers of Americans could afford a car like the Model T in the 1910s. Yet the pent-up demand that fueled the early car boom also speaks of a harsher side of American life at the time. Making up almost 60 percent of the population in the early twentieth century, rural Americans constituted the largest group among Ford's customers. Life on the farm may not have been short, but it was often brutish and nasty, involving long hours as well as much drudgery. Domestic amenities in the countryside remained primitive, as most households still lacked the central heating and indoor plumbing that city dwellers increasingly took for granted. Rural Americans embraced the Model T as a vehicle with many functions, turning Ford's creation into a versatile tool that did far more than move people and goods. Fixing a belt to the rear axle or crankshaft, farmers employed their cars to drive grindstones, pumps, saws, butter churners, and more. Given the vehicle's ability to keep going in difficult terrain, it also served as a precursor of the tractor. "In the fields," recalled a historian whose family had used a Ford on their farm, "the Model Ts pulled hay rakes, mowers, grain binders, harrows, and hay loaders."[14]

First and foremost, the car recommended itself as a source of individual mobility that alleviated the social isolation of America's rural population. Beyond rendering chores such as buying supplies and selling produce in rural towns far easier, it opened up new opportunities for sociability. With a Model T, country folk could visit relatives and acquaintances as well as drive to town for dances and movies much more quickly and frequently than by horse-drawn buggy. Men were by no means the sole beneficiaries of this new mode of transport. Women greeted the expansion of their social horizons with a mixture of relief and enthusiasm. "Your car lifted us out of the mud. It brought joy into our lives," the wife of a farmer wrote in a personal letter to Henry Ford in 1918. By 1920, 53 percent of farm households in the Midwest owned a car, while the rate stood at 42 percent in the Far West. Ford himself took pride in the popularity of his car among country folk—and not just because this development generated great profits for his company. Having grown up on a Michigan farm, he had firsthand knowledge of the harshness of country life.[15]

Middle-class urban and suburban Americans also embraced the Model T, albeit on slightly different grounds. Since they had readier access to public transport than country folk, urban car owners put their automobiles mainly to recreational use on weekends and during vacations. As a result, city residents made up the bulk of the clientele for the many touring accessories ranging from awnings to tents to collapsible beds that soon became available and converted many a Model T into "Hotel Ford." Irrespective of the car's strong appeal as a leisure item, urban demand did not keep pace with sales in rural areas during the Model T's first decade. Some professionals, including doctors and lawyers, bought cars for work-related reasons, but on the whole, city dwellers found fewer uses for the Model T as a work tool than did farmers. Compared with rural residents, they also faced higher outlay for garages, as well as maintenance, because they tended to be less adept at fixing their vehicles.[16]

Whatever their place of residence, drivers quickly established close, highly emotional bonds with their automobiles. "In its original state," a journalist commented in 1915, "a Ford" was not well suited to "express your individuality . . . for Fords are all alike." Such uniformity, of course, derived from a product policy encapsulated in Ford's famous quip that "any customer can have a car painted any color that he wants so long as it is black." Yet as suppliers flooded the market with a plethora of technical extras and makeover accessories as diverse as turn indicators, seat covers, fenders, stylish radiator hoods, and much more, owners modified their cars in accordance with their aesthetic preferences and practical needs. The numerous personal names and nicknames Americans bestowed on their cars show that these alterations as well as daily use turned a standardized technical object into a unique, personal possession for which owners developed great affection. It wasn't long before "Tin Lizzie" and "flivver" entered the lexicon as collective terms of endearment, signaling the Model T's ascent to national icon—a trajectory many cars for the wider population were to follow in subsequent years.[17]

As the Model T generated seemingly inexhaustible demand, Ford Motor Company embarked on a search for a cost-effective way of mass-producing this highly complex artifact composed of numerous technical components. The solution devised in Dearborn was to prove profoundly influential throughout the twentieth century. Beyond expanding its workforce from 450 in 1908 to 32,679 in 1921, the firm initiated pathbreaking organizational changes in manufacturing routines by taking the division of labor to unprecedented heights. To be sure, the advantages of arranging manufacturing processes into sequences of individual, specialized tasks had ranked prominently among the world's economic truisms for well over a century by the time Ford began to assemble the Model T. The famous opening chapter of Adam Smith's *Inquiry into the Nature and Causes of the Wealth of Nations* extolled as early as 1776 how a small pin-making workshop raised its output more than a thousand times by splitting production procedures into "eighteen distinct operations."[18]

Ford and his managers, however, were inspired by a far more advanced form of labor division that was widely commented upon in the late nineteenth century. Manufacturers of intricate mechanical objects such as sewing machines and bicycles assembled their products in carefully coordinated work sequences out of interchangeable components. Most notably, they employed specialized machine tools like punching machines and electric welders that were easy to operate, could be entrusted to largely unskilled workers, and hence decreased the need to rely on highly trained artisans. Pioneered in the United States, this production mode became known as the "American system" of manufacture and attracted note in several respects. In essence, the use of mechanized tools transferred manufacturing skills from man to machine and thus "deskilled" the human labor required in the factory. This new form of mechanization was particularly advantageous in the United States because it compensated for the chronic shortage of skilled labor that persisted despite large immigration waves. The advanced mechanization of production put businesses in a position to increase output while lowering prices. The "American system" manufactured large runs of identical commodities because its procedures relied on interchangeable components and uniform labor routines. In short, the development of the "American system" opened the door for the cost-effective mass production of complex, highly standardized mechanical commodities with the help of largely unskilled labor.[19]

Although the early American motor industry was aware of the benefits of these novel manufacturing methods, it paid them little attention as long as car makers operated in a small luxury market. Artisanal forms of production, in which teams of skilled men assembled large parts of a car, such as the chassis, from beginning to end, prevailed in the budding auto industry. Ford, however, developed the Model T with an eye toward a much wider sales territory. As the order volume increased, the management at Ford began to draw on the mechanized production methods of the "American system." In-house engineers, for instance, developed and installed a plethora of specialized machine tools to manufacture interchangeable components for the Model T. The speed, precision, and consistent quality of these machines made the work of the most dexterous craftsmen pale in comparison. In 1915, a visiting engineer marveled at the mechanization at Ford by describing a "special semi-automatic" machine that drilled "holes in a cylinder casting from four sides at once, forty-nine holes in one operation." This efficiency, however, came at a price. While artisanal car production had easily accommodated special requests from customers, Ford's most famous car emerged as a standardized commodity. Offering technical variations in the Model T (beyond the type of car body) would have required expensive modifications of procedures and equipment that would have driven up the price. As he drew on the lessons of the "American system," Henry Ford oversaw the birth of an inflexible variant of mass production.[20]

Maintaining a steady production flow involving thousands of workers and millions of parts confronted the management with a particularly tricky task. Compared with sewing machines and bicycles, mass-producing automobiles involved far more complex logistics and resulted in numerous bottlenecks. Ford's managers found the solution to their work-flow problems on the killing floors of Chicago's slaughterhouses, where hundreds of thousands of carcasses, which were suspended from ceiling rails and transported at a set pace along specialized butchering stations, were dressed at great speeds by a small number of meat packers.[21]

After several visits to Illinois, the management at Ford transferred the principle behind these gory "disassembly lines" to the plant in Highland Park in 1913. Mechanically conveying a work piece past a series of highly specialized production and assembly points had several advantages from the management's perspective. What soon came to be known as the "assembly line" cut production time drastically, and thus laid the ground for the persistent price falls that were a precondition for the Model T's commercial success. "Production nearly doubled every year for a decade after 1913, while the price of a Model T dropped by two thirds," one scholar has summarized.[22] The assembly line also extended managerial control on the factory floor, where workers had so far designed numerous stalling tactics to slow down manufacturing routines. As the pace of the conveyor belt defined the work rhythm, laborers had no choice but to adapt their speed to the standards imposed by the management. The assembly line allowed the executives at Ford to replace skilled workers with unskilled laborers who could be trained "within a few hours or a few days," as *My Life and Work* boasted. In 1913, the company began to hire numerous foreign-born immigrants, especially from southern and eastern Europe. Within a year, these recent arrivals made up over 70 percent of the factory's employees. America's first iconic automobile was built by people many of whom spoke barely a word of English.[23]

For the workforce, the arrival of the assembly line meant a drastic deterioration of labor conditions. The monotony and the physical strain of life on the assembly line took a harsh toll. Immediately after the line's introduction in 1913, the annual staff turnover rate shot up to over 370 percent as thousands of employees left the company each month. Many simply found employment at Ford intolerable. One worker recalled how, after a first day at Highland Park in the early 1920s, a friend returned home in a state of utter exhaustion: "He would sit in a chair and didn't care whether he ate dinner or not. . . . He was just so tired, and his body ached so that he didn't care whether he moved or not." Two days later, the new recruit quit. As Ford emerged as the first in a long line of auto manufacturers notorious for labor routines that subjected employees to the rhythm of the machine, it was no coincidence that a much-loved automobile like the Model T owed its existence to backbreaking, monotonous work routines. Only advanced mechanization lowered production costs sufficiently to bring the automobile within the financial reach of the broader population.[24]

If the assembly line became symbolic of workplace alienation in the car factory, Ford's response to the initial hemorrhage of personnel set an example that other car manufacturers copied almost as frequently as his manufacturing routines. In January 1914, Ford Motor Company announced that it would double workers' daily wages to $5 for an eight-hour shift. The day the new remuneration plan went into practice, applicants laid siege to the recruitment office amid riotous scenes. Even the fire department, which turned its hoses on the throng in the near-zero temperatures of the Michigan winter, could not disperse the crowd clamoring for one of the well-paid jobs.[25] Only after the proclamation of the $5 day did demand for employment at Ford pick up. Beyond its pay policy, the company strove to consolidate its workforce through welfare measures, establishing hospitals, a savings and loans association, sports facilities, and night schools offering English lessons and trade apprenticeships. A deep paternalism underpinned Ford's managerial ethos. In addition to abstaining from all forms of trade-union activism, employees had to subject themselves to strict controls at work and at home. Investigators from the company's "sociological department" kept thousands of files on the private lives of individual workers, interviewing family members and neighbors to uncover evidence of gambling, drinking, or sexual dalliance. To qualify for the company's material and educational offers (including the $5 day), Ford workers were expected to lead a wholesome life of thrift, and those found wanting risked wage reductions or worse. The immigrant labor force paid a steep price for the moderate prosperity that came with manufacturing Tin Lizzie.[26]

Meanwhile, Henry Ford's reputation grew with his business. Having honed a marketing talent in the early days of the automobile to drum up publicity for his vehicles, he now employed his promotional skills to fashion himself as a "millionaire folk hero." In writings and interviews, Ford accentuated a public persona as a plainspoken man of common sense, employing simple examples to illustrate how seemingly inconspicuous measures yielded spectacular effects: "Save ten steps a day for each of twelve thousand employees and you will have saved fifty miles of wasted motion and misspent energy." Eliminating waste, or the search for "efficiency," to put it in slightly different diction, figured as a major leitmotif among Ford's entrepreneurial principles. In keeping with the ethos of productivity, Ford insisted that the $5 day was not a form of largesse but a just reward for the profits dedicated workers had helped create. Under capable entrepreneurial leadership, Ford argued, capitalism delivered the goods not only in the form of commodities but also high wages. Hundreds of thousands of visitors traveled to the gigantic factories at Highland Park and, from 1919, River Rouge to inspect Ford's expanding empire as the company turned into a vertically integrated concern that purchased iron ore and coal mines, forests, and a railway line before embarking on a disastrous adventure to maintain its own rubber plantation in Brazil.[27]

Fame did not shield Henry Ford from public attacks. In addition to the working conditions at Highland Park and his harsh stance on trade unions, Ford's anti-Semitism, which he aired in more than one publication, attracted censure in the United States. Neither controversial management practices nor his anti-Jewish views undermined Ford's national stardom, however. His elevated position survived the economic difficulties that befell his company in the mid-twenties, when it failed to notice that consumers developed novel preferences after the first wave of mass motorization. As Ford continued to pursue a one-model policy based on the Model T, American drivers demanded wider product choice and more comfortable vehicles. Ford lost its position as market leader to General Motors, which offered a range of brands with the aim of encouraging customers gradually to "trade up." GM's Chevrolets proved particularly popular among American drivers in the second half of the twenties. Despite these problems, however, Ford's status as a venerated celebrity remained largely intact. In Tin Lizzie, he had created the first automobile that fully qualified as a car for the general population. In the American imagination, this feat ranked far above its creator's anti-Semitism, or any managerial shortcomings or persistent complaints about inhumane working conditions.[28]

Commercial triumph propelled Henry Ford's name far beyond the American national pantheon. In addition to his expanding business, the best seller penned in Ford's name played an important role in securing and consolidating his unmatched global reputation. The German translation of *My Life and Work* sold over two hundred thousand copies within little over a year of its publication in 1923. In Germany, the book met with avid interest because Ford's success story epitomized the dynamism that allowed the American economy to outperform Western Europe by widening margins. In light of the Weimar Republic's manifold economic ills, many Germans picking up a copy of *My Life and Work* hoped to hold in their hands a prescription book offering a cure for their country.[29]

In the early twenties, no commentator courted controversy by describing Germany's economic situation as abysmal. Within less than a decade after the outbreak of World War I, Germany had gone from a position as Europe's fastest-growing industrial power to what many contemporaries regarded as a basket case. Military defeat, a tumultuous revolution resulting in the Weimar Republic's precarious democratic constitution, and the Versailles Treaty with its unspecified demands for German reparations complicated attempts to restore a stable political order and economic growth after 1918. Although the notorious reparations issue proved less of an economic liability in the medium term than German officials had initially feared, the country had to cope with serious economic disruption. As a result of Europe's territorial reorganization after 1918, the country lost one-third of its coal reserves and three-quarters of its iron ore mines to France and Poland. Political and economic unrest repeatedly dominated the headlines in the early postwar years, culminating in a bout of hyperinflation in 1923 that was as financially costly as it was psychologically unsettling for many Germans. Only a radical currency reform and a renegotiation of reparations, as well as the influx of American business loans, established a modicum of political and economic order to Germany, initiating the Weimar Republic's phase of brittle stability between 1924 and 1929.[30]

Against this backdrop, Ford figured as a crucial reference point in a passionate and deeply controversial debate about the United States, which the Weimar public increasingly regarded as a "cipher of unbridled and unconditional modernity." No matter whether they discussed Hollywood movies, life in New York and Chicago, or American women's comparatively liberated mores, commentators saw in American affairs positive and negative aspects of a future society characterized by industry and commerce.[31] Like America in general, Ford exerted a broad yet divisive appeal as Germans hailed and condemned the American entrepreneur in keeping with varying ideological agendas. Adherents of *Fordismus* among the German business community, for instance, emphasized that the productivity gains generated by mass-production methods would provide an indispensable instrument to raise the efficiency of the Weimar economy. German trade unionists, irrespective of Ford's profound aversion to organized labor, commended the American entrepreneur's high-wage policies, pointing out that in real terms American workers' incomes stood at twice the level of pay in Germany. Next to these acknowledgments of prosperity stood a chorus of skeptics that painted Ford as a "fake messiah" with a pernicious gospel. While social reformers focused their criticism on workplace alienation, businessmen warned that Fordist mass production would undermine German industry's reputation as a purveyor of high-quality products. Cultural doomsayers advanced the most comprehensive dystopias, thundering against a future consumer culture dominated by uniform, standardized commodities that would signal nothing less than a "mechanization of work, thought, and life itself."[32]

As pro- and anti-American commentary turned Henry Ford into a deeply controversial household name in Germany, a stark absence marked the Weimar debate about Fordism: it occurred to hardly any German observer to ask whether the "universal motorcar"—the product upon which Ford's empire rested—would anytime soon provide a realistic prospect for German consumers. Germans of the Weimar era looked upon Ford as a mass manufacturer rather than a car manufacturer, an indicator of the contrast between economic conditions on both sides of the Atlantic. German visitors encountered America's auto culture with incredulity. Arriving in Detroit, engineer Franz Westermann, who took pride in having maintained his composure in the bustle of Manhattan, was reduced to disbelief when he beheld the unending procession of automobiles in Michigan's motor city. "I laughed," he recalled, "at the fact that Detroit is the home of approximately twice as many cars as there are in all of Germany . . . , that hundreds of auto dealerships [feature] showrooms as large as our small car factories, that there are almost as many cars in the streets as pedestrians." At the end of the day, as he stepped to the window of his hotel room, he could not avert his gaze from the "hundreds of cars, one next to the other, as far as the eye could see."[33]

Westermann's incredulity illustrates how distant a prospect most Germans considered a "universal motorcar" for their own country. As late as the twenties, motor vehicles still remained a rare sight in many parts of Germany. To be sure, after the restoration of political and economic stability in the mid-twenties, the number of passenger cars on German roads more than quadrupled, from 80,937 in 1922 to 422,612 in 1928. Yet this was a numerical expansion within strict limits—and not just when compared with the 15.4 million cars in the United States in 1925. France, a country with merely two-thirds of Germany's population, witnessed a rise in vehicle registrations from 242,358 to 757,668 during the same period. Even in Berlin, which featured the highest concentration of German automobiles by a wide margin, the volume of motorized road traffic remained modest into the mid-twenties. After the Great War, the German capital may have gained an international reputation as an exciting center of cultural experimentation for its vibrant nightlife, provocative art scene, and dynamic press landscape, but it proved unnecessary to install traffic lights in this modern metropolis before 1925.[34]

Many factors were responsible for the automobile's slow proliferation in the Weimar Republic. For German car manufacturers, Fordism remained an elusive goal as long as the sector was dominated by comparatively small companies unable to shoulder the investments necessary for mass production. In 1927, no fewer than twenty-seven automakers operated in Germany. Daimler and Benz, both of which produced limousines for affluent drivers, each had over two thousand workers on its payroll, but made on average a mere 4.4 and 5.3 vehicles per day respectively in 1924. Labor-intensive production routines and high material costs left both companies with crippling debts. Even after a merger that resulted in the formation of Daimler-Benz AG in 1926, the new company found it impossible to finance assembly lines or the numerous specialized machine tools that were prerequisites for mass-production methods. Production costs at Daimler-Benz remained far higher and output considerably lower than at comparable companies in the United States.[35]

Only Opel, by far the largest German car producer, with twelve thousand employees in the late twenties, stood out for introducing Fordist production modes to reduce costs and increase capacity in 1924. "Gigantic eight-story buildings, high towers, and rising smokestacks" lent the factory in Rüsselsheim the appearance of a "fantastic city," a richly illustrated company history gushed. Manufacturing a fourteen-horsepower and a forty-horsepower model on assembly lines, Opel established itself as Germany's market leader, turning out about half the country's vehicles in 1928. The "miraculous workings of the 6,000 machines" in the assembly halls may have regularly impressed visitors, but rationalization had clear limits in Rüsselsheim. In contrast to American practice, Opel did not assemble its models with interchangeable parts. Moreover, skilled manual workers made up two-thirds of the firm's employees, a ratio that left unskilled laborers, who dominated car production in the United States, in a minority in Rüsselsheim. Even at Germany's most advanced automaker, mechanization had gone nowhere as far as in the United States.[36]

In light of the limited capacity of domestic firms, foreign producers viewed Germany as a promising future market. To circumvent high import tariffs, American companies established assembly plants in Germany to manufacture vehicles from components shipped across the Atlantic, allowing them to capture a quarter of the German market by 1928. Ford and General Motors, who became the most influential foreign players in Weimar Germany, made the decision to preempt future protectionist measures by establishing full-scale production sites in Germany. While Ford erected a new factory in Cologne that opened its gates in 1931, General Motors purchased no other company than Opel in 1929. The German press reacted to the takeover of Germany's largest car manufacturer with apprehension, considering the prospect of an invasion of the country's roads by cars made by foreign corporations as an alarming indicator of national weakness. At the time, few observers would have predicted that German companies would emerge at the forefront of auto manufacturing in the second half of the twentieth century.[37]

Rather than primarily the result of poor management, the precarious position of the German car sector in the twenties reflected far larger, fundamental problems that lay beyond the industry's control. Simply put, the German automakers suffered from a chronic lack of domestic demand. To some extent, social geography hampered the automobile's proliferation in the Weimar Republic. A densely populated country with a well-established public transport system like Germany required motorized individual transport less urgently than the United States, with its vast stretches of sparsely peopled areas.[38]

The most important impediment, however, was the state of the economy. After a decade of war, political upheaval, and hyperinflation, the earnings of the vast majority of the population in Weimar Germany remained severely depressed. Around 1925, a survey concluded that only half of the country's industrial workers brought home the annual wage of 1,000 reichsmarks ($250) widely considered the minimum necessary to ensure basic lodgings and food for a worker's family with two children. In nominal terms, an annual pay packet at this level amounted to merely a quarter of the wage levels of Ford workers after the introduction of the $5 day. Small wonder, then, that German trade unionists drew attention to Ford's remuneration policies. Most German farmers found themselves in similarly strained material circumstances. Over three-quarters of rural landowners in Weimar Germany cultivated plots smaller than five hectares (12.5 acres), eking out incomes below subsistence levels. In 1928, average rural earnings stood at a mere 1,105 reichsmarks ($263) per annum.[39] In pronounced contrast to the United States, where the rural population had proved pivotal for making the Model T a commercial success, a German farmer behind the steering wheel was a rare sight indeed. If farmers and workers contemplated motorized individual transport, they thought of the motorcycle rather than the automobile. While a motorbike offered no protection from the elements and exposed riders to greater physical dangers than an automobile, it compensated for these disadvantages with far lower purchase prices and maintenance costs. From 1923 to 1929, motorbikes outsold cars by one-third, leading to an increase in registrations from 59,389 to 608,342.[40]

As long as farmers and workers, who made up over 75 percent of the population, could consider car ownership merely a distant aspiration, the majority of drivers belonged to Weimar Germany's troubled middle class. While members of the affluent elite such as entrepreneurs, bankers, and high-level managers had succeeded in protecting their assets throughout the war and hyperinflation, the bulk of the German middle class—shopkeepers, tradesmen, salaried employees, and civil servants, among others—reeled from a series of financial blows, including the loss of investments in war loans and the evaporation of savings in repeated inflation waves between 1914 and 1923. Next to their counterparts in Great Britain and France, the German middle class cut a shabby figure, taking fewer holidays, owning fewer radios and home-movie cameras—and, predictably, driving fewer cars. In 1928, a commercial survey calculated that the country could absorb no more than a further 220,000 vehicles, roughly the number of carless, well-to-do professionals like medical doctors and lawyers, as well as high-ranking white-collar employees and civil servants. Although market research cannot be taken at face value, the study shows that even at the end of the Weimar Republic's "golden years," any call for more than a modest social expansion of car ownership beyond the ranks of the solid middle class possessed a deeply unrealistic ring.[41]

Given the constraints imposed by the national economy, small and medium-size cars with engines developing less than forty horsepower accounted for over 75 percent of automobiles in operation in 1926. With a production run of around 120,000 from 1924 to 1931, the "small Opel" proved the most popular model by a wide margin. Nicknamed "tree frog" for its green coat of paint, the fourteen-horsepower vehicle came as a two- or a four-seater and reached top speeds just shy of thirty-eight miles per hour. Although Opel had copied the design of Citroën's 5 CV and thus saved the development costs for the vehicle, the car retailed between 2,300 and 3,200 reichsmarks ($550 and $760) in early 1929—far more than the Model T had cost almost a decade earlier. The sales tag, however, did not stop a test driver from hailing the "tree frog" in 1930 as "Germany's most popular car that withstood the critical eyes of the layman, the expert, and the connoisseur."[42]

Drivers who found the Opel too expensive could turn to the Dixi, a limousine with a four-stroke, fifteen-horsepower engine that could travel at forty miles per hour. Based on the British Austin Seven, the Dixi was another vehicle with a foreign pedigree. Made by motorcycle and aero engine manufacturer BMW, the car cost around 2,500 reichsmarks ($595) in 1929 and sold between five thousand and six thousand models annually. Technical problems were not responsible for the failure

A driver showing off his Hanomag microcar, a two-seater affectionately known to Weimar Germans as the "rolling bread loaf" because of its rounded front and rear sections. Postcard, author's collection.

of BMW's first automotive offering to match Opel's popularity.[43] Rather, the car compared unfavorably with its main competitor in terms of size. An advertisement may have assured prospective buyers in 1929 that the Dixi "transported three people and their luggage" or "two grown-ups and two children" with ease, but BMW found it necessary to relaunch the vehicle with a larger body within three years. While the automotive press breathed a sigh of relief that the car was no longer a "baby," the increased price of 2,825 reichsmarks ($672) disappointed journalists.[44]

The only domestic design to gain a notable following was the ten-horsepower, two-seater produced by Hanomag from 1925 to 1928. Soon christened the "rolling bread loaf" because of its compact rounded body section at the front and back, this car attracted almost sixteen thousand buyers and retailed at around

2,000 reichsmarks ($475) toward the end of its production run. Although owners insisted that a Hanomag "offered excellent comfort," the vehicle became the target of numerous jokes inspired by its basic technology. "A piece of tin and paint from a jar is all that makes a Hanomag car," a popular rhyme declared. Similar to the transmission of a motorcycle, a chain rather than a driveshaft ran between the engine and the rear axle. While this design solution kept production costs low, it proved unreliable given the wear and tear involved in propelling a seven-hundred-pound vehicle rather than a far lighter motorbike. Most curiously, however, driver and passenger could enter the Hanomag only from one side, because a second door would have unbalanced the small car. When Hanomag went into the red in 1928, it discontinued production—much to the regret of devoted fans, who retained fond memories of the "undemanding, indestructible, and ever-ready . . . little one-cylinder vehicle."[45]

Beyond the comparatively high price and technical limitations of its cars, prohibitive maintenance costs inhibited Germany's transformation into a nation of drivers. Despite falling expenses for tires, gasoline, and engine oil during the twenties, keeping a BMW Dixi on the road required around 1,200 reichsmarks, one owner estimated. German motorists spent roughly three times as much on gasoline and up to seven times as much on road tax as Americans, triggering polemical comparisons in the press that cast the motorist as a "fat milk cow" sucked dry by a state allegedly prioritizing investment in publicly owned railways. The largest single expense car owners faced, however, had nothing to do with the state. Since it was generally agreed that automobiles required shelter from the elements, urban owners, most of whom resided in apartments, had to hire a garage at an annual cost of between 400 and 700 reichsmarks—roughly the equivalent to the cost of a two-bedroom apartment.[46]

The need for shelter, in turn, reflected yet another obstacle in the path of the automobile's proliferation in Germany. Compared with American models, many German automobiles proved unreliable technological artifacts that demanded much tender loving care. A technical primer from 1925 contained a seemingly endless list of reminders, starting with the admonition "never to leave" a new car "standing in the sun" because fresh paint easily threw bubbles. "Dust and dirt from the street," the manual continued, had to be removed after each journey to "prolong the car's life." Beyond providing cosmetic advice, the work emphasized the importance of regular checkups on mechanical components: "Every six months, the springs need to be greased with a mixture of oil and graphite. To this end, the springs are relaxed and the mixture described above is filled between the spring bearings. The ball bearings of the wheels have to be lubricated with good engine grease every 900 miles. It is advisable to clean the ball bearings thoroughly every three months." For those who had neither the time nor the skills to perform these procedures themselves, frequent and costly trips to a service station were inevitable to ensure a vehicle's smooth operation. Under these conditions, winter was the season automobile owners faced with particular trepidation, because keeping a car in good working order often entailed numerous chores during that season. These typical maintenance routines made a mockery of the advertisements that overtly played on prevalent notions of female technical incompetence by recommending particular models to "lady drivers" for their supposedly low service requirements. In everyday life, the work involved in maintaining most vehicles during the Weimar Republic placed the motorcar fairly and squarely in the male domain.[47]

High costs and technical deficiencies, however, did little to diminish the high esteem in which drivers held their automotive possessions. While the very wealthy kept one or more cars as luxury items, the majority of drivers came from middle-class backgrounds and had an automobile for work-related reasons. For medical doctors, lawyers, and traveling salesmen, as well as tradesmen, a contemporary observed, "the small car is primarily a professional vehicle that makes its proprietor independent of the railway, the tram, and other means of public transport." Only as "an indispensable tool" that extended its owner's professional sphere and thereby promised to boost earnings did the automobile justify the substantial financial outlay and the personal labor required for its upkeep. Most Germans who bought a car during the Weimar Republic were motivated by tangible material benefits.[48]

At the same time, owners insisted that they regarded their cars as far more than prosaic work instruments. Despite its professional benefits, they pointed out, the automobile revealed its true appeal in private life outside working hours. As a matter of fact, many drivers praised the car as an effective antidote to the pressures of professional life as well as workaday tedium. "At the end of a workday," a car owner enthused, "there is probably no more enjoyable pleasure for the harassed and stressed-out city dweller than fleeing from the urban sea of houses with a little, swift car." While many car owners counted themselves among "the modest, yet respectable middle class" and therefore did not experience the harsh forms of alienation widely associated with, say, manual factory labor, they nonetheless emphasized that genuine self-fulfillment could be found only outside professional contexts.[49]

After working hours and during weekends and holidays, the car morphed from a professional tool into a means of escape that literally allowed people to get away from it all. Like no other form of transport had done before, the automobile enhanced the liberating effect of spare-time activities. As one advocate of the car complained, railways "tortured" passengers by "squeezing" them into "overcrowded" carriages that traveled according to set timetables and tracks. By contrast, motor vehicles offered far more freedom, flexibility, and comfort, giving drivers a chance to choose their fellow travelers as well as routes and speeds while taking breaks along the journey. As they gained "complete independence," car owners praised their vehicles for making them feel like "lords" and "pashas." The automobile liberated free time, thereby fostering a sense of personal autonomy that middle-class drivers missed in everyday life. The automobile's capacity to enrich private life functioned within strict limits, however. Most small vehicles accommodated two or three people at best. Drivers, therefore, found it difficult to praise the motorcar's potential to enhance family life during the Weimar Republic. Although small automobiles undoubtedly strengthened a sense of personal autonomy, they were still a far cry from qualifying as fully fledged family vehicles.[50]

Like ramblers on foot, Weimar Germans going for a Sunday drive visited picturesque ruins of churches and castles, stopped at lakes for a swim, napped in meadows, and took in scenic vistas. Given the absence of alcohol limits, they also took advantage of restaurants, pubs, and taverns situated along their itineraries, often beginning their excursion with a breakfast of "broth and beer," followed by lunch as well as coffee and cake in the afternoon before completing the trip with a dinner accompanied by "wheat beer and petroleum," the latter a high-octane spirit. Some outings turned into booze cruises, but most accounts convey that automobilists, rather than employing the car for cultural experimentation, were in pursuit of profoundly conventional and idyllic pleasures. As such, the motorcar promised an escape not only from workaday life but also from the political, social, and economic turmoil that characterized the Weimar Republic. Above all, a yearning for an unspectacular normality fueled middle-class desire for the automobile in the twenties.[51]

Nonetheless, the automobile's appeal was by no means limited to the quiet pleasures enjoyed by middle-class car ramblers. While numerous commentators had denounced motor vehicles as a novel hazard in Germany (and elsewhere) before 1914, new traffic laws and a process of accommodation between drivers and public had long undermined fundamental opposition to the automobile by the 1920s.[52] As fundamental animosity faded, the automobile consolidated its status as an object of veneration far beyond the circles of car owners. The virtually boundless enthusiasm the car elicited in Weimar Germany manifested itself most visibly among the mass audiences at the nation's racetracks. Despite deep political crisis and economic hardship in May 1932, the German Grand Prix on Berlin's AVUS circuit attracted over three hundred thousand spectators. "Berlin was in the grip of a previously unknown racing mania," a journalist gasped. "Whoever could somehow find the money invested it in an AVUS ticket," including "the unemployed [who] went hungry to afford a standing-room ticket." Commentators of all political stripes agreed that spectators were richly rewarded with "a gigantic struggle" as Manfred von Brauchitsch and Rudolf Caraciola, Germany's most popular racing drivers "chased each other" at "terrifying speeds" in excess of 140 mph.[53]

The media storms that accompanied events like the 1932 AVUS race did more than celebrate auto racers as daredevils. The thrill of the racetrack aligned the motorcar with numerous other technologies whose velocity and power had long fascinated the public in Germany and elsewhere. Railways, ships, airplanes, airships, motorbikes—all these and other "technological wonders" fueled the public imagination as they covered growing distances in shorter amounts of time, prompting numerous observers to extol the "conquest of nature." Ever since the 1850s, technological exploits had been greeted by a swelling chorus that took mankind's expanding control over the natural world as evidence for the advent of a novel era whose characteristics differed from all previous historical epochs: "modern times." In the eyes of many contemporaries, the motorcar counted among the technologies set to propel mankind into a new and exciting "modern age."[54]

That this appealing technology remained beyond most people's reach frustrated the citizens of the Weimar Republic. Regardless of political affiliation, the press lamented the absence of an affordable automobile. When Hanomag unveiled its small vehicle at the Berlin auto show in 1924, the liberal *Berliner Tageblatt* welcomed it as "an option for all social circles." The leading Social Democrat daily *Vorwärts,* however, disagreed with this assessment in its review symptomatically entitled "the absent people's automobile." In addition to being "much too expensive," the left-leaning daily explained, the Hanomag two-seater with its one-cylinder engine fell short on several technical counts. "An automobile for the people needs to be designed as a four-seater, feature a water-cooled four-cylinder engine, and possess detachable wheels and rims. Electrical lights and an automatic starter ought to be a matter of course. The sales price would have to become as cheap as imaginable," the paper concluded as it articulated the need for an inexpensive family vehicle that was easy to maintain and repair. When General Motors took over Opel five years later, the issue was still debated in similar terms. The *Berliner Tageblatt* commented that "a small car with a low purchase price and cheap maintenance costs" represented an indispensable prerequisite for the social extension of car ownership. Although calls for an affordable family vehicle surfaced right to the Weimar Republic's end, the topic never made it to the forefront of the public agenda.[55]

A car for the wider population remained a secondary public concern not only because the Weimar Republic confronted far more pressing problems as a consequence of chronic economic and political instability. Rather, the issue itself was a nonstarter. As long as incomes remained low and automobile prices high, the "people's car" remained an amorphous concept that circulated in the background but never took center stage in the public sphere despite a profound affection for the automobile in wider society. To be sure, Henry Ford had demonstrated how to mass-produce a basic, cheap, and reliable automobile, but his triumph also revealed the economic gulf that separated the United States and the Weimar Republic. As long as depressed wages and salaries restricted demand, firms had few incentives to invest in the mechanized manufacturing methods that held the key to lowering auto prices. Conversely, as long as cars remained expensive, only a small number of middle-class customers, who venerated the car as a source of professional gain and private pleasures, could contemplate buying a vehicle. Although German society displayed considerable desire for the automobile, it failed to produce substantial commercial demand. The German auto industry and its potential customers were thus jointly caught in a deadlock that appeared impossible to break with the help of a "people's car" within the foreseeable future. In fact, when the German economy collapsed in the wake of the crash on Wall Street in 1929, the prospect of a car for all became an even more distant prospect than it had been for much of the preceding decade, highlighting the material limits of Weimar's social modernity.

2

A Symbol of the National Socialist People's Community?

On July 30, 1938, Henry Ford celebrated his seventy-fifth birthday amid much pomp and circumstance in Detroit, beginning in the morning with a party involving eight thousand children who sang "Happy Birthday" and ending in the evening with a banquet for fifteen hundred diners. Among the very few well-wishers whom the renowned industrialist received in person that day was Karl Kapp, the German consul in Cleveland. Kapp owed the privilege of meeting Ford face to face to a special gift he bore on Adolf Hitler's behalf. Citing Ford's "pioneering work in motorization and in making cars available to the masses," the consul awarded the American entrepreneur the Grand Cross of the Order of the German Eagle, the highest honor the Nazi state could bestow on a foreigner.[1]

This decoration, which provoked immediate anger among the American Jewish community, reflected the high esteem in which Hitler had long held Ford. Although the head of the Nazi regime regarded the United States with deep ambivalence for its democratic constitution, the global appeal of its popular culture, and its supposed materialism, as well as its geopolitical power, he was unwavering in his admiration for the tycoon from Detroit.

As early as 1922, Ford's portrait adorned Hitler's private office in Munich.[2] Like many other Europeans, Hitler marveled at Ford's commercial success, viewing him as a social benefactor for developing an affordable car as well as for doubling workers' wages. Yet Hitler's admiration extended far beyond the car maker's entrepreneurial activities. In particular, the Nazi leader was taken by the infamous anti-Semitic tracts entitled *The International Jew* that appeared under Ford's name in the early twenties. Hitler not only agreed with Ford's denunciation of "the Jew" as the "grasper after world control"; in the context of Germany's political and economic turmoil in the immediate postwar years, he was one of many on the German right to share the American's arbitrary charge that "the main source of the sickness of the German national body" was "the influence of the Jews." In the early twenties, Hitler, who was then establishing himself as the leader of the Nazi Party, went so far as to place *The International Jew* prominently on a list of mostly anti-Semitic "books that every National Socialist must know." While Ford made a halfhearted attempt to distance himself from the anti-Jewish tirades published in his name at the close of the twenties, he was not prepared to turn down an award from the radically anti-Semitic regime that had directly sought his company's advice more than once in the previous two years.[3]

Neither the German nor the American acknowledged that the medal was also a reward for the counsel the Nazi government had recently received from Ford's company. Starting in 1936, several German delegations had consulted the Ford works as the Nazi leadership pursued an ambitious plan for mass motorization. In light of the economic conditions that had hampered the proliferation of the automobile after World War I, Germany made a highly unlikely candidate for such an enterprise, not least since the depression of the late twenties and early thirties had further impoverished large sections of the population. Tackling seemingly insurmountable obstacles, however, did not deter the National Socialists, or so they at least claimed on numerous occasions. On the contrary, Hitler relished striking public poses that cast him and his followers as launching daring initiatives that others did not even contemplate. In his address at the Berlin auto show in 1937, the führer invoked his own career trajectory as an illustration of National Socialism's defiant fanaticism, incidentally offering a glimpse of the energies behind the regime's cumulative radicalization that would set the world alight in 1939: "I need not assure you that a man who has managed to rise from the rank of an unknown soldier . . . to the leadership of a nation will also manage to solve the coming problems. No one shall doubt my determination to implement the plans I have made whatever it takes."[4]

If the National Socialists staged themselves as a force drawn to audacious missions, it was anything but self-evident why they included mass motorization among their high-profile initiatives. After all, attempts to render the motorcar accessible to the wider population possessed only tenuous connections with the aggressive expansionist agenda that represented Nazi Germany's ultimate policy aim. However, for all its radicalism, racism, militarism, and political repressiveness, the regime was by no means impervious to the mood of the German people. While it remains unclear to what extent the government's increasingly radical political and racist measures met with approval and support in German society at large, policies catering to private aspirations, such as the provision of cheap holiday trips, proved deeply popular. Given the strong appeal the automobile had exerted in the twenties—especially on the middle class, which formed a solid core of Nazi Party voters—populism provided an important motive for the regime's push ahead on the development of a car for all. As the regime pledged to render accessible a consumer good that had hitherto remained altogether unaffordable for the vast majority of the population, it sought to portray Nazi Germany as an attractive country that had much to offer to its people. The "people's car" was only one of many consumer items that the dictatorship included in visions of an affluent National Socialist future. The plan to design and market a car for the wider population possessed intimate links to National Socialism's ideological preoccupations and thus went beyond a pragmatic initiative to bring a highly desirable consumer durable within the reach of more consumers. In fact, the quest for a "people's car" corresponded to and aimed to enhance wider attempts on the Nazi government's part to remodel German society. Efforts to turn Germany into a country of car owners were closely tied to other policies promoting the automobile in general and coincided with measures to establish a new culture of driving in keeping with a broader vision of a modern National Socialist nation. As much as opportunism, ideology drove the Nazi project for a "people's car."[5]

When the National Socialists joined the government on January 30, 1933, Hitler and his followers moved with unexpected speed to push aside the conservative majority in the cabinet and consolidate their power through "a combination of pseudo-legal measures, terror, manipulation—and willing collaboration," as Ian Kershaw has summarized.[6] After the dramatic economic slump since 1929, which had seen unemployment in Germany rise to six million, and the widespread political turmoil during the Weimar Republic's final years that the Nazis had fanned through street terror, only a comprehensive "national revolution" could save the country, proclaimed Hitler's followers. Casting themselves as right-wing revolutionaries, the National Socialists sought to rejuvenate the nation. In order to combat the allegedly pernicious effects of Marxism, liberalism, democracy, and political pluralism, as well as a form of capitalism that placed class interests before those of the nation, the Nazis advanced a vision of a novel powerful Germany.

According to the conglomerate of ideas that constituted Nazi ideology, a racially homogeneous, yet socially hierarchical "people's community" *(Volksgemeinschaft)* of "Aryans" provided the cornerstone for "Germany's rebirth." As a best seller penned by the party's press secretary in 1934 stated, the Third Reich aimed to restore "the immutable values of the Nordic race" that rest "deep in the German soul."[7] Nazi ideology revolved around a quest to recover and reinvigorate the racial foundations of the *Volk,* a term only imperfectly translated by the English word "people," with its lack of racial overtones. Without a renewal of the "people," the argument ran, it would prove impossible for the nation "to maintain itself and rise to new greatness" in a Darwinist international environment. The construction of the "people's community" not only motivated the persecution of political opponents as well as the progressive exclusion of the Jews alongside other so-called "community aliens" *(Gemeinschaftsfremde);* it also fueled the imperial quest for "living space" *(Lebensraum)* through wars of aggression in eastern Europe, for which the Nazis, immediately upon gaining power, began to prepare for through a vast rearmament drive.[8]

The rhetoric of racial regeneration lent National Socialism a deeply atavistic quality. While Soviet socialists of the thirties can be compared to engineers who forged an altogether new type of socialist person as a first step toward communism, the Nazis are better understood as archaeologists in pursuit of racial restoration. National Socialism's emphasis on purportedly immutable racial traits implied that it aimed to excavate and reawaken qualities in the German people that lay dormant beneath the debris of misdirected historical change.[9] Despite this atavistic core of its "blood and soil" ideology, National Socialism was not an antimodern movement that wished to turn back the historical clock. The Nazis not only enlisted numerous scientists, including medical doctors, eugenicists, biologists, and psychiatrists, in their racial project; they also repeatedly claimed that the nation could only hope to assert itself economically and militarily with the help of a highly productive industrial sector. Beyond promoting productivist notions, the Nazis urged the "people's community" to follow Hitler, who frequently traveled by car and airplane, in adopting "a thoroughly modern lifestyle using the newest technological instruments." In short, the Nazis sought to create a highly technicized environment permeated by a spirit of modernity in which unalterable racial characteristics of the German people would flourish powerfully.[10]

The National Socialists' fascination with technology helps to explain why Hitler, who had been in office less than two weeks, took the time to open the International Automobile and Motorcycle Exhibition in Berlin on February 11, 1933. While the Reich chancellors of the Weimar Republic had declined similar invitations, the dictator accepted gladly, a gesture the show's organizers acknowledged with a donation of 100,000 reichsmarks to his campaign fund. Hitler, in fact, was a long-standing car enthusiast who never learned to drive, but thoroughly enjoyed being chauffeured in the high-end Mercedes limousines the party had put at his disposal since the mid-twenties. Speaking in front of Germany's leading car managers in Berlin, he declared that the automobile ranked "alongside the airplane" as "mankind's most marvelous means of transport." His government, the leader continued, would break with previous policies that had allegedly "done heavy damage to German car manufacturing" and instead "promote this most important industry in the future." From his earliest days in office, Hitler launched car-friendly policies that set the ideological context for the quest for the "people's car."[11]

The "encouragement of sporting events" was the first pledge Hitler held out to Germany's auto managers in February 1933.

Between 1933 and 1939, the national government paid over five million reichsmarks in subsidies to the racing teams maintained by Daimler-Benz and Auto Union. This sum covered less than a quarter of the costs for auto racing, but lent both teams a competitive advantage and contributed to their unprecedented supremacy on Europe's tracks. From 1934 to 1937, German cars won no fewer than nineteen of the twenty-three Grand Prix competitions and set numerous speed records in excess of 250 mph. Held in front of hundreds of thousands of spectators, these races possessed a significance that went far beyond their character as mass spectacles of speed. As a motoring journal gushed in 1936, the contests between Auto Union's V-16, 5.8-liter, 450-horsepower cars and Mercedes-Benz's "silver arrows," with their eight-cylinder, 4.2-liter engines developing 420 horsepower, demonstrated that the "modern racing car" made in Germany "absolutely dominates this era of motor sport." The party press meanwhile interpreted triumphs on the racetrack as illustrations of "the importance of German technology and Germany's will to rise to power." After 1933, the press repeatedly narrated victories as evidence of swift national regeneration. When German competitors occupied all three places on the podium in Monaco in 1936, a reporter celebrated this feat with a turn of phrase that, in light of later developments, gains a profoundly ominous quality: "Germany has won a hot battle in superior manner—now we advance to further battles."[12]

If the promotion of automotive competitions reflected National Socialism's aggressive desire for international supremacy, the second initiative Hitler announced in Berlin in February 1933 promised comprehensive domestic reconstruction. According to the dictator, "the implementation of a generous road-building program" presented a task of high national priority. Drawing on proposals from the Weimar Republic, the regime unveiled an ambitious scheme in June 1933 to complete six thousand kilometers of four-lane, long-distance *autobahn*—or highways—within five years, thereby initiating roadworks on a scale unmatched anywhere else at the time. In keeping with the proposal's ambitious nature, the autobahn attracted propaganda efforts that were exceptional even by the Third Reich's extravagant standards. After a steady stream of articles, pamphlets, photographs, and newsreels had celebrated the project's early progress, Hitler opened the first stretch between Darmstadt and Frankfurt in May 1935 in a choreographed mass ceremony involving over six hundred thousand people. Fritz Todt, the scheme's coordinating engineer, used the occasion to praise "Adolf Hitler's roads" as the "symbol of the new Germany."[13]

The new highways, the official press stressed, revealed their significance on several levels. In cultural terms, the Nazis regarded the highways as proof of Germany's creative vitality. Soon dubbed the "pyramids of the Reich," the autobahns, a glossy party publication declared, would etch the Third Reich "into the book of world history." As they hoped to muscle their way into the world's cultural heritage by road building, the Nazis tied this infrastructure project to the "people's community." Beyond offering future commercial advantages, the new, comfortable roads would "link the people" across Germany, thereby strengthening cohesion among the population, a party official predicted in 1935.[14] Proponents also celebrated the autobahn as a remedy for industrialization's effects. "We live in a technological age," a brochure explained, "and the more we take possession of it, the more we have a desire to return to nature. As the car serves to cover large distances quickly, it also serves as a bridge to nature. Meanwhile the autobahn is the technologically most advanced road, a mediator between man and landscape." Connecting urban areas with the countryside that, according to the gospel of "blood and soil," provided the fount of regeneration, the highways would help sustain the racial foundations in an industrialized Germany, this argument implied. Last, but not least, the autobahns were expected to contribute to the restoration of the "people's community" through a huge public-works program. Promising to recruit over three hundred thousand unemployed workers and inject over five billion reichsmarks into the national economy, Hitler boasted in 1933 that the autobahn marked "a milestone for the construction of the German *Volksgemeinschaft*."[15]

Although the fanfare surrounding the autobahn underpinned assumptions that road building supported the country's recovery and reduced unemployment, the program fell drastically short of the regime's promises.[16] Never employing more than 124,000 workers, the autobahn resembled other public job-creation programs in Nazi Germany in making "little if any contribution to the ongoing reduction in unemployment." Joblessness dropped in Nazi Germany because of a wider economic upturn that had set in during the summer of 1932—half a year before the Nazis took power. Once the regime was in power, its economic stimulus derived from a debt-fueled armament drive that disrupted public finances and caused acute shortages of construction material and manpower, thus delaying road building. Although no other country possessed roads like Nazi Germany's thirty-nine hundred kilometers of four-lane highways in 1942, the autobahn network remained patchy and years behind schedule. In economic and infrastructural terms, the highway program was a spectacular failure. Nonetheless, the emphasis that propaganda placed on the autobahn's capacity to enhance labor markets, national cohesion, and racial regeneration established a firm link between motorization policy and the invigoration of the "people's community."[17]

Other initiatives pointed toward a similar ideological target. If motorization policies were to contribute to German society's remaking in the image of the National Socialist "people's community," the argument within the government ran, far more than an improved infrastructure was required. In addition to launching a construction program, the regime sought to turn Germany's expanding roads into public arenas that put the *Volksgemeinschaft* and its behavioral ethos on display. To bring the conduct of road users in line with the notion of the "people's community," the government undertook a far-reaching attempt to reshape Germany's quotidian traffic culture and passed a new Reich Highway Code in May 1934. This new framework replaced the previous, uneven system of regional road regulations in which speed limits had differed significantly across Germany. Beyond providing an example of the "coordination" *(Gleichschaltung)* through which the new regime consolidated its power, this initiative strove to infuse the country's driving culture with new behavioral standards. Paragraph 25 of the code stated that "every participant in public traffic has to behave in a manner that neither endangers nor obstructs nor impairs anyone more than is altogether inevitable in a given situation." This stipulation gave drivers extensive leeway in determining their road conduct because they could do just about anything short of imperiling each other. "As long as the road is free," a journalist commented in 1934, "one can drive according to one's wishes," adding explicitly that, in the absence of oncoming traffic on a two-lane road, drivers did *not* have to remain on the right. Furthermore, the code abolished all speed restrictions. While cars and other vehicles had previously been barred from going faster than twenty to twenty-five miles per hour within built-up areas, the Nazi government left the determination of appropriate speeds to those behind the wheel. In a move that appears at odds with the regime's dictatorial character, the new traffic code curbed state control, placed considerable trust in drivers, and granted them unprecedented latitude.[18]

To some extent, the highway code was a response to calls from the automobile lobby, which had opposed speed limits for years. Moreover, the new guidelines fit in with a wider European movement against the supposed overregulation of the road, which, for instance, had led British legislators to repeal all speed limits in 1930. In the United Kingdom, lawmakers abolished speed limits because they credited the middle-class drivers, who made up the majority of British car owners, with the discipline to follow "informal, gentlemanly codes of behavior," thereby minimizing recklessness.[19] While the German government also placed its faith in the power of informal conventions to guide road conduct, the British gentleman was not the inspiration behind the Nazis' move. Rather, the Reich traffic code amounted to an enactment of "National Socialist ideas," a party handbook entitled *People to the Gun* insisted in 1934. As it granted drivers more autonomy and removed speed restrictions, the highway code and its paragraph 25 did anything but promote anarchy. On the contrary, it ranked "common good above individual good," the publication explained, because a sense of mutual obligation was supposed to control road conduct. In a deeply ironic twist to its title, *People to the Gun* called upon drivers to subscribe to the maxim "consideration of everybody for everybody." It was by no means an isolated appeal for prudent behavior on the road in Nazi Germany. As the party paper *Völkischer Beobachter* clarified, steering a large car "with strong horsepower . . . does not increase [a driver's] rights, but his duties to treat others with consideration." As late as 1939, Hitler threw his weight behind this principle, denouncing those who "treat other people's comrades [*Volksgenossen*] inconsiderately [as] fundamentally un–National Socialist."[20]

Party officials and the press declared "chivalry" and "discipline" as the virtues fostering driving styles that would prevent mayhem and maintain "order" in the absence of speed limits.[21] Extending to all road users from pedestrians and bicyclists to motorcyclists and drivers, the ethos of mutual consideration was supposed to establish a new "traffic community" *(Verkehrsgemeinschaft)*. Frequently invoked by party members, transport experts, and legal commentators, the "traffic community" was intended to offer "a mirror image of a people's community," according to the party handbook cited above. A primer for car owners from 1938 hit a similar note: "Driver, set an example of comradeship and chivalry behind the wheel. The traffic community is a piece of the people's community."[22] Although the "traffic community" was in principle open to both sexes, the emphasis repeatedly placed on "chivalry" reveals that, in keeping with National Socialism's deeply ingrained, hierarchical gender notions, the regime expected drivers to be primarily men.

The formula of the "traffic community" took up several core concerns of Nazi ideology. For everyone from lowly pedestrians to bicyclists to motorcycle and car owners, it was intended to "bridge existing class antagonisms," as *People to the Gun* claimed. At the same time, the regime by no means aimed to level the hierarchy between various road users, but promoted a code that acknowledged and moderated the differences between a highly powered minority of drivers and a majority that commanded less forceful machines or none at all. Responsibility for the smooth operation of the "traffic community" primarily rested with its "members," the road users themselves, among whom automobilists carried a particular responsibility, given their position at the apex of the hierarchy. The authorities took a background role in this novel form of community. Rather than tightly control day-to-day traffic affairs, state bodies, including the police, received directives to focus on driver education. Egregious offenders causing accidents faced punishment, but, as a legal commentator remarked, an "exceptionally generous" spirit characterized the new traffic order, trusting Germans on the road to find a modus vivendi on the basis of mutual respect.[23]

Collectivistic and individualistic elements, then, characterized the concept of the "traffic community." After all, participation in traffic could give rise to a sense of community only if drivers and other road users developed sufficient individual initiative to fill their respective roles. Rather than merely put masses of Germans into cars, the Nazis quite literally envisaged the mobilization of individuals on a broad scale. In addition to supporting their quest for a "traffic community," this acknowledgment of the individual was also designed to enhance the emancipatory character with which leading Nazis credited the automobile. In his speech at the Berlin auto show in February 1933, Hitler illustrated the car's liberating effects by contrasting it with the railroad, whose set routes and schedules had allegedly brought "individual liberty in transport to an end." The advent of the motorcar, however, gave "mankind a mode of transport that obeyed one's own orders. . . . Not the timetable, but man's will" determined the car journey. It was thus only consistent for the regime, a legal expert pointed out, to support the automobile's liberating properties through legislation that guaranteed "maximum freedom in traffic." Irrespective of its celebration of collectivism on countless occasions, the regime cast itself as a champion of individualism in its motorization policies.[24]

Of course, the National Socialist "traffic community" was predicated on a deeply illiberal, discriminatory variant of individualism. In keeping with its ideological proximity to the racially homogeneous "people's community," it predictably accorded no place to Germany's Jewish population. While the country's leading car club, ADAC, expelled its Jewish members as early as May 1933, the state banned Jewish drivers from the nation's roads in the context of anti-Jewish measures following the violence during so-called Kristallnacht (or Night of Broken Glass) on November 9, 1938, which resulted in the death of ninety-one Jews and the vandalization of 267 synagogues as well as about 7,500 businesses. Over the following month, the government barred Jews from access to cultural institutions, deprived them of welfare payments, and inaugurated a wave of expropriations. On December 3, 1938, Heinrich Himmler, chief of police and head of the SS, prohibited all German Jews from driving and owning automobiles. Victor Klemperer, a fifty-seven-year-old professor of Jewish descent who had become a passionate driver since passing his test three years earlier, paraphrased Himmler's decree accurately: "The Jews were 'unreliable,' therefore could not be permitted to sit behind the steering wheel. Moreover, their driving activities insulted the German traffic community, not least since they dared to use the autobahns built by German workers' hands. This ban hits us exceptionally hard."[25] The individualistic elements of the "traffic community" were thus closely correlated with the regime's racist preoccupations. In the Third Reich, the promotion of the automobile went hand in hand with anti-Semitism.

The "traffic community" did not live up to the Nazis' high hopes. The repeal of speed limits as well as comparatively lax traffic supervision left the authorities with few tools to counter the extraordinarily high proportion of lethal traffic accidents. From 1933 to 1939, between 6,500 and 8,000 Germans perished on the road annually, marking the Reich as Europe's most dangerous traffic environment. In 1939, Hitler railed against this recalcitrant trend. "The men," he thundered, "who deprive the nation of 7,000 dead and 30,000 to 40,000 injured are a plague on the people [*Volk*]. Their actions are irresponsible, and their punishment is therefore a matter of course." By identifying careless driving and "speeding" as the main causes of accidents, the dictator indirectly admitted that the regime's policies of regulation had failed. As a result, the administration modified the highway code in November 1937 and allowed the police to prosecute not only drivers directly responsible for crashes but also those who endangered traffic without causing accidents. Since death rates remained high, the regime went one step further in May 1939 and reimposed speed limits at roughly forty miles per hour in towns and sixty on the open road. By the government's own admission, too many road users failed to adopt circumspect behavior behind the wheel. The figure of the caring, considerate Nazi driver, on which the administration's regulatory policies rested, remained an ideological fiction.[26]

More than ruthless behavior behind the wheel, however, it was the flagging proliferation of the automobile in German society that provided the most important impediment to the development of a fully fledged "traffic community." If the regime considered the car "the most modern means of transport," as the führer proclaimed in 1934, a community of road users dominated by pedestrians and bicyclists fell well short of National Socialism's vision of a Germany at the forefront of technological developments. To be sure, auto sales increased steadily after the Nazis came to power. At unemployment's peak in 1932, the car market had been practically lifeless. That year, owners decommissioned almost one-third of all German cars to save money, and dealerships sold a mere forty-eight thousand vehicles. In the following year, sales of new cars rose to ninety-four thousand. In part, this upward trend reflected tax exemptions the regime granted in 1933 to reduce maintenance expenses by up to 15 percent. The government also lowered the cost for driver's licenses in late December 1933. The expansion of the auto market continued until the outbreak of World War II. In 1937, the number of registered cars exceeded 1 million for the first time and reached 1.3 million in 1939. Beyond the government's financial alleviations, the wider economic recovery supported sales, as did falling manufacturing costs. Opel, for instance, cut the price of its twenty-three-horsepower, four-seater P4 model from 1,990 reichsmarks in 1934 to 1,450 reichsmarks in 1936, a move that helped the General Motors subsidiary consolidate its dominant position in Germany with a market share of over 40 percent. Although the rate of automobile ownership rose from 1 in 135 to 1 in 61 persons from 1932 to 1937, this trend did little to close the gap between Germany and Western European rivals. After all, Great Britain and France had reached similar levels a decade earlier.[27]

Roughly a year after coming to power, the regime added an ambitious call for mass motorization to its vision of a technologically modern, highly motorized "people's community." Returning to the auto show in Berlin in March 1934 after about a year in power, Hitler painted a gloomy picture of the nation's automotive affairs. To match existing British and American levels, Germany's car pool had to increase from its current half million to three million or twelve million vehicles respectively. "The statement that [the present state of affairs] reflects our people's general living standard or its economic and technical capacity is utter nonsense," Hitler lectured the nation's auto managers, swiping aside the sector's long-standing commercial difficulties. In addition to raising issues of national prestige, the comparative scarcity of automobiles on Germany's roads posed a social problem in Hitler's eyes. That "millions of decent, industrious, and hardworking fellow citizens" could not contemplate purchasing a car allegedly left the führer with a "bitter feeling." It was high time, he went on, for the automobile to lose its "class-based and, as a sad consequence, class-dividing character" by affording "ever greater masses of the people the opportunity" to buy a vehicle. Bringing the car within the reach of broader sections of society, Hitler pointed out, would broaden access to a "useful" object and tap a rich "source of joyful happiness on Sundays and during holidays," which currently remained the preserve of the social elite. In short, the government aimed to transform the automobile's status within German society from a tool of social exclusion into one of cohesion.

Responsibility for this transformation rested on the shoulders of the car manufacturers, as Hitler saw "the German car industry's most important task [as] pushing ahead the construction of the car that will inevitably open up a group of new buyers that runs in the millions." In particular, such a vehicle had to "adapt its price to the spending power of millions of potential buyers." As he called on the motor industry to design a cheap car for Germany's comprehensive motorization, Hitler outlined how he anticipated his extravagant demand to be implemented. The car sector, he urged, should model its approach on the radio industry, which, between May and August 1933, had formed a consortium under the auspices of Joseph Goebbels's propaganda ministry to develop and produce a basic radio set for the low-end market.[28] Officially named the "people's receiver" *(Volksempfänger)*, this simple radio cost 76 reichsmarks—roughly a quarter less than conventional sets. The Volksempfänger sold like hotcakes, attracting around half a million orders within four months of its launch, in part because an installment plan allowed customers to spread payments. While this wireless remained altogether uncompetitive in international markets dominated by far more sophisticated, similarly priced American radios, it outsold rival models in Germany by a wide margin. As it boosted the regime's ability to beam its propaganda messages into the home, the "people's receiver" brought the radio within reach of Germans who had previously regarded this consumer item as an unaffordable luxury.[29]

By referring to the Volksempfänger in a speech on an economical car, Hitler did more than recommend to the auto industry a collaborative approach to design and production that had proved successful in a different context. He also aligned the cheap automobile with a growing number of consumer products through which the regime sought to demonstrate how the emerging "people's community" would raise the average German's living standard. Over the years, reports about a whole range of "people's products" appeared in the press, all of them, propagandists never tired of repeating, proof of National Socialism's commitment to consigning an economy characterized by shortages to the dustbin of history. The "people's fridge," the "people's television set," the "people's apartment," the "people's tractor," and other projected commodities, the government promised, would follow the "people's receiver" and usher the members of the *Volksgemeinschaft* into a novel era of affluence and technological modernity. The "people's car" thus formed part of a wider vision of a National Socialist consumer society. Set against the background of the recent, profound slump, with mass unemployment and a marked fall in incomes, the prospect of a host of "people's products" clad the regime in a mantle of economic boldness whose utopianism was thoroughly in keeping with the Nazis' self-styled image as national revolutionaries. Even while most people had been confined to watching cars pass by in the street or admiring them at racetracks, exhibitions, and in glossy magazines, the regime promised to turn the average German from a spectator into a practitioner of modernity by widening automobile ownership.[30]

In his speech at the 1934 Berlin auto show, Hitler had avoided the term *Volkswagen,* or "people's car," but many knew that this was what he had meant. "Who will build the people's car?" asked car magazine *Motor und Sport* once the exhibition closed.[31] This question immediately gave the car industry a serious headache, which became even bigger when it learned of the specifications for the vehicle drawn up by the Reich transport ministry. By the next summer, officials demanded to see the prototype of a car for four to five passengers that cost no more than 1,000 reichsmarks, traveled at a top speed of fifty miles per hour, consumed between four and five liters of fuel over 100 kilometers, and could easily be converted for military use by mounting a machine gun on it.

The government signaled unmistakably that, in technical terms, the "people's car" was intended to differ fundamentally from the small, cramped vehicles at the bottom end of the Weimar auto market, which, while traveling at far lower speeds, had found it hard to accommodate four passengers. Rather, the regime aimed at bringing out a family car at an exceptionally attractive price.

Germany's car manufacturers may have been delighted at the government's general pro-car stance as well as the tax cuts enacted in 1933, but the call for this affordable vehicle struck the business community as fundamentally unworkable. While the request for military adaptability could be accommodated by designing an alternative body, several other stipulations put the auto industry in a quandary. No one in the trade deemed the regime's deadline realistic. Moreover, new small cars at the time started with dealership prices of around 2,000 reichsmarks. Of course, German auto executives understood that low purchase and maintenance costs were a precondition for a mass market for automobiles, but how a car selling at half the current rate could possibly generate profits exceeded their imaginative capacities. As to how such a vehicle was to be manufactured on a scale untried in Germany, and who would finance such a vast investment—those were issues the industry did not even begin to contemplate in 1934. In short, the automobile entrepreneurs found Hitler's request for a "people's car" altogether unrealistic.[32]

Rather than risk Hitler's ire by voicing their concerns openly, the country's leading manufacturers played for time. They instructed their lobbying body, the Reich Confederation of the German Auto Industry (RDA), to endow a research society with 500,000 reichsmarks and passed the task of developing a prototype to an engineering consultancy. A transparent hope lay behind this move. In light of Hitler's ambitious request, an independent expert would confirm the brief's impracticality and prompt the regime to modify or abandon its quest for a "people's car." RDA expected the consulting process to reach a foregone conclusion, not least because of the vehicle's low stipulated price. In this, the auto industry seriously underestimated the dynamism the project was to gain over the next few years. As a matter of fact, the manufacturers inadvertently played a decisive role in lending the project a material foundation because they approached an engineer who had no intention of entering the history books as the undertaker of the "people's car."[33]

A native Austrian born in 1875 and who, like Hitler, pursued his high-flying ambitions in Germany, Ferdinand Porsche was an aging man in a hurry when RDA contacted his consultancy firm in Stuttgart in 1934. Before 1918, he had won his automotive spurs with designs for luxury vehicles and a tractor for heavy field artillery for the Austrian army. After the collapse of the Austro-Hungarian Empire, he moved to the Daimler works in Stuttgart, where he oversaw the development of the supercharged SSK Mercedes that aroused attention on Europe's racetracks in the late twenties and early thirties. Because of severe cost overruns and technical glitches in experiments for a small car, he lost his position as engineering director at Daimler-Benz in 1929. Porsche was undoubtedly a resourceful engineer, but he was also known for his explosive temperament and disregard of financial discipline. While the latter qualities left him, at the age of fifty-four, with a reputation in industrial circles as an unemployable perfectionist, his dedication to building technically sophisticated cars inspired the loyal team of imaginative auto engineers that he assembled in his freelance consultancy firm in Stuttgart in the years after 1931.[34]

Although RDA's strategy initially appeared to bear fruit, it failed to derail the "people's car." Porsche and his team could draw on previous studies they had conducted on inexpensive economy cars, for motorcycle manufacturers Zündapp and NSU, but when they presented a proposal conceived around a two-stroke, one-liter engine in June 1935, they could not conceal that this vehicle would retail at 1,400 to 1,450 reichsmarks. A technical commission convened by the German auto industry concluded that Porsche had failed to meet the target price, which remained at 1,000 reichsmarks. This setback, however, did little damage to the project or Porsche's role in it, because by the middle of 1935 the engineer was well on his way to becoming Hitler's confidant in technical matters. Porsche stood high in Hitler's esteem for rewarding the dictator's trust in May 1933, when the engineer was among a delegation from Auto Union that lobbied the new head of government for a subsidy for auto racing. The Auto Union car Porsche's development team masterminded with the funds won competitions within a few months, thus laying a foundation for the jubilations in the press over Germany's supposed revitalization that accompanied successes on Europe's racetracks.[35]

Upon his installation as chief designer of the "people's car," Porsche took great care to cultivate cordial relations with the führer. He reported enthusiastically on the project, studiously avoiding mention of the numerous technical problems that, for instance, beset his team's efforts to design a lightweight, robust, economical engine. While behind closed doors the skeptical representatives of the leading car manufacturers ridiculed the "people's car" as "the führer's pet idea," Porsche must have struck Hitler as the epitome of the "man of action" whom the National Socialist press venerated for his energetic persistence in pushing aside daunting obstacles. In addition to a selective approach to the truth, Porsche possessed an assured instinct for impressing the leader. On July 11, 1936, the engineer landed a particular coup when he secretly took two advanced prototypes to Hitler's Bavarian mountain retreat at Obersalzberg for a display to a select circle of party grandees including Hermann Göring and head of the autobahn scheme Fritz Todt. The occasion left a lasting impression on Hitler, who still reminisced almost six years later, in the middle of World War II, about how "these people's cars [had] zipped up and down the Obersalzberg, whizzing around his large Mercedes automobile like bumblebees."[36]

RDA, meanwhile, grew increasingly alarmed as Porsche demanded ever more funds for a vehicle that, as the German car industry gradually realized, had the potential of turning into a technologically sophisticated competitor in the small-car market. When the federation of German car manufacturers eventually extracted itself from its contract with Porsche in 1938, it had subsidized the project with no less than 1.75 million reichsmarks, a princely sum that allowed Hitler's engineer-of-choice to put his previously struggling consultancy on a financially sound footing. RDA, which had attempted to torpedo Porsche with several critical memos, realized it had no choice but to instead accommodate him with virtually open funds so as not to exacerbate tensions with Hitler. In 1936, the führer returned to the Berlin auto show to vent his frustration at the lack of enthusiasm among the nation's car managers. "I have given orders to pursue the preparations for the creation of the German people's car with relentless determination," he declared. "And I will bring them to a conclusion, and that, gentlemen, will be a successful conclusion."[37]

Two important, intimately linked issues, however, presented seemingly intractable problems. None of Porsche's experimental cars could be produced and distributed for the 1,000 reichsmarks on which Hitler insisted with an eye toward securing broad popular appeal. This key obstacle rendered it profoundly unclear which concern would manufacture the automobile. After all, the project promised nothing but losses. The resolution to both questions was characteristic of Hitler's disregard for economic constraints in general. In the euphoria that followed the demonstration of the two prototypes at Obersalzberg on July 11, 1936, the dictator decreed a retail price of 990 reichsmarks—a highly symbolic amount, which may have signaled the vehicle's affordability, but lacked all financial viability. On the same date, the regime's inner circle also decided to exclude existing automakers from the production of the "people's car." Instead, they planned to erect a new factory with an annual capacity for three hundred thousand automobiles that would start delivery by early 1938.

Rather than alert Hitler to the financially unrealistic nature of this plan, Porsche sensed an opportunity to pursue a personal ambition that extended far beyond his task as the prototype's designer. "I cannot help gaining the impression that Porsche is nurturing the dream of becoming the technical director of a big specialized works for the construction of the people's car," Robert Allmers, the head of Daimler-Benz, had observed several months earlier. Hitler's favorite engineer, in other words, had made no secret of his hope that the leader would create a new factory for the "people's car." As the project's most loyal supporter, Porsche clearly had good chances to run this facility. In the autumn of 1936, he traveled to Michigan for an inspection tour of the world's most advanced car factories. The Ford works at River Rouge attracted his particular curiosity.[38]

The decision to build a vast factory for the "people's car" reveals how, in their ambition and joint defiance of economic fundamentals, Hitler and Porsche complemented each other. Porsche's prototype assured Hitler that the vision of a mass motorized "people's community" lay within the regime's reach. Meanwhile, by offering Porsche a dual role as the designer of the "people's car" and the manager of its production site, the Nazi government presented the engineer with the career prospect of a lifetime: he could turn himself into Germany's Henry Ford. While political obsessions fueled Hitler, Porsche was primarily driven

A celebratory portrait of Ferdinand Porsche (1875–1951), who oversaw the creation of the "Strength-through-Joy Car" in the Third Reich. Porsche became one of the most prominent engineers in Nazi Germany. Bundesarchiv, image 183-2005-1017-525.

by technical aspirations. As he took great care to set himself aside from Nazi officials by always donning a three-piece suit rather than a party uniform in public situations, Porsche cultivated the image of the apolitical expert. Irrespective of his exterior appearance, he was by no means a politically neutral engineer whom a racist government manipulated to secure his collaboration. On the contrary, Porsche became an influential player within the regime itself, actively seeking Hitler's support as he pushed the "people's car" ahead because he was fully aware that the dictatorship offered him opportunities that private enterprise could not.[39]

With Hitler's backing, Porsche shrewdly navigated between the Third Reich's numerous power centers and placed his consultancy in a position to develop a vehicle that closely resembled the later Volkswagen. Although this car has come to be most closely associated with Ferdinand Porsche's name, it owes its existence to numerous specialists at Porsche's consultancy who were experts in axles, transmissions, suspensions, engine design, and other areas. Erwin Komenda, for instance, focused on the car's distinctive rounded body. The Stuttgart-based engineers did not work in a conceptual vacuum. They borrowed heavily from Béla Barényi, a twenty-seven-year-old unemployed engineer of Hungarian extraction, whose sketches for a small, rotund economy automobile with an air-cooled back engine had appeared in the French automotive press in 1934. The air-cooled engine and streamlined body design owed crucial intellectual debts to the Tatra T97, first manufactured in Czechoslovakia in 1937. Meanwhile Hitler had little direct influence on the car's design. His initial objections to a rear engine left no imprint on the final product, and the prototype's silhouette bore no resemblance to a pencil sketch the dictator probably made during a meeting with Porsche in March 1934.[40]

The vehicle that emerged from the workshop in Stuttgart was a technically sophisticated economy car. Its heart was a four-cylinder, one-liter boxer engine in the rear, readily accessible and easily exchangeable by loosening two fastening screws. Developing up to twenty-three horsepower, it was, by the standards of the day, a powerful aggregate for a small vehicle measuring a little over four meters in length. Placing the engine in the back delivered torque directly to the rear axle and ensured that sufficient pressure rested on the rear wheels to improve grip. The compact boxer design with its short, horizontally opposed cylinders allowed pistons to move at comparatively low speeds. In addition to preserving precious space, this arrangement reduced wear and tear, thereby guaranteeing longevity as well as low running costs. The engineers had opted for an air-cooled design to limit the car's weight to around 1,000 kilograms and enhance its fuel efficiency, which stood at around seven liters per 100 kilometers. The air-cooled technical solution also meant that the coolant could neither boil nor freeze, allowing the engine to function under extreme weather conditions while diminishing the need to shelter the vehicle in subzero temperatures. A car with this type of engine, which promised sturdy performance along with simple upkeep, undoubtedly recommended itself for subsequent military conversion. First and foremost, however, the vehicle's technical features acknowledged complaints about cumbersome maintenance requirements, as well as the financial strictures faced by the wider population, both factors impeding Germany's motorization in the twenties. In addition to minimizing operating costs, an air-cooled automobile could be parked outside in the winter, an important consideration in a country where garages were in notoriously short supply. Funded lavishly in its design stages, the "people's car" responded to the culture of material scarcity that pervaded interwar Germany.[41]

Although the brainchild of Porsche's engineers heeded the need for low running costs, it could by no means be dismissed as a cheap construction. While high ground clearance and a closed, sturdy underbody allowed the car to traverse difficult terrain, its individual wheel suspension, which featured torsion bars at the front, promised good performance on the road. Equipped with four gears, the creation of Porsche's consultancy held out the prospect of smoother rides than conventional economy vehicles restricted to two or three gears. Thanks to its ability to maintain top speeds over sixty miles per hour over long distances, the "people's car" was ready to hit the autobahn. The engineering team had also incorporated a heating system as a standard feature, a costly extra for virtually all other European automobiles of the thirties. In terms of exterior appearance, the all-steel body—by no means yet the international norm, even for upmarket American automobiles in the thirties—was the vehicle's most striking characteristic. Designed to accommodate up to five people, the rounded, aerodynamically tested hull lent the small automobile flowing, streamlined contours that took up aesthetic elements of the international art deco style.[42]

In the summer of 1937, Porsche arranged for the most extensive and expensive road trial a German automobile had undergone up to this point. Conducted with thirty prototypes made by Daimler-Benz in Stuttgart, the tests covered over 2.5 million kilometers, cost roughly half a million reichsmarks, and confirmed that Porsche's team had assembled a technically and aesthetically ambitious product. Compared with the boxy vehicles that dominated the small-car segment during the Third Reich, the projected "people's car" boasted a far more modern look, achieved higher speeds, and rested on an exceptionally solid construction. Next to other European economy automobiles that were either available or close to reaching the market in the late thirties, the German "people's car" stood out, too. It promised considerably more space and comfort than the two-seater Fiat 500 with its water-cooled half-liter front engine, which the Italian population nicknamed *Topolino* or "little mouse" soon after its launch in 1936. And the 2CV that Citroën was developing especially for French rural drivers since 1937 featured a two-cylinder front engine with a three-gear transmission to propel a vehicle weighing less than four hundred kilograms at a top speed of around thirty miles per hour—an approach that prioritized lightness over the robustness favored by the German engineers from Stuttgart. While Fiat and Citroën opted to build economy cars that incorporated pared-down automotive technology, the engineering that went into Porsche's prototype bore few traces of economy. As they aimed to keep running costs for drivers low, the technicians of the "people's car" had adopted a far more technically ambitious solution than their Italian and French colleagues. Of course, this approach to automotive engineering also provided one reason why Porsche's design proved impossible to produce at 1,000 reichsmarks.[43]

Since Hitler delayed a decision about a production facility throughout 1936, it remained unclear where and how the car would be built until February 1937, when Robert Ley, the head of the German Labor Front (Deutsche Arbeitsfront, or DAF), brought his organization into play. Founded as an authoritarian replacement to the banned trade unions in May 1933, DAF focused on labor management issues and became the party's largest agency, in which virtually the entire "Aryan" workforce was required to enroll. With its vast fee-paying membership as well as properties confiscated from the trade unions, the German Labor Front was one of few institutions that could afford to be involved in the project for the "people's car." In the course of 1937, Ley oversaw several key decisions. The plant, he stipulated, was to be built in the deep countryside about forty-five miles east of Hanover, close to Fallersleben in the Lüneburger Heide, on a site with good transport links because of its proximity to the Mittellandkanal (the central German ship canal) and a railroad line connecting the Ruhr area and Berlin. He also dispatched Porsche along with several DAF officials on a second tour of the United States, where the visiting party inspected car factories, met Henry Ford, and hired several experienced engineers of German extraction from the Ford works. Ley played a central role in approving the 170 million reichsmarks required to build a gigantic plant. Featuring its own power station, foundry, rolling and casting facilities, mechanical workshop, body shop, stamping presses, tool shops, and an administration building, the works were planned for a yearly output rising from 150,000 vehicles in 1939 to 450,000 within five years, eventually aiming for an annual target of no fewer than 1.5 million cars. DAF envisioned a vertically integrated factory of truly Fordist proportions, whose capacity for mass production not only would make the works near Fallersleben Europe's largest auto plant by a wide margin, but also aimed to exceed the capacity of the River Rouge works by

The "Strength-through-Joy Car" on display in Berlin in 1939. Developed by Ferdinand Porsche with support from the Nazi regime, this automobile incorporated most of the features that were to make the Volkswagen a commercial success after World War II. Postcard, collection of Achim Bade, Munich.

which it was inspired. Since this new manufacturing center was situated in a rural, almost unpopulated area, the regime also commissioned an architectural firm in March 1938 to develop a model city for a population of thirty thousand to sixty thousand inhabitants.[44]

In addition to raising the public profile of Ley's organization, the factory scheme corresponded well with DAF's stated purpose to transform industrial relations in the workplace. Erecting a vast manufacturing site from scratch offered a rare opportunity to show how the "plant communities" *(Betriebsgemeinschaften)* held up as the new ideal form of labor relations would foster a collaborative spirit between management and workers. The new factory, DAF speculated, would demonstrate how "workers of the fist and workers of the head" producing the "people's car" could bring about a hierarchical, yet harmonious and cohesive "people's community" in the workplace. The German Labor

Front also took the *Volkswagen* project under its wing because the automotive scheme sat well with the organization's involvement in matters of consumption. Through its programmatically named division "Strength through Joy" (Kraft durch Freude, or KdF), DAF sponsored countless spare-time activities targeted at employees of limited means. Low-priced outings and holidays made up KdF's most successful offerings, designed to reenergize workers with the ultimate aim of raising their productivity. Before the war, KdF developed into Germany's largest travel agency, organizing trips for a total of fifty-four million Germans, including a number of highly publicized cruises to the Mediterranean, Norway, and Madeira. Assuming a prominent role in the motorization program, which promised to enhance the leisure time of millions of Germans, thus corresponded with KdF's self-proclaimed commitment to "social justice" or "real socialism." In the second half of 1937, KdF officials, therefore, began to take the lead within the German Labor Front in pushing ahead the "people's car" as a "social project."[45]

On May 26, 1938, a bit more than two months after German troops had marched into Austria, the Third Reich's leadership assembled on a forest clearing near Fallersleben to lay the foundation stone for the factory. Fifty thousand spectators, most of whom had been transported to the deep countryside by special trains, set the stage for the hour-long ceremony broadcast live on national radio. When Hitler arrived amid "clarion calls and shouts of *heil* . . . the SS had much difficulty" in controlling the crowd, since "everybody streamed and pushed ahead" to catch a glimpse, an excited local schoolgirl wrote. In the cordoned-off area reserved for Hitler and his entourage, three models of the "people's car"—a standard limousine, a limousine with a retractable canvas roof, and a convertible—gleamed in the sunshine, strategically placed in front of the wooden grandstand draped with fresh forest green from which the party grandees delivered their speeches. Robert Ley, the first speaker, ended on a theme that became an omnipresent propaganda trope, calling "the factory for the people's car one of [Hitler's] favorite works." After a KdF official had outlined the vehicle's technical characteristics, the führer took to the stage and emphasized his longstanding dedication, claiming in the early stages his idea had been routinely dismissed as "impossible." Buoyed by his recent foreign political success and preparing to lay the foundation stone for the factory, he triumphantly declared: "I hate the word impossible." In recognition of Strength through Joy's support, the dictator named the vehicle to be manufactured at the site "KdF-Wagen" (KdF Car) before expressing the hope that the entire project would shine as "a symbol of the National Socialist people's community." To bring proceedings to a close, Hitler was driven through cheering crowds to the railway station in the passenger seat of the convertible next to Ferdinand Porsche's son at the steering wheel, while the chief engineer rode in the back.[46]

Over the next year, Strength through Joy unleashed a public-relations storm to promote the car. In addition to issuing a special stamp and inviting journalists for test drives, KdF displayed prototypes at offices, factories, exhibitions, party gatherings, and automotive competitions. The party organization also courted the international media, for instance securing favorable coverage in the *New York Times,* which showed itself impressed by the prospect of autobahns crawling with "thousands and thousands of shiny little beetles," thereby incidentally offering the earliest use of the moniker that was to emerge as the car's endearing nickname after World War II. Most important, however, to drum up demand, a stream of writings assured the German population of the car's impeccable quality. Issued with an initial print run of half a million copies and costing a mere 20 pfennigs, the brochure entitled *Your KdF Car* stressed that many of the vehicle's design features, including the engine's location in the back and

Hitler delivering his propagandistic speech on May 26, 1938, when he laid the foundation stone for the factory intended to produce the "people's car." Ferdinand Porsche is on the right. Fresh tree branches decorate the stage above, which holds a marching band in front of a huge swastika. Bundesarchiv, image 183–H06734.

The German post office issued this special stamp as part of the regime's wider propaganda campaign to popularize the KdF Car. The image emphasizes the spaciousness, speed, and comfort of the vehicle, which is shown cruising on the autobahn. Collection of Achim Bade, Munich.

the suspension system with torsion bars, were "almost exact replicas" of those Porsche had used in his Auto Union racing cars. In the Third Reich's firmly controlled media environment, publication upon publication raved about the KdF Car's excellent performance on the road, agreeing that the vehicle amounted to a "technological miracle."[47]

The automobile's technical quality, several columnists emphasized, reflected the regime's pledge to offer an affordable car because only a reliable product kept maintenance costs at a minimum. Due to its extensive testing phase, the vehicle would suffer none of the "infant's diseases" that had so frequently frustrated

Die Deutsche Arbeitsfront

KdF-WAGEN-SPARKARTE

NR. 1/606. Voraussichtliches Lieferjahr

Wagenführ Karl

Vor- und Zuname (bei Frauen auch Geburtsname)

Wohnort: Mchn. Poststation: Mchn.

Straße: Adalbertstr. Nr. 82/2.

Geboren am: 11.1.10 in: Mchn.

Genaue Berufsangabe: Vermessungsassessor

Besitzt Führerschein: ja Klasse: 3 b.

31. März 1939

Diese Karte ist ausgestellt am:

von der Kreis-Dienststelle:

Gau: Gebietsdienststelle München-Nord Augustenstraße 18

(Unterschrift des Ausstellers) (Dienststempel)

Volkswagen-Werk

Every German wishing to acquire the KdF Car had to save up for it in advance. The credit of each saver was documented in a booklet into which would-be owners glued little red stamps, one of which can be seen in the bottom right corner. Each stamp was worth five reichsmarks. Savings book, collection of Achim Bade, Munich.

drivers in the twenties, the argument ran. The regime's commitment to low-cost motorization manifested itself in the car's purchase price of 990 reichsmarks, as well as the installment plan the administration had devised. To acquire a "people's car," Germans joined a savings scheme, paying in 5 reichsmarks weekly until they reached the delivery price. While this plan was highly advantageous for Strength through Joy because it raised capital before production began, Ley stated his organization's intentions in altogether different terms: "In Germany, there is not supposed to be anything the German worker cannot partake of."[48]

The vehicle itself struck propagandists as ideally suited to complement Strength through Joy's profile as a leisure organization. As *Arbeitertum,* the German Labor Front's official magazine, explained, the "KdF Car [was] destined to offer every German member of the people's community relaxation and pleasure in their free time." Another publication praised the automobile because it would bring "strength, happiness, and pleasure to millions of people who have up to now been forced to do without them." Numerous illustrations emphasized this theme. Color drawings set the automobile in front of a background of mountain scenes to hint at possible outings to Alpine regions. Public-relations photographs, meanwhile, encompassed scenes of family and friends relaxing in forest clearings or next to their tent by the lakeshore, enjoying the countryside thanks to the KdF Car.[49]

Advertisements along these lines not only took up deeply conventional motifs that were ubiquitous in the promotion of holiday destinations as well as in prospectuses for automobiles and motorcycles; they also portrayed car-related leisure as altogether apolitical and hence free from the regime's ideological preoccupations. Although the "people's car" owed its existence to deeply ideological impulses tied to racial regeneration, promotional material consistently played up its apolitical benefits. Driving held the key to a host of private pleasures, the regime's

The party repeatedly praised the "Strength-through-Joy Car" as a source of recreational pleasure. This promotional photo staged the car beside a lake, with a couple resting on their air mattresses in front of a small tent. Bundesarchiv, image 146-1988-019-16.

propaganda insisted. Nonetheless, militarist and other political motifs occasionally crept into ads composed along seemingly apolitical lines. A drawing in a KdF prospectus, for instance, staged a black "people's car" in front of a detached house, whose male proprietor tends to his flower garden while a little boy in the foreground gazes longingly at the automobile. Irrespective of its domestic subject matter, this scene of middle-class bliss bears a bellicose trace. At first sight, the small boy, possibly the car owner's son, appears a picture of innocence, but by pulling along a little toy cannon he nonetheless betrays Nazi Germany's militaristic culture.[50]

While it is notoriously difficult to gauge what the state of public opinion was in Nazi Germany, owing to the regime's tight control of the public sphere, plenty of evidence exists that the German population responded to the vehicle with strong and

The Third Reich launched an extensive publicity campaign to promote the "people's car." This illustration from a contemporary prospectus celebrates the "limousine" as an integral part of middle-class domestic bliss. Courtesy of Volkswagen Aktiengesellschaft.

positive interest. The son of a worker recalled his enthusiastic reaction to the car's announcement in the following manner: "I was totally fascinated. You know, that everybody, every family could afford a car over time, that's great what they are planning." In a secret report, a critical Social Democrat went so far as to attribute to German society a "KdF Car psychosis. For a long time, the KdF Car was the main topic of conversation among all sections of the population." Soon after Hitler had laid the foundation stone for the factory, *Motor und Sport,* a leading car magazine, begged readers to stop inquiring about the KdF Car because editorial staff could no longer cope with the deluge of letters. Between early August 1938 and the end of 1939, about 270,000 people opened a savings account with KdF—a considerable number in a country that commanded only 1.1 million registered automobiles in early 1938.[51]

While many joined the savings scheme for work-related reasons, hoping to increase their earning power through a car, others were undoubtedly drawn to the automobile as a source of enjoyment and fun. After all, the automobile had lost none of its appeal as an elusive instrument of leisure since the end of the Weimar Republic. A student from Munich, who had unexpectedly been taken along for a car ride in the countryside by a wealthy friend in 1935, could hardly contain his elation in a letter to his parents: "We returned to Munich at 8 pm. We are completely thrilled by the drive. The things we have seen! I could never have visited all these places by rail. In one afternoon, we have enjoyed the entire beauty of . . . upper Bavaria in a relaxed, completely untaxing drive." For Germans with limited disposable incomes, yet good prospects like this young man, an automobile for 1,000 reichsmarks offered a deeply alluring aspiration.[52]

Although the KdF Car attracted over a quarter million savers in less than eighteen months, the initiative was ultimately an economic failure. Since the Nazi leadership had disregarded basic accounting principles, a provisional calculation from 1939 concluded that a production run based on orders placed so far would generate a ruinous loss of 1,080 reichsmarks per vehicle—more than the automobile's projected retail price! The German population generated nowhere near the demand required for the volume of production that would have brought manufacturing costs even remotely within the proximity of the official price.[53]

Irrespective of strong interest, order numbers remained low in part because many Germans who could afford a car lacked faith in the regime. While some potential customers hesitated to pay in advance for a car whose production facility had not yet begun operations, others regarded the Third Reich's increasingly aggressively foreign policy with growing concern. And when war broke out, applications for KdF Cars collapsed almost completely.

Between 1940 and 1945, a mere seventy thousand new savers enrolled in the scheme.[54] Above all, however, the vast majority of the German people were simply too poor for a car. Despite full employment thanks to a wider economic recovery driven by re-armament, incomes did not increase significantly in comparison to the Weimar years. In the mid-thirties, when 83 percent of the population earned less than 200 reichsmarks per month, few blue-collar families with dual incomes commanded more than 225 reichsmarks. While urban workers struggled materially, living standards in the countryside recovered from the slump's immediate effects, with help from rising agricultural prices as well as sweeping government-sponsored programs to write off debts. These gains, however, did not raise farming incomes to a level that would have allowed Germany's rural population to play the pivotal role their counterparts had assumed in the mass motorization of the United States. Rather than save for a car, people on lower incomes were far more likely to ogle one of the motorcycles, whose numbers increased from 894,000 in 1934 to 1,582,872 in 1939. On the eve of World War II, most of these machines had engines below 200 cc, cost less than 500 reichsmarks, and attracted an overwhelmingly blue-collar clientele. The motorbike thus qualified far more as a "people's vehicle" than any automobile. Above all, however, Germany remained a country of bicycles. In 1939, there were about twenty million of them.[55]

Even the middle class on monthly salaries between 300 and 360 reichsmarks, who emerged as the core clientele of the KdF scheme, could only keep up their contributions by cutting back in other areas. As a result, the project widely hailed to provide a "people's car" turned into a program directed almost exclusively at an upper-middle-class minority that lived in relative material comfort. Tellingly, a third of those who saved for a KdF Car already owned an automobile. The failure of the plan to broaden access to the automobile became clear almost immediately. In 1938, the secret report on the mood in German society compiled by the Security Service of the SS did not mince its words: "Orders for the KdF Car have so far not met with expectations. As a result, the first annual production could not yet be sold. The participation of workers ranges between 3 to 4 percent."[56]

Those Germans who paid into the scheme were to be disappointed. Before the Third Reich collapsed, the company founded under the name Volkswagenwerk in September 1938 produced no more than 630 KdF Cars in its vast factory, most of which were delivered to prominent regime leaders. That the savers were to be disappointed was not due to a lack of dedication on the part of Ferdinand Porsche, who assumed a critical managerial role during the factory's construction phase in 1938 and 1939, pushing the works toward completion through his high-level contacts within the political administration. While the regime channeled material resources and manpower toward sectors of direct military importance, Volkswagenwerk's relevance to the coming war effort was a bone of contention. Amid chronic currency shortages, it proved difficult to import the specialized equipment required for a highly rationalized auto plant. Construction material including cement, steel, and glass were also in short supply. When the military commandeered the majority of workers erecting the factory to accelerate the completion of fortifications on Germany's western border, construction near Fallersleben could proceed only because the German Labor Front secured the approval of Mussolini's regime to hire six thousand Italian laborers. Visitors to the site in late summer 1939 would have been impressed by the factory's huge exterior front measuring 1.3 kilometers (0.8 mile) in length, but the halls behind the redbrick façade still lacked much of the necessary manufacturing equipment. The rudimentary state of the adjacent settlement provisionally named "City of the KdF Car" revealed, particularly dramatically, the severe overstretch of the German economy in the run-up to the war. In 1939–1940, some 80 percent of its inhabitants lived in wooden barracks because no more than 10 percent of the program for residential dwellings could be completed before building activities ceased in 1941.[57]

When Nazi Germany invaded Poland in September 1939, the Volkswagenwerk management jettisoned plans to manufacture the KdF Car and shifted its focus to preserving the plant's independence within the war economy. In the hope of reconverting the factory to its civilian purpose after the war, Ferdinand Porsche relied heavily on family members to safeguard the operation of the factory during the war. While his son Ferry retained a visible presence, his son-in-law Anton Piëch played the most important role, taking over operational command at the works in May 1941. During the war, the factory attracted a variety of armament contracts, which, motor vehicles aside, ranged from the production of stoves, mines, and antitank weapons to the manufacture of wings for the medium-range Ju 88 bomber. Having selected the factory as a repair shop, the Luftwaffe also chose Volkswagenwerk as the main production site for the V-1 flying bomb, thousands of which Germany launched as "vengeance weapons" against Britain in the later stages of the war.[58]

To guarantee steady output, Porsche and Piëch, like virtually all German industrial employers in war production, lobbied the administration for forced labor. As the number of workers rose from 6,582 to 17,365 from December 1940 to April 1944, the rate of foreign nationals stood at 60 percent, a proportion about twice as high as in the rest of the weapons industry. Their working conditions reflected their place within the Nazi racial hierarchy. While Danish and Dutch workers frequently arrived on regular work contracts, POWs, political prisoners, and forced laborers from eastern Europe made up the bulk of those on the lines. Eastern Europeans and political prisoners were particularly likely to lead lives characterized by long hours, dangerous work, insufficient nutrition, inadequate clothing, overcrowded housing in cold, drafty wooden barracks, negligent medical care, and incarceration in camps. They not only experienced random violent outbursts by guards but suffered from acts of casual sadism like the ones perpetrated by a canteen cook who laced kitchen leftovers with glass shards so undernourished inmates injured themselves while rummaging for food. Polish national Julian Banaś, who arrived at Volkswagenwerk against his will in October 1942, soon gained clarity about his position: "I understood then that I am a slave. . . . The first days in KdF City made me understand that I was an object. An object that can work." Decades after his ordeal, "the feeling of being in a situation over which one has no influence whatsoever" still haunted him.[59]

To be sure, the aim of operations around the factory was industrial production rather than political and racial persecution, but chronic neglect in conjunction with arbitrary mistreatment had murderous consequences. Ranking low in National Socialism's racial hierarchy, eastern European prisoners counted among those with the worst experiences. The workforce included over one thousand young Polish and Belorussian women, who arrived undernourished, often lacking even shoes and winter clothing. Most of them had fallen into German hands in waves of random arrests by the SS in their home countries. Some female forced workers either reached KdF City with undetected pregnancies or became pregnant in the camps, where male and female prisoners could visit each other. From the middle of 1944, eastern European women were forced to give up their children soon after birth to a nursery in the nearby village of Rühen. The conditions in the ward for newborns at Rühen defied belief, a British prosecutor explained in 1946: "At night the bugs came out of the walls of those barracks and literally covered the children's faces and bodies. It has been described by one witness as a living ant hill. . . . Some children had as many as thirty to forty boils or carbuncles on their bodies." Undernourished and surrounded by vermin, all of the roughly 300 to 350 newborns who arrived in Rühen between July 1944 and April 1945 quickly died of infections. Next to Rühen, the concentration camp in nearby Laagberg stood out for a sadistic regime maintained by a commander who systematically worked some of the six hundred to eight hundred inmates, mainly political prisoners, to death. If Volkswagenwerk survived the war as an operational manufacturing site, it did so only because an overwhelmingly foreign-born workforce of forced labor maintained production amid countless human rights abuses.[60]

Military vehicles featured prominently among the wartime products made by the enslaved workforce. Before the outbreak of World War II, experiments with a military conversion of the KdF Car had stalled because of tensions within the army's procurement service, as well as technical problems. In the fall of 1939, Porsche's team resumed work on a new Wehrmacht commission that passed a series of official tests in January 1940. Compared with the KdF Car, the military vehicle possessed higher ground clearance, a more powerful transmission, wider wheel gauge, and rims suitable for broader, more robust tires. Most notably, it was fitted with a lightweight angular body with a retractable canvas roof. After it had become evident that the Wehrmacht was woefully underequipped with light automobiles, the military command placed over sixty-five thousand orders for the modified vehicle, along with an amphibious version fitted with a propeller that could traverse rivers. Attracting the popular moniker *Kübelwagen* (bucket vehicle) for the shape of its seats, this car established a prominent presence in numerous war theaters.[61]

Producing the Kübelwagen played a significant role in the survival of Volkswagenwerk as a going concern during the war. Demand from the Wehrmacht gave management the chance to invest in the plant, complementing the existing machine park by installing a paint shop, additional conveyor belts, as well as new automated lathes, drills, and presses. By 1942, the works had turned into a plant that partly implemented rationalized manufacturing procedures while producing an automobile based on the KdF Car. As thousands of forced laborers turned out tens of thousands of Kübelwagens, the factory demonstrated its potential in a first and tentative phase of Fordist mass production under wartime conditions. Crucially, this initial stage drew to a close only with the arrival of American troops in the "City of the Strength-through-Joy Car" on April 11, 1945. Although Allied air raids had killed dozens of workers in the summer of 1944, car production continued until the very end of the war. The aerial attacks damaged numerous buildings, but they destroyed only 7 percent of the production equipment. While the management dispersed the manufacture of most weapons to other locations, car assembly remained at the main site, having mostly moved to the lower floor to guard against the effects of bomb attacks. Even in the first three months of 1945, when public order collapsed all over Germany, this arrangement ensured steady production and put the plant in a position to deliver over four thousand Kübelwagens.[62]

In the many theaters of war in which the Wehrmacht operated, this military vehicle proved the maneuverability, reliability, and sturdiness of the civilian design on which it was based. Since it was common knowledge that the Kübelwagen's body sat on the slightly modified chassis of the KdF Car, its presence during the war provided a potent reminder of the regime's pledge to mass motorize Nazi Germany. In the war's early phase, propaganda took up this aspect and articulated soldierly fantasies of

A Kübelwagen in North Africa in 1942. The three Wehrmacht soldiers in the vehicle survey the expansive desert battlefield; a smoke column rises in the background. German service members venerated the Kübel for its robust performance even in extreme environments. Bundesarchiv, image 101-784-0228-29a.

car ownership in the wake of a German triumph. "After the war, one of these little devils [that is, a Kübelwagen] will belong to me. It will be prettier than this: with shiny paint, a nice body, retractable roof, an ashtray and a flower vase, but this I know: its heart will remain the same," a motor journalist turned soldier wrote in 1941. A year later, Goebbels published a newspaper article reminding the German population what they were fighting for. Predictably, his vision of a National Socialist future included the prospect of "a happy people in a country full of blossoming beauty, traversed by the silver ribbons of wide roads, which are open to the modest car for the small man."[63]

Still, as the Wehrmacht's fortunes turned, propaganda visions of postwar affluence receded into the background. As Hitler's armies found themselves on the defensive, the Kübelwagen may well have sparked associations among the military rank and file that were far more in keeping with the theme of a leaflet the Allies dropped over German lines in 1942. Under the headline "Kraft durch Freude!" it contrasted a photo of an enthusiastic Hitler inspecting a miniature model of the KdF Car in 1938 with the picture of two German soldiers' twisted corpses on the Libyan desert ground next to a Kübelwagen.[64] While visually invoking the Nazis' full-mouthed promise of prosperity, this montage staged the "people's car" as the symbol of a deadly betrayal the German population had suffered at National Socialism's hands. Toward the war's end, the "people's car" conveyed more than one message, drawing attention to the regime's pledge to promote mass motorization in an affluent National Socialist Germany and to its failure to deliver on this promise as a result of policies that ended in murderous defeat and genocide.

The quest for a "people's car" formed part of National Socialism's policies that claimed a prominent place in visions of Germany as an energetic, racially homogeneous, highly militarized, and technologically modern "people's community" that brought previously unaffordable consumer goods within the reach of "ordinary" citizens. While auto racing signaled Germany's international competitiveness, autobahn construction, propagandists emphasized, demonstrated the regime's commitment to the "people's community" through a scheme that ensured racial regeneration and national cohesion in the long run. A new highway code granted motorists unprecedented latitude and enhanced the automobile as an instrument of individual liberty. The "people's car" thus fit squarely into the regime's vision of a "traffic community" whose modernity was predicated on mass motorization.

Although motorization policies commanded prominence and attracted considerable resources despite the regime's focus on rapid rearmament, it is striking how many of them ended in failure—and not just because Nazi Germany lost the war. The

British war propaganda leaflet dropped over German lines in Africa aimed to alert Wehrmacht soldiers to the Third Reich's broken promises. In the top image, Porsche (left) and Hitler inspect a model of the "people's car" in 1938. The bottom image, taken four years later, depicts two dead soldiers in the desert next to a Kübelwagen, the military vehicle based on Porsche's design. Postcard, collection of Achim Bade, Munich.

new traffic code required several modifications, including the reintroduction of speed limits in 1939 to reduce a consistently high number of fatalities. The highway construction program did not come close to offering employment on the scale promised and resulted in an incomplete, patchy network. And the "people's car" was founded on wishful thinking because German society did not even remotely command the economic resources to support mass motorization.

Simply dismissing the Third Reich's motorization policy as a fiasco risks obscuring its important postwar legacies, however. The modified highway code of 1937 remained largely in force until 1971. After 1945, the West German autobahn network built on the foundations laid in Hitler's days. Regarding the "people's car," National Socialism left behind a technically sophisticated, extensively tested prototype as well as a vast, debt-free production facility that operated throughout the war, not least turning out military conversions. The regime's propaganda campaigns had also envisioned a future in which mass motorization took prominence, emphasizing the automobile's capacity to enhance manifestly apolitical forms of leisure, including weekend outings and holidays. The enthusiastic reception that large parts of German society accorded to the prospect of universal car ownership affirmed that the National Socialists had successfully tapped a strong and broad social aspiration. While they never oversaw its production, the Nazis thus launched the "people's car" as an enticing and affordable object of private desire.

At the same time, the "people's car" emerged from the Third Reich with a profoundly ambivalent legacy, since it owed its existence to a murderous regime that had placed a shining future before the German population but left postwar society with a ruined country as well as the moral burden of genocide and countless war crimes. Volubly touted in Nazi propaganda as a "symbol of the people's community" yet never produced before 1945, the vehicle could easily have epitomized Germany's defeat and monumental moral bankruptcy. Developments at the factory threatened to deepen the imprint of the Third Reich's criminal nature on the car. Wartime operations at the works near Fallersleben relied on forced laborers from across Europe whose sufferings ranged from malnutrition and inadequate housing to arbitrary violence to the deaths of hundreds of war workers' children. To be sure, the forced laborers did not produce the "people's car." Nonetheless, their activities maintained military production right to the end and thus involuntarily helped preserve the works that had originally been built for the "people's car." Both its ideological lineage as Hitler's prestige project and the inhumane management practices at the factory during the war lent the vehicle a deeply compromised pedigree by 1945.

Yet the "people's car" did not perish with National Socialism. After Germany's catastrophic collapse, the vehicle owed its survival to a shifting international constellation that, within a few years of the war's end, proved conducive to rapid economic reconstruction in those parts of Germany under the control of the Western Allies. International historical forces had already shaped the car in important conceptual and social respects before 1945. German dreams of a *Volkswagen* as well as the organizational arrangements at the factory erected for its production gained crucial impulses from examples set by Henry Ford across the Atlantic. And the manufacturing site near Fallersleben—itself completed with the help of a sizable contingent of Italian workers—had operated in the war with a heterogeneous labor force from all over Europe. While the "people's car" thus possessed substantial American and European dimensions by the war's end, the immediate postwar years added an altogether new transnational facet. In keeping with Germany's unconditional surrender in 1945, German authorities lost operational control of the factory. It was thus the British occupiers, in whose zone the factory came to be situated, who initiated the civilian production of Ferdinand Porsche's design in 1945 amid numerous practical difficulties in a devastated country.

3

"We Should Make No Demands"

When news reached Anton Piëch at the Volkswagen works on April 10, 1945, that the Americans were marching toward the "City of the Strength-through-Joy Car," the head of operations suspended the production of Kübelwagens and fled south, joining his father-in-law Ferdinand Porsche on the Austrian family estate in Zell am See. All German armed forces simultaneously abandoned the town. A day after Piëch's hasty retreat, the American army took KdF City without firing a single shot. Since the settlement was quite literally not on their map, the American soldiers paid no further attention to it. Rather than explore the factory and station an occupying detachment in the town, the U.S. forces followed in pursuit of their German opponents, who were withdrawing eastward in chaos.[1]

In contrast to the orgy of violence and destruction in most of Germany, the Third Reich imploded almost silently in KdF City. Nonetheless, in the absence of all state authority, public order soon collapsed when the malnourished and badly clothed forced workers freed themselves from their camps and turned their pent-up resentment against the German population. As one German resident recalled: "They shouted slogans and then it started.

Toward the evening, they broke out . . . they broke down front doors and then there were several deaths. A butcher and a few others. They emptied out the apartments and destroyed everything." Not all of this violence was random, however: "They did look for certain people," targeting those who stood out for abusing foreign workers, the German witness acknowledged. To avoid getting caught up in reprisals, some Germans went into hiding in the large forests surrounding KdF City. Only when American troops formally occupied the town on April 15 did a modicum of stability and calm return.[2]

After the Allied victors had finalized their plans for Germany's division into four occupation zones in June 1945, KdF City came to be located on the eastern fringe of the British sector, a mere five miles from the Soviet zone. A host of practical challenges awaited the British authorities in the town counting around seventeen thousand residents, half of whom were of non-German nationality. In addition to securing public peace, British officials had to ensure sufficient provisions, help manage the return of thousands of foreign workers to their home countries, and oversee the allocation of scarce housing resources to thousands of refugees soon flooding the town. Since the occupiers could devote only limited manpower to these and other tasks, they quickly delegated numerous responsibilities to German officials. To this end, they established a new German local administration whose structures and procedures broke with the practices of the Nazi era. A visible sign of the new beginning was the British approval of the new name, taken from a nearby medieval castle, that the German mayor selected for the town in June 1945. KdF City was henceforth to be known as "Wolfsburg."[3]

Mass motorization—a Nazi dream and Wolfsburg's raison d'être—was, of course, an issue no one wasted even a thought on in 1945. At the time, the disastrous consequences of National Socialism's violent attempt to implement its vision of a modern

Germany dominated everyday life throughout the ruined country. Dramatic shortages of accommodation in destroyed cities, a lack of food, clothing, and fuel, a devalued currency, disruptions to the transport network, and stringent controls over business created harsh economic realities in which many Germans experienced daily life as an existential struggle to make ends meet. Initial Allied schemes such as the Morgenthau Plan of 1944, in which the American secretary of the treasury envisioned a demilitarized, deindustrialized, and largely agrarian Germany, held out little promise of bringing about the economic and social conditions conducive to wide car ownership. Under these circumstances, the future of the car factory as well as that of Wolfsburg—both tainted by their Third Reich origins—was anything but clear. If, in 1945, the inhabitants of most German cities asked themselves in trepidation what the coming years would bring, the insecurity surrounding Wolfsburg was particularly pronounced. The catastrophic fall of the Nazi regime and the devastation of Germany had removed the foundations of the town's very existence. Yet by the time the British occupation drew to a close in 1949, both Wolfsburg's and its automobile works' future was secure, as the factory celebrated the assembly that May of the fifty thousandth car since the end of the war. By December, the figure stood at eighty-five thousand.[4] And the vehicle that left the factory gates with increasing frequency was none other than a slightly modified version of the model Ferdinand Porsche had designed for the National Socialist regime.

It was thus during the occupation period that Wolfsburg emerged as a site of civilian automobile production. Why the British decided to manufacture a car with a Nazi pedigree and how this decision ensured Volkswagen's survival are not the only important issues in the car's immediate postwar history. Despite a highly fragmented press landscape at the time, local developments soon attracted the curiosity of newspapers, magazines, and newsreels across the western sectors. Reporters not only increasingly took note of the growing economic activities in Wolfsburg; in light of a stunning victory of right-wing extremists in local elections in late 1948, some commentators also asked whether the town five miles from the Soviet zone was turning into a hotbed of "neo-fascism." The occupation years thus highlight the many economic, social, and political difficulties that beset auto production in Wolfsburg in the immediate wake of the war and explain why, in the case of the small, rounded car, leaving behind the legacy of the Third Reich involved a protracted and complicated process that did not come to an end with the departure of the British authorities.[5]

The British military authorities who assumed responsibility for large swaths of northwestern Germany in June 1945 viewed it as their prime task to maintain public order in their sector while international negotiations took place to determine Germany's fate. To consolidate control, the occupiers formulated a discrete and comparatively modest objective. Unlike the Americans who arrived in their zone in the southwest with ambitious plans to stamp out National Socialism's cultural roots among the German population through comprehensive reeducation, the British aimed to ensure their own safety from German hostility. According to British reasoning, this aim was best served by expelling Nazi officials from key administrative positions and replacing them with Germans who had kept their distance from the fallen regime. At the same time, the British occupiers initially displayed little interest in transforming the political views of the wider population. After all, Germany's prominence in starting two bloody global wars within less than three decades suggested to influential British observers the existence of a deeply aggressive streak in the German "national character" that, the argument ran, would take decades to eradicate. Moreover, the established practice of indirect rule in the far-flung colonial empire, where a minuscule number of British officials often exerted control over large territories by governing through co-opted members of the local elite, shaped the occupational regime in postwar Germany. Rather than seek to initiate a comprehensive cultural transformation, the British regime in Germany was designed to avert open security challenges to occupational authority with the help of local officials who appeared reliable.[6]

Beyond maintaining public order in their zone, the British occupiers faced the task of feeding and housing a population made up of ethnic Germans as well as millions of other Europeans stranded on German soil amid pervasive destruction. Since the Allies had developed no comprehensive plan for postwar Germany before the cessation of hostilities, it was unclear how the country's occupation was to be managed. With respect to the economy, the Allies agreed that German industry should lose its military potential, but beyond this basic point the country's future productive capacity remained a topic of contention. While the Allied plan of March 1946 for the "Level of Industry" aimed to lower the capacity of German heavy industry (coal, iron, and steel) to half of the output achieved in 1938, the British remained doubtful about this punitive approach. To be sure, British officials agreed that Germany should pay for the damage and suffering it had inflicted on the world, for instance through dismantling and transferring production facilities, but they warned against strangling Germany economically. A stilted German economy, their reasoning went, would drive up the need for imports of food and other vital commodities, thereby increasing occupation costs. Since Britain emerged from World War II deeply indebted, an expensive occupation regime threatened to stretch the country's limited resources beyond the breaking point. British observers also cautioned against potential political repercussions of stifling the German economy. An industrially emasculated Germany would not only leave a strategically undesirable power vacuum in central Europe; persistent poverty and social discontent could also lead to a resurgence of the German radical Right, a scenario deeply reminiscent of the Weimar Republic, whose instability had partly derived from the harsh economic regulations initially imposed by the Versailles Treaty. In light of these economic and political risks, "constructive pragmatism" came to characterize British economic policy after July 1945.[7]

It was far too early to envision the Volkswagen works, whose capacity as a production site for civilian automobiles was yet unproven, as a potential pillar of postwar German stability amid the large-scale devastation and chaos that characterized the country in 1945. Nonetheless, the vast plant aroused strong interest as a potential reparation target. To secure the site, the British authorities dispatched Ivan Hirst, a twenty-nine-year-old army major in the Royal Electrical and Mechanical Engineers Corps with a degree from the Manchester Institute of Technology, to Wolfsburg. When he arrived in early August 1945, Hirst found that despite the piles of rubble from Allied air attacks, 70 percent of the factory's buildings and over 90 percent of its machinery had survived. This made the Volkswagen works the only German automobile plant that, in principle, could be set to production immediately after the war. Hirst also uncovered sizable stocks of various materials the German management had hoarded toward the end of the war. Prompted by a superior officer who recalled the KdF Car from a visit to the Berlin Automobile Exhibition in 1939, Hirst sent one of the few models with a civilian body built during the war to the British army headquarters, where it proved its functionality.[8]

During August 1945, the British military lobbied the Allied authorities for permission to produce the KdF Car to alleviate

Ivan Hirst, the British officer who oversaw the factory in Wolfsburg during the occupation period and initiated Beetle production after the war. Courtesy of Volkswagen Aktiengesellschaft.

the vehicle shortage that hampered efforts to run the occupied territory. As a report from July 1946 stated, the British zone, which counted over twenty-two million residents, featured no more than sixty-one thousand cars, 65 percent of which were classified as "worn out."[9] Cars manufactured in Wolfsburg, British administrators reasoned, would not only support day-to-day operations in their sector; such vehicles would also provide an alternative to expensive automotive imports from Britain, which threatened to raise the cost of running their zone. To render their request attractive to the fellow occupying powers, the British successfully held out the possibility of making cars produced at the Volkswagen works under their aegis available to the American, French, and Soviet administrations. On August 22, 1945, the British military government could issue an order to Hirst to manufacture and deliver twenty thousand cars by July 1946. While this plan was wildly optimistic, it won the Volkswagen works a crucial reprieve. Although the factory was not formally removed from the list of potential reparations, it was temporarily exempted.[10]

Establishing and maintaining production proved highly complicated tasks that required a manager with an exceptional combination of skills, and in Ivan Hirst the British had found a man who was ideally suited for the job. With a passion for automobiles, an extraordinary talent for improvisation, and a winning, modest manner that abstained from triumphant gestures *vis-à-vis* the German population, he stressed that "we . . . regarded ourselves as temporary trustees of the factory." By invoking "trusteeship," the British head of the Volkswagen works signaled that he understood his role to extend beyond the short-term preservation and management of the industrial installation. Having gained prominence in the context of British imperialism in Africa during the interwar years, the idea of trusteeship denoted an approach to foreign rule that was motivated by a moral desire to act for the benefit of populations under British control. While the philanthropic effects of trusteeship in Africa have been hotly contested in the postcolonial world, for the Volkswagen factory the method proved deeply advantageous. As Wolfsburg's trustee, Hirst played a central role in laying the foundation for a successful business.[11]

In the first months, removing rubble from the workshops, making basic repairs to the sewage system, securing vital materials, establishing contacts with suppliers, and drawing up production plans absorbed Hirst's energies. Delays and complications led to numerous setbacks during the fall of 1945. By the end of the year, the factory had turned out no more than fifty-eight vehicles, because it took until late November to ready the plant for assembly. To fulfill the overall production quota, the British authorities expected monthly output in Wolfsburg to reach four thousand in January 1946—an altogether utopian target.[12]

In part, Hirst could not meet his superiors' expectations because he faced intractable difficulties in creating the stable, reliable, and competent workforce indispensable for orderly, large-scale industrial production. While numerous Germans who had filled key posts at the plant during the Third Reich remained in Wolfsburg, their presence proved a mixed blessing. Hirst knew that he could only operate the huge factory with experienced German staff, but his desire to retain them clashed with denazification efforts to prevent a return of Nazis and active regime supporters to leading jobs in postwar society. In the summer of 1945, an early wave of arrests led to the prosecution of individuals in key positions who had committed substantial crimes. As part of this early judicial drive in June 1946, a British military court ordered the execution of the medical doctor under whose authority the children born to eastern European forced laborers at the Volkswagen works had died in the camp in Rühen. That autumn, the Allies in the western zones turned their attention to the wider population. While the Americans required all adult Germans in their sector to complete a questionnaire containing 131 queries about an individual's party affiliations, career, military service, and more, the British took a more focused approach. In the hope of controlling the mountain of files inevitably generated by such an exercise, they instructed 10 percent of the German people who had held prominent social, administrative, and economic jobs during the Third Reich to fill in a shorter form.[13]

Although he had no sympathy for former Nazis, Hirst regarded the procedure as a "headache" and initially attempted to minimize the impact of denazification on the factory because the removal of qualified personnel threatened to disrupt production. Yet when it emerged that only a very few factory employees had been dismissed by January 1946, both Germans familiar with events at the works prior to 1945 and senior administrators stationed elsewhere in the British zone protested, pressing for a repetition of the exercise. As a result, local denazification committees ordered the dismissal of 228 individuals in June 1946, a group that included the head of operations, the technical director, four department heads, and several foremen. Output at the factory immediately fell by 60 percent.[14]

Although the British authorities declared denazification complete in July 1946, the issue dragged on and poisoned the atmosphere in Wolfsburg for months, because those who had lost their jobs launched appeals. In February 1947, 138 of them had their verdicts overturned. Apart from relief, these individuals felt deeply aggrieved at having experienced what they considered an unjust ordeal. Meanwhile, Germans who had maintained a critical distance from the Third Reich became increasingly disillusioned by the numerous revisions of initial decisions. Despite its shortcomings, denazification exposed erstwhile adherents of the regime to an unsettling shock, "furthering their willingness to adapt to new political conditions" in most parts of West Germany. In Wolfsburg, by contrast, lingering discontent with denazification added to a politically explosive constellation that was to result in a spectacular resurgence of right-wing extremists in late 1948.[15]

Despite the complications surrounding denazification, Hirst managed to recruit a sizable labor force, which rose from roughly 6,000 employees in January 1946 to 8,383 in December 1947. Annual turnover rates, however, stood at 50 percent. As a result, the management was permanently forced to integrate new, inexperienced workers into production routines, a process that lowered productivity significantly. Many workers at the Volkswagen works left in search of more stable jobs, regarding the Wolfsburg plant as merely a temporary employer because it remained on the list of potential reparations. Moreover, a dismal housing situation did little to encourage residents to stay. Since the National Socialist regime had never built the model town outlined in its master plan, the camps that had accommodated forced workers until April 1945 provided the most common dwelling available to factory employees. Beyond primitive sanitation and heating provisions, overcrowding and a lack of privacy characterized life in the camps. Several families frequently shared one room, separating their areas by suspending blankets from the ceiling.[16]

Although most forced workers departed in 1945, this extremely basic housing stock remained oversubscribed because, with its dozens of barracks from the Third Reich, Wolfsburg quickly emerged as a center attracting many of the refugees drifting around Germany in their millions after the war's end. Demobilized German soldiers and released POWs made their way to Wolfsburg in considerable numbers in the hope of finding employment there. These arrivals often brought with them violent wartime manners, as a resident recalled. Faced with a hesitant official, one former soldier resorted to open physical threats to secure residence permits for himself and his comrades: "He took the man [the official] by his tie and said: 'Now listen. In my life, I have learned nothing but how to kill. Whether I kill one more man makes no difference to me. So, if you pigs don't give us an address right now, I will ram you right into the ground.'" This type of rough conduct unsurprisingly prompted residents to leave Wolfsburg for places promising more civilized forms of behavior.[17]

Local overcrowding also resulted from an influx of thousands of Germans who had fled the Soviet sector, as well as those sent to Wolfsburg by British and German authorities keen to steer migrants away from bombed-out urban areas. The latter included ethnic Germans whom governments in Poland, Czechoslovakia, Hungary, and elsewhere ejected in their millions from their territories in 1945. Several thousand of these so-called "expellees" ended up in Wolfsburg, in many cases reaching its camps with nothing but the clothes on their backs. Wolfsburg also became home to several hundred "displaced persons"—a shorthand term for civilians from all over Europe unable or unwilling to return to their countries of origin at war's end. The high turnover rate at the Volkswagen factory was directly linked to this transient local labor pool. Many deracinated refugees caught their breath, worked a few days at the plant, and then continued their search for a future elsewhere.[18]

Meanwhile, many of those laborers who signed up for regular employment over a longer period proved themselves unreliable. Monthly absentee rates could reach 40 percent until the middle of 1948 and further undermined efforts to raise productivity. Numerous workers missed work at the factory because food shortages forced them to look for provisions elsewhere. In a society operating on the brink of, or below, subsistence levels, frequent absences were part of many workers' survival strategies. In 1945, the British occupiers set individual food allocations at a mere 1,000 calories per day before raising them to 1,100 calories in March 1946. A typical daily ration consisted of just two slices of bread, a little margarine, a spoonful of watery soup, and two small potatoes.[19] As a consequence, "everybody was only after food. There was much talk about food," a refugee remembered. In this situation, workers at the Volkswagen plant had to complement their official allocations by other means. In addition to cultivating small plots of land, they undertook foraging trips to the countryside to acquire food from farmers. Desperate demand for virtually everything fueled a burgeoning and illegal black market, in which cigarettes supplanted the largely worthless reichsmark as the lead currency. In Wolfsburg, conditions were particularly desperate because the refugees making up a high proportion of the population had lost practically all their belongings and thus had nothing to trade. Since securing food

German workers protesting against material shortages during the occupation period while posing next to the ten thousandth vehicle produced in Wolfsburg. The drawing in the foreground depicts a beer mug, a sausage, a cigar, and other basic consumer goods that were in short supply in the immediate postwar years. Courtesy of Volkswagen Aktiengesellschaft.

and other basic staples during the so-called "hunger years" took up considerable amounts of time, it was no surprise that many workers could not afford to show up at the factory every day. Nothing illustrates the mood among the German workforce better than a photograph taken in October 1946 as the plant made its ten thousandth vehicle: "10,000 cars, nothing in our stomachs, who can take it?" read one of the signs the workers had drawn; another sketch in front of the auto body showed a mug of beer, a sausage, and a smoking cigar—all treasures well beyond the average Wolfsburger's reach.[20]

The ubiquitous shortages also directly affected production. Although the British administration gave the plant priority access to supplies, securing adequate monthly steel and coal quotas proved a major challenge because of Germany's disrupted infrastructure and a pervasive scarcity of raw materials. The low quality of the steel that did reach the plant often caused concern. Germany's division into occupational sectors presented another obstacle because it cut the plant off from manufacturers outside the British zone that produced components such as carburetors, headlights, and other electrical parts. The management had to design specialized machine tools to work materials of inferior quality. All these bottlenecks rendered it virtually impossible to establish reliable manufacturing routines.[21] When an exceptionally harsh winter struck Europe in November 1946, no amount of improvisation could maintain production. As the ports froze over and the railway network collapsed, the distribution of raw materials and food came to a virtual standstill all over the country. Inside the factory, temperatures slumped to 19 degrees Fahrenheit (-7 degrees Celsius) in early December 1946. Out of concern for the health of the workforce and amid low stocks as well as mounting problems with malfunctioning equipment, management suspended production until March 10, 1947. Like other Germans around the country, the inhabitants of Wolfsburg spent the winter of 1946–1947 suffering severely from the bitter cold and lack of food.[22]

It comes as no surprise that automobiles produced under these inauspicious circumstances exhibited numerous faults. As the factory lay quiet in December 1946, Hirst issued a memo to German managers in which he expressed his hope that "upon resuming production, 60 percent or more of the cars that come off the line will no longer require repairs or an engine change." This high rate of defects reflected the manufacturing problems that resulted from inadequate materials and unreliable machinery. Fault upon fault dogged the cars made in Wolfsburg, internal reports lamented. Hoods and doors did not close properly, headlights cracked, cylinders possessed "too short a lifespan," and the steering "was heavy going in many cases." Many finished cars had to be completely repainted before final delivery because they were often stored in halls with war damages in which wind and rain corroded their military matte green paint. In terms of reliability and workmanship, the automobiles that left Wolfsburg in the immediate postwar years were still a far cry from the Volkswagens that were to establish themselves as synonyms for quality and dependability in the fifties.[23]

As Hirst struggled to maintain production, the question of whether the works would be dismantled as a reparation continued to hover in the air. In the course of high-level discussions about a potential transfer of the production site to Great Britain, officials in London asked British auto manufacturers for an evaluation of the vehicle as well as the plant. Several visits to Wolfsburg and car tests by British auto engineers in 1945 and 1946 yielded a mixed verdict. The plant's size and equipment impressed British visitors, as did the vehicle's potential for Fordist mass production, since its design eliminated "hand work to an unusual degree."[24] British experts commended the car for its "excellent road holding quality" that derived from the torsion bar suspension. At the same time, test drivers noted serious problems. Smelly fumes in the car's interior, poor brake performance, an "underpowered" engine, and deafening noise counted among the main complaints.[25] Despite several technical virtues, the tenor went, the automobile required "considerable modification . . . to conform to the standard" the British public expected from a commercially available vehicle. As a result, the British car industry saw little advantage in transferring the Volkswagen production site to the United Kingdom, especially since such a step would also have thrown the landscape of motor manufacturing across the Channel into turmoil, given the vast proportions of the Wolfsburg plant. British industry by no means easily dismissed the erstwhile KdF Car. Rather, the factory's size and the vehicle's deficiencies during the first year of production explain why British manufacturers developed no plans to produce the car in the United Kingdom. With the international economy in ruins, no one could anticipate that a smelly, loud, uncomfortable automobile would soon sell like hotcakes.[26]

The decisive factor that secured Wolfsburg's survival was the new economic policy the Western Allies formulated in response to both concrete problems in Germany and a rapidly changing global political constellation. The "hunger winter" of 1946–1947 had demonstrated that German society could not sustain itself under the existing draconian regime. Apart from facing the specter of popular revolts, the Western Allies began to fear that persistent material discontent would render the West German population susceptible to political overtures from the Soviets, whose attempts to expand their European sphere of influence the Americans and British regarded with growing suspicion. The Americans, who soon emerged as the dominant force in the Western camp, realized that Germany afforded a key political battleground amid increasingly palpable tensions between East and West. Since several international conferences failed to reach an agreement on Germany's political future, the Soviets and the Western Allies took steps to consolidate their occupation zones politically and economically, thereby further stoking mutual mistrust. In October 1946, the U.S. administration abandoned its restrictive economic policy vis-à-vis German industry, a step signaling the onset of energetic reconstruction policies in the former enemy country. In 1948, this change of strategy culminated in the inclusion of Germany's western zones in the Marshall Plan, the famous loan program through which the United States helped revitalize Western European economies.[27]

The Volkswagen works was one of the earliest beneficiaries of new Western economic policies. In the spring of 1947, the American and British authorities ordered the production of 160,000 vehicles from West German plants, thereby "virtually exempting the German car industry from dismantling." In light of these new circumstances, the British authorities embarked on a search for an energetic German chief executive with the skills to oversee the large production facility in Wolfsburg. Ideally, this new "general director" would possess a background combining proven managerial experience with technical expertise in the auto sector that would allow him to set the plant on a course of expansion. A record of comparative detachment from the Nazi regime and a relatively uncompromised political past were crucial requirements for a candidate seeking to run a factory that owed its existence to a National Socialist prestige project and was thus prone to future scrutiny. Given the German auto industry's international backwardness before 1939 and the central part business had played in a war economy that relied on millions of forced laborers, the occupiers were looking for a manager with rare qualifications indeed. Ferdinand Porsche, who was imprisoned for eighteen months in France after the war for having requisitioned machinery from French automakers, was altogether unsuitable for this position. While Porsche's design company in Stuttgart survived the immediate postwar era and subsequently flourished under his son, the creator of the "people's car" was to visit the Volkswagen plant only one more time before his death in 1951.[28]

When the British authorities learned in the fall of 1947 that Heinrich Nordhoff (1899–1968) lived in Hamburg and was looking for a permanent position, they did not hesitate to invite him to Wolfsburg. A graduate in mechanical engineering from the Technical University in Berlin, Nordhoff had joined Opel in 1929 just after the company was taken over by General Motors. He had filled a leading position in the service department, which involved several visits to the United States; recognizing his negotiating skills, the Opel board had sent him in the late thirties to

Berlin, where he procured contracts from the Nazi government. As an executive at Germany's largest producer of passenger cars during the Third Reich, he had also counted among the managers who had vainly tried to undermine the regime's plans for a "people's car." At the end of the war, Nordhoff ran Opel's truck factory in Brandenburg, the largest installation of its type in Europe. By no means meteoric, Nordhoff's career nonetheless made him stand out among Germany's executives for his first-hand knowledge of American business practices in the auto industry, as well as his experience with mass production.[29]

Although he wished to continue at the General Motors subsidiary, Nordhoff had been denied a work permit by American occupation authorities and lost his position at GM owing to the vagaries of the denazification process—and despite support from his employers in Detroit. While his denazification panel made up of German jurors had recommended him to be cleared, the American military government, which retained vetoing rights, banned him from future employment in their sector. The Americans reached this decision because the Nazi regime had honored Nordhoff as a "leader of the war economy," an award received by a heterogeneous group of four hundred businessmen that included, but was by no means exclusively composed of, die-hard supporters of National Socialism. Nordhoff, who had never joined the Nazi Party, primarily owed this award to his managerial contributions to the war effort rather than to his political allegiance. While the Americans considered all such leaders ideologically suspicious and hence unsuitable for senior posts, the British had fewer qualms. Rejected by the Americans while simultaneously courted by the British, Nordhoff considered himself a prime example of "how questionable the whole procedure" of denazification was.[30]

However inconsistent his denazification examination may have been, Nordhoff did not emerge from the Third Reich free from moral blemish. During the war, the truck factory under his management in Brandenburg had employed between sixteen hundred and twenty-one hundred foreign workers. Nordhoff, a practicing Catholic, appears to have secured sufficient food rations, clothing, and accommodation for the forced workers, but he could not prevent their mistreatment at the hands of SA guards whose violence upheld the system of compulsory labor. "To get the works through the war intact," Nordhoff's biographer has written, the manager "arranged himself with the system of forced labor." To be sure, he was not one of the numerous German entrepreneurs and managers who treated forced labor with indifference or brutality. Yet despite his Catholic creed, his detachment from the Nazi Party, and his attempts to limit war workers' suffering, Nordhoff became complicit with the National Socialist terror regime by supporting its war effort.[31]

If Nordhoff's endeavors to preserve his personal moral integrity during the war set him apart from many of his colleagues, his promotion to "general director" at the Volkswagen works after a period of compulsory unemployment fits into a broader pattern among the West German economic elite. As initiatives to rebuild the economy gathered momentum during the incipient Cold War, the Western Allies saw no alternative to rehabilitating numerous managers with politically compromised careers. As a result, a cohort in their forties with experience in influential, albeit second-tier, appointments in the war economy assumed positions at the helm of the postwar West German economy. Since many established directors also resumed their posts, West German business circles displayed a pronounced continuity in personnel across the political watershed of 1945.[32]

When Nordhoff took up his new post on January 1, 1948, the survival of the Volkswagen works constituted the main focus of his activities. After visits to Wolfsburg the previous fall, he harbored no illusions about the difficulties he faced. So far, the factory had mainly produced vehicles for military authorities, but the prospect of competing in a civilian market was an altogether different proposition. The new general director unleashed a frenzy of activity, regularly working seventeen-hour days to put the factory on a stable footing.[33] In the early months of 1948, Nordhoff established a hierarchical management structure to tighten his control of the plant, a step supported by the British authorities. Drawing on his experience at Opel's service department, he expanded the dealership network that the British had started in October 1946. He negotiated several export deals to Switzerland, Sweden, and Belgium, which brought much-needed U.S. dollars into the company. Above all, Nordhoff focused on the factory and its product. His initial survey revealed that the plant manufactured in far too inefficient and expensive a manner, and as a result he demanded that individual productivity rise by 30 percent. Improving the car's quality became a key objective, and the new chief executive intensified efforts to eradicate the vehicle's technical deficiencies that threatened its competitiveness.[34]

That it proved impossible to raise monthly output substantially over one thousand vehicles in the first half of 1948 cannot be regarded as a failure on Nordhoff's part but rather reflected Germany' dysfunctional economy. In 1947, the Western Allies began to acknowledge that German business activities would emerge from hibernation only after serious reforms. Allied experts as well as a group of liberal German economists identified two closely related problems. To begin with, the rationing system severely curtailed supplies, thereby fueling a black market that could never function as a site for sustained growth given its illegal nature. Since wartime inflation had eroded the value of the reichsmark, the black market in turn ran on cumbersome barter arrangements that frequently forced traders to assemble complicated chains of exchange to secure a particular commodity. In short, Allied scarcity planning and a worthless currency combined to fuel a shadow economy that was not only illegal but inefficient and therefore a serious impediment to growth.[35]

As Nordhoff sought to stabilize operations in Wolfsburg, rumors abounded that official planning and rationing would be severely curtailed in the summer of 1948. This implied that the market forces of supply and demand would soon determine commodity flows and prices to a far larger extent than had been the case in Germany since the mid-thirties, when the National Socialists had intensified their state-directed armament drive. The introduction of market mechanisms, in turn, required a currency reform to rebalance the volume of money in circulation with existing levels of economic activity. It was on June 20, 1948, that the authorities launched a new currency in the western sectors—the deutsche mark (DM).[36]

Over the years, German collective memory came to associate this date with shop windows that overnight became filled with previously unavailable goods, retrospectively acting as a harbinger of future prosperity. At the time, however, most contemporaries reacted with profound shock at the draconian stipulations that Ludwig Erhard, the economic director in the British-American bizone and West Germany's future economics minister, implemented to restrict the amount of DMs in circulation. While every citizen received DM60 in two installments, private savings accounts were effectively cut by 93.5 percent. Salaries, wages, rents, and public pensions retained their nominal value, but all other regular payments, including private pensions, were reduced by 90 percent. The business world did not escape unscathed either, as companies had to write off 90 percent of their financial reserves.[37]

Erhard coupled the currency reform with a deregulation of prices on most goods, thereby largely abolishing the rationing system and forcing the black market out of business. While acknowledging the financial losses his policies involved for many average citizens, Erhard hoped his measures would spur economic growth in several respects. "Whereas before June 20 money had been plentiful but in practice hardly worth earning," a scholar has observed, "now goods were appearing in far greater quantities but money was scarce. This gave the German people a tremendous incentive to work."[38] Erhard and his supporters thus gambled that his reform would drive the population to their jobs and counteract the high rates of absenteeism that afflicted workplaces all over Germany. The reformers also calculated that businesses with drastically reduced reserves would need the new, scarce currency to keep their operations going, thereby literally forcing them to make money by producing and selling in a competitive marketplace. The coming months came to show that the currency reform and the introduction of free markets acted as vital stimuli to the economic reinvigoration of Germany's western sectors. Nonetheless, the road to growth was paved with many obstacles.

After the currency reform, the Volkswagen works soon showed signs of improvement. Fluctuation and absenteeism among the workforce plunged from late June 1948 onward, since employees now valued their jobs as a source of regular, meaningful income. A steady workforce facilitated production planning, as did the more predictable flow of supplies that developed thereafter. As a consequence, individual productivity increased, stoking an increase in output. In July 1948, eighteen hundred automobiles left the factory in Wolfsburg; in June 1949, the figure stood at over thirty-eight hundred cars. The factory's rising production curve reflected expanding demand from a clientele that underwent a major change in 1948. While the occupiers had received almost three-quarters of the plant's output in 1947, a year later 60 percent of the automobiles made in Wolfsburg went to domestic customers, among whom businessmen received the lion's share. Preferential deliveries for "industry," a newsletter explained, occurred in the hope of aiding a wider economic recovery. Improved sales and productivity put the management in a position to announce a 15 percent pay rise in October 1948, which took the average hourly wage to DM1.34, a rate significantly above the remuneration levels of most German workers outside coal mining. The budding West German press noted these encouraging economic trends and began to cite Wolfsburg as an example of the upturn in the territory covered by the deutsche mark.[39]

As sales improved at home, Nordhoff never lost sight of the international market and redoubled efforts to develop a model suitable for export. To ensure the vehicle's attractiveness to customers abroad, he invited foreign car dealers to the factory in early October 1948 and asked them to evaluate five versions of the Volkswagen car with varying interior and exterior designs. "We know that our cars have deficiencies. We do everything in our power to remove these deficiencies," the general director explained before underlining the importance of the traders' feedback: "We want to and have to leave the basic construction unaltered. We have changed the exterior appearance, and that is not negligible. Please be critical. Please be thorough. Your opinion will have decisive consequences," he assured the salesmen. The car's basic technical characteristics, Nordhoff suggested, were sound, but the automobile required substantial aesthetic modifications.[40]

While we know nothing of the dealers' reactions to these experimental models, the management launched a new "export version" in June 1949. Designed for more comfort than Porsche's original, it still ran with an unsynchronized transmission and remained equipped with mechanical brakes. The improved model, however, featured more-effective soundproofing. Above all, in-house designers had turned their attention to the car's spartan interior, embellishing the dashboard with a chromium strip, and fitting a sunshield as well as adjustable upholstered seats covered with sturdy, color-coordinated fabrics. The vehicle also sported a durable coat of glossy synthetic resin paint that increased the body's longevity and enhanced its visual appeal. The journalists at the launch may have regarded the new automobile's shiny exterior primarily as a commercially motivated cosmetic alteration, but Nordhoff elevated this aspect to one of the car's defining features. On offer in "pastel green," "medium brown," and "Bordeaux red," the "export limousine," the general director explained, had received a "paint job absolutely characteristic of peacetime." In the context of a vehicle with a Nazi pedigree, this attempt to link glossy paint with peace was less curious than it may initially appear. The automobile of 1949, Nordhoff implied, differed from the "people's car" commissioned by the militaristic Nazi dictatorship as well as the matte green vehicles produced under British management. As he highlighted the export model's new aesthetic features, Nordhoff strove to dissociate it from its compromised recent past and recast it as a postwar commodity.[41]

But divesting the automobile made in Wolfsburg of its Nazi sheen would take more than shiny new paint. After all, numerous material characteristics, including the car's distinctive silhouette, its air-cooled rear engine, and its torsion-bar suspension, hailed directly from the Third Reich and thus evoked the artifact that the National Socialists had trumpeted as a symbol of the *Volksgemeinschaft* only a few years earlier. Beyond memories of the dictatorship, political developments in Wolfsburg themselves were liable to compound the automobile's image problems. In 1948 and 1949, the town made the national news not only for auto manufacturing but also for a singular upswing of right-wing extremism that turned nationalist enemies of democracy into the strongest force on the city council. In January 1949, a reporter for a Berlin daily went so far as to warn against "neo-fascism in Wolfsburg."[42]

Two months earlier, a splinter group under the name of the German Justice Party (Deutsche Rechtspartei, or DRP) had secured 64 percent of the local vote, ousting the Social Democrats from the city council. Composed overwhelmingly of former members of the Nazi Party, the DRP fought the election in November 1948 on a profoundly revanchist platform that included demands for an immediate end to denazification and the restitution of full sovereignty to a unified, remilitarized Germany in its prewar borders. This nationalistic outbreak in late 1948 might have been dismissed as an episodic protest, but further ballots in May and August 1949 saw the extreme Right gain 48 and 40 percent in Wolfsburg respectively. On these occasions, some voters demanded on their ballot papers: "We want Adolf Hitler." Only an unprecedented coalition between Conservative Christian Democrats, Social Democrats, *and* Communists supported by the British authorities kept the nationalists out of local office. Similar leanings became manifest in elections of workers' representatives at VW, which saw a former Luftwaffe major and founding member of several radical right-wing organizations win the highest number of votes into the early fifties. Swastika graffiti repeatedly appeared on walls of the plant. No other city in the western sectors displayed even remotely similar levels of support for revanchist tendencies as Wolfsburg.[43]

To some extent, the DRP profited from a wider groundswell of nostalgia for the Nazis as well as burgeoning frustration with denazification. As polls in the British sector revealed in May 1948, fewer than 30 percent of the public deemed National Socialism a "bad idea," while a solid majority thought it had been "a good idea badly carried out." Those considering denazification a success numbered below 30 percent. These figures partly reflected the inconsistency with which denazification panels passed judgments, but they also illustrate an outright refusal among large parts of the German population to confront the "question of guilt" *(Schuldfrage),* as philosopher Karl Jaspers termed it in 1946.[44]

The Allies' increasingly lenient line vis-à-vis war criminals only reinforced this tendency. In 1946, the British authorities had not hesitated to execute the doctor responsible for the deaths at the children's camp at Rühen, but two years later they acquitted local party leader Ernst Lütge, who claimed to have shot dead an unarmed Ukrainian war worker in self-defense during an interrogation at the Volkswagen plant in November 1943. Lütge based his plea on prominent character testimonials, including a lengthy statement by Ferdinand Porsche, who praised the party official for his supposedly humanitarian conduct. Lütge, Porsche asserted, had secured plentiful food rations for forced workers and "insisted on treating foreigners correctly and decently. . . . I recall him as an immaculate and humane individual."[45] That the Ukrainian victim had attracted Lütge's attention for illicitly boiling a potato reveals Porsche's claim about the supposedly plentiful rations as a threadbare, cynical attempt at self-exoneration. Moreover, the party official, the prosecution uncovered, had already been sentenced by a German court in 1944 after shooting and beating to death another forced war worker who had stolen a piece of fruit.[46]

In an environment in which mendacity could gain the upper hand and convicted killers like Lütge walked free, Wolfsburg proved a particularly fertile ground for right-wing extremists. To a degree, the Right thrived because the DRP focused its campaign almost exclusively on Wolfsburg. Moreover, as a historian has observed, the town "lacked seasoned politicians who could have gathered experience in the Weimar Republic" to oppose the Right effectively. Wolfsburg contained its fair share of people with little desire to let go of the recent past. Among this group that amplified right-wing electoral campaigns, a journalist observed, were "former officers, professional soldiers, and privates [*Landser*]," as well as people "who had been forced to abandon their livelihoods for political reasons" after 1945. The latter comprised erstwhile SS men and local Nazi dignitaries, including Wolfsburg's former lord mayor and city treasurer, who both had lost their jobs in the town administration but secured positions at the plant. In effect, the journalist concluded, Wolfsburg acted as "a reception camp for failed people" who preserved an oppressive "military esprit de corps."[47]

Meanwhile, the German Justice Party's pledge to campaign for national unity in prewar borders exerted a deep appeal on the thousands of ethnic German refugees and expellees that made up around half of Wolfsburg's population—more than twice the regional average. Despite their employment at the Volkswagen works, their lives showed few signs of improvement in late 1948. While housing shortages persisted, the introduction of the deutsche mark earlier that year triggered a substantial price rise for many basic staples, causing the population to take to the streets in numerous cities. All over West Germany, refugees and expellees counted among those hardest hit, since they lacked material reserves. In Wolfsburg, with its disproportionate share of utterly impoverished recent arrivals, soaring discontent at seemingly unceasing deprivation played a major role in the exceptional success of the Right.[48]

Reports about Wolfsburg being a stronghold of the radical Right posed a substantial commercial danger for the VW works, threatening to recontaminate the car with its ideological origins. After the German Justice Party's electoral triumph in late November 1948, Nordhoff, who had initially refrained from addressing the workforce on the Nazi past for fear of alienating employees, abandoned his reticence in an impassioned speech to the entire factory. "That prices are high, that all goods are scarce and frequently of inferior quality," the general director pointed out, "is neither the fault of the [German bizonal] administration, nor can it be changed by joining an organization or . . . by banging the drum in the street—tempting as that may be." Rather than expect "the living standard of 1938," he continued, the citizens of Wolfsburg had to acknowledge the causes and scale of the disastrous present:

> Past and present are separated by a war that we started and that we lost, by the death of millions of men in their best years, by the loss of tens of thousands of valuable machines and factories, of unimaginable treasures of raw materials; they are separated by the destruction of our currency and savings, by the emaciation of our fields and the terrible breakup of our country into bits and pieces that no one yet knows how to put together again.[49]

Beyond defending the recent currency reform, Nordhoff unequivocally called upon his compatriots to accept the burden of defeat, as well as Germans' liability for their own plight. Nordhoff clearly rejected the notion that National Socialism had been a "good idea badly carried out." He positioned himself against a nascent culture of collective victimhood that cast Germans as passive casualties of Allied bombings and forced expulsion, amid silence about their active involvement with the Nazi regime. Nordhoff did not deny German suffering, but from his perspective Germans had to accept responsibility for their difficulties. As he read his employees the riot act, Nordhoff stayed within clear limits. For instance, he mentioned neither the genocide of the Jews nor other crimes committed in Germany's name, and thus steered well clear of the "question of guilt." Within a climate that overwhelmingly cast Germans as passive sufferers, his appeal to accept the war's material consequences was as exceptional as his silence about suffering among Jews and non-Germans was conventional.[50]

Above all, the head of the Volkswagen factory called upon employees to seize the initiative in accordance with the following maxim: "We should make no demands, but quietly do the only thing that will lead us from our deepest misery, and that is to work." He returned to this theme throughout 1948, demanding that the workforce "achieve" *(leisten)*. Having stated in June that the currency reform would "reveal who achieves something," he took up this idea a few months later in a reminder that "one thing alone gets us ahead: work and achievement." By reiterating the need to "achieve" in the face of obstacles, Nordhoff invoked a figure of speech that carried deeply familiar overtones for most employees, given that numerous Nazi orators had resorted to the identical formulation during the war. In May 1942, for instance, Hitler declared the war "a battle of achievement for German enterprises" that would end in "final victory." Even politically well-intentioned speakers such as Nordhoff found it impossible to move beyond the recent past in historically uncontaminated language.[51]

Given the long shadow the Third Reich cast over Germany in the second half of the forties, the Volkswagen car did not yet push its Nazi past to the sidelines and fashion a new public image. When Nordhoff presented the first "export" version to the public in June 1949, his cautious attempt to relate the vehicle to a postwar era of peace was bound to prove largely ineffective. For one thing, West Germany's fractured media landscape restricted the circulation of stories offering novel interpretations of the car made in Wolfsburg. More important, however, the inchoate nature of Germany's political and social contours impeded the formation of an alternative image for the car made in Wolfsburg. Attempts to associate the vehicle with peace could only prove largely ineffectual as long as Wolfsburg made headlines for a resurgent Right that profited from local radicals and the plight of large numbers of refugees. As a result, the erstwhile KdF Car began its postwar career in a state of cultural limbo between the collapsed Third Reich and the emerging Federal Republic of Germany.

The enduring basic question of who actually owned the vast factory after the Third Reich's end exemplified the uncertainty surrounding the car. When the British authorities lifted many economic restrictions after the foundation of the West German state in 1949, they could not identify a proprietor of the plant. After months of wrangling, the British occupiers and the new federal government in Bonn signed a clumsy compromise that placed the works under the joint ownership of the Federal Republic and the state of Lower Saxony. The negotiators themselves recognized this legal construction, which subjected a company operating in a market economy to twofold state supervision, as an undesirable economic hybrid and agreed on the need for a future federal law to settle the issue of ownership. It was thus under merely provisional proprietorship that the company—which increasing numbers of contemporaries now began to call "Volkswagen"—expanded production at the beginning of the Federal Republic's history.

Despite the insecurity that surrounded both car and production site in 1949, the occupation period laid important foundations for the factory's subsequent economic good fortunes. The pragmatic decision of the British to alleviate a vehicle shortage in their sector by putting Ferdinand Porsche's prototype into production proved crucial for the future of the Volkswagen works because it removed them from the list of potential reparations. Amid tremendously difficult conditions between the summer of 1945 and late 1947, Ivan Hirst initiated serial manufacture of the civilian vehicle, began to build a national dealership network, hired a sizable workforce, and prompted technical improvements on the rudimentary Volkswagen car. Hirst prepared the ground for Heinrich Nordhoff's initiatives that aimed to put the Volkswagen works on a sound commercial footing from

1948. The former Opel manager not only launched a drive to increase productivity but made the development of a vehicle suitable for international markets one of his priorities. Steering the factory through the difficult period of the currency reform, the new manager oversaw an incipient expansion of production that boded well for the future.

In fact, the occupation period signaled a shift in the hierarchy among German auto firms. While operations in Wolfsburg began in late 1945, it took until the second half of 1947 for Opel, Ford, and Mercedes-Benz to resume production. All these companies had suffered far more destruction in air raids than Volkswagen, which the Allied authorities privileged immediately after the war. This early period of activity worked to Wolfsburg's commercial advantage. By the beginning of 1949, the Volkswagen factory had established a significant early lead over competitors, turning out 19,127 vehicles in the previous year while output at Opel, Germany's dominant car producer before the war, stood at just 5,762.[52] It was still altogether unclear whether these figures heralded a lasting transformation in West Germany's car landscape. Nonetheless, a new automotive player had appeared on the scene. With its vast, debt-free, and Fordist production facilities and its well-tested, Porsche-designed automobile, Volkswagen enjoyed significant long-term advantages as it began commercial operations in the late forties. In the next decade, the car was to step out of the Third Reich's shadow and establish itself as the undisputed collective symbol of the young Federal Republic. Volkswagen's subsequent economic strength and the car's cultural prominence, however, owed an immense debt to developments in Wolfsburg during the immediate postwar years when the British began the process that turned Europe's largest automobile factory from a Nazi prestige project and military production site into a civilian auto works.

4

Icon of the Early Federal Republic

"Last Saturday, newsreel reporters and television journalists had before their lenses an extravaganza in honor of the economic miracle." Thus began the weekly magazine *Der Spiegel*'s detailed article on the lavish festivities organized by the Volkswagen works in August 1955 to celebrate the one millionth car produced in Wolfsburg. After religious services in the morning, one hundred thousand people gathered in a temporary stadium erected for the occasion and watched a spectacle of "international attractions" that had the reporter rubbing his eyes incredulously: "Scantily clad ladies from the world famous Moulin Rouge swung their legs, South African negro choirs sang spirituals, 32 Scottish female highland dancers stomped around to the sound of bagpipes, Swiss flag bearers twirled their standards." After three hours of dazzling music and dance, general director Heinrich Nordhoff, recently awarded a prestigious Starred Federal Service Cross, took to the stage to the tunes of twelve marching bands playing Johann Strauss and thanked the labor force for its hard work in a short speech that closed with a characteristic clarion call: "Ahead to the second million!" After his address, Nordhoff switched into the role of a game show

Workers surround the one millionth VW, which rolled off the line in 1955 painted all in gold. The company staged a weekend of sumptuous celebrations in honor of this symbol of the "economic miracle." Courtesy of Volkswagen Aktiengesellschaft.

host, overseeing a lottery that distributed fifty-one Volkswagens among the gathered employees. Those desiring further diversion could proceed to a fun fair with merry-go-rounds, a roller coaster, and bumper car rides or watch soccer matches throughout the weekend. "It was a royal festival lasting three days," the journalist concluded.[1]

Everyone, including the twelve hundred press representatives who had accepted VW's invitation to attend the celebration, knew that the company's ability to hold court so sumptuously only ten years after Germany's ignominious collapse rested upon roaring demand for its main product. In Wolfsburg, the car was the star, and VW acknowledged this fact by painting its one millionth model in gold and encrusting its bumper with rhinestones. Beyond evoking local pride, the vehicle developed into a veritable golden calf all over West Germany in the course of the fifties. Rather than remain a neutral, functional object, the Volkswagen turned into a prominent collective symbol inseparable from West Germany's rapid recovery, which soon became known ubiquitously as the "economic miracle." By the end of the fifties, *Der Spiegel* labeled the car "the German miracle's favorite child." VW pursued an active public relations policy to encourage such flattering readings of its main product. No matter whether the company gave a press conference, celebrated a production jubilee, launched a modified model of its best-selling vehicle, or opened a stand at an auto show, Nordhoff insisted on lavish settings, impeccable hospitality, and generous reimbursement arrangements for hundreds and at times thousands of reporters and guests. In addition to wooing the press, the company not only issued numerous booklets, brochures, and pamphlets about its product but, in 1954, produced a seventy-five-minute factual color film entitled *Under Our Own Steam* distributed across the country to propagate news of Volkswagen's steady rise. VW's PR targeted drivers with *Gute Fahrt* (Safe Journey), a monthly magazine that reached a six-figure circulation by the mid-fifties.[2]

The Volkswagen's elevation to iconic status in West Germany cannot be primarily attributed to its manufacturer's promotional initiatives. Even the most expensive public relations strategy would have fallen flat had it not resonated with contemporaries. In the eyes of many citizens of the early Federal Republic, the Volkswagen appeared as the embodiment of core values that defined the new country they were in the process of building. Although burdened by its origins in the Third Reich, the Volkswagen's ambivalent legacy presented far fewer difficulties for the vehicle's iconic ascent than one might expect. The car's appeal derived in part from the opportunities it offered to recount Germany's recent and highly compromised history along favorable lines. These tales embraced extremely selective interpretations of the immediate past, foregrounding certain motifs while marginalizing those that would have cast postwar Germans in a morally questionable light. Rather than straightforward "suppression," it was a particular mode of recounting Germany's immediate past that helped the Volkswagen pass through a "historical carwash" after 1945.[3]

As the vehicle emerged as a collective marker, much of its symbolic power sprang from its ubiquity. With annual sales figures reaching six-figure territory by the mid-fifties, the VW's central role in the Federal Republic's mass motorization ensured that the car was virtually omnipresent in West German society. The vehicle's proliferation left a deep imprint on the word "Volkswagen." Initially a vague term denoting an affordable car for the wider population—an idea the Nazis had taken up with much clamor—"Volkswagen" turned into a thoroughly familiar brand name for a successful company, as well as its most successful product.

With its high public profile and its firm roots in the private realm, the Volkswagen functioned as one of West Germany's few largely uncontested collective symbols. Germany's recent descent into barbarism, and the country's subsequent split into two separate states as a result of the Cold War, complicated virtually all attempts to provide the foundations for national collective identities. In the political domain, the search for collective identification proved particularly difficult, since nationalism had been thoroughly discredited by National Socialism. Moreover, in light of the country's recent division, few contemporaries viewed the Federal Republic as a "nation." In consequence, for the time being, the new democratic institutions of the Federal Republic elicited little enthusiasm. Bonn, West Germany's capital, was ironically dismissed by many contemporaries as a "federal village" for its small size of roughly one hundred thousand inhabitants. It even took the new country's political elite several passionate conflicts before reaching an agreement about a national anthem in 1952.[4]

The dearth of federal symbols and the antagonistic debates surrounding them, however, could not conceal the wider population's desire for reaffirming collective icons. When the West German team surprised everybody by returning from the soccer world tournament in Switzerland in 1954 as champions, they were greeted by a wave of enthusiasm. This victory offered a "collective experience of success that gained almost mythical symbolic power." While the ensuing festivities, during which team members paraded through Munich in Volkswagen cabriolets, revealed a widespread desire for collective reference points among the West German population, the so-called "miracle of Berne" remained an isolated and altogether exceptional episode. In this respect, the triumph on the soccer field differed fundamentally from the Volkswagen, which gained a steady and thematically broad iconic presence in West Germany owing to its prominence in everyday life.[5]

As the Volkswagen developed into an unprecedented commercial success, public attention did not remain focused solely on the vehicle but frequently turned to the production site to which it owed its existence. In the fifties and early sixties, the VW works and its adjacent town acted like a magnet for scores of reporters and writers because developments in Wolfsburg provided shining examples of the reconstruction processes then transforming the Federal Republic from an impoverished, rubble-strewn country into an affluent industrial society with full employment. Albeit a geographically marginal, medium-size municipality less than ten miles from the Iron Curtain, Wolfsburg assumed a conspicuous place in the West German imagination and helped shape central contemporary assumptions about the nature of the postwar recovery. Founded with much pomp and circumstance in the Third Reich, the city had failed to live up to the regime's promises, yet its economic trajectory after the war demonstrated how West Germany had moved beyond the recent past, directing the public gaze toward a sector that had historically played only a minor part in the German economy. As Volkswagen rose to corporate prominence, a slew of articles and books about the factory and its benefits for Wolfsburg's inhabitants added crucial elements to the car's iconography in West Germany, linking it closely to a success story that unfolded in the realm of industrial production.

The transformation of the Volkswagen works from an economically unrealistic initiative to prominent, successful big business was intimately tied to the economic boom that underpinned the Federal Republic's political and social consolidation in the fifties. After the outbreak of the Korean War in 1950, the West German economy grew annually by 8 percent over the next ten years and prospered at an average annual rate of 6.5 percent throughout the sixties. Far more than coal mines and iron and steel works, it was the rapid expansion of chemical and electrical companies, mechanical engineering firms, and car producers that fueled West Germany's economic rebirth, a trend that signaled a gradual structural shift in the country's manufacturing base away from the traditionally dominant sectors of the Ruhr. At the same time, the Federal Republic gained an increasingly prominent presence in the global economy. Between 1950 and 1960, the country's share of overall global exports rose from 3 percent to 10 percent. Responsible for roughly one-fifth of the Federal Republic's GDP by 1960, success in export markets—in conjunction with domestic economic reconstruction—laid a crucial foundation for a considerable improvement in living standards. The completion of five million new dwellings in the fifties, more than half being affordable public housing, helped ease the postwar shortage of residences, while a fall in unemployment from

13.5 percent in 1950 to just 1 percent at the end of the decade went hand in hand with significant pay increases at a time of low inflation. As a result, average real incomes doubled in the 1950s. Most of these upward trends continued until the oil crisis of the early seventies, making the period between 1950 and the oil shock the longest and strongest boom in German history.[6]

Economic historians have advanced several explanations to account for this unique upswing. One school of thought has emphasized that the West German economy succeeded in unleashing a productive potential in the fifties previously stifled by the political turmoil of the interwar years and subsequently by the belligerent economic policies of the Nazis. The immediate postwar years amounted to a reconstruction phase for an economy that already ranked among the most highly developed in the world before the boom began. Another line of interpretation has drawn attention to the positive effects resulting from the enactment of Ludwig Erhard's vision of a "social market economy" during his tenure as economics minister. According to Erhard, the state's role was to establish a dynamic economic order that safeguarded free enterprise and competition, encouraged entrepreneurship with a sense of social responsibility, expanded property ownership among the wider population, and put in place social safeguards for the weak. Although only partially implemented, the "social market economy" successfully limited economic state interventionism and curtailed a widespread inclination among big business to avert competition through cartels and trusts. Erhard pursued an economic policy designed to bring "prosperity through competition," as the English title of his programmatic best seller put it. Yet West Germany's boom would have been inconceivable without the fundamental reconstruction of the international economy that occurred under American auspices after World War II. The Bretton Woods system of fixed exchange rates and the American-led push toward the liberalization of international trade through the General Agreement on Tariffs and Trade provided a crucial context in which West German companies returned to the world market. U.S. support of Western European initiatives for supranational economic and political cooperation achieved similar effects, since these initiatives led to the foundation of the European Economic Community in 1957 and gradually opened Western European markets for West German companies. A combination of the Federal Republic's own growth potential, a new international institutional framework, and German economic policy, then, were responsible for West Germany's economic expansion.[7]

In few places did the boom manifest itself as potently as in Wolfsburg. During the fifties, Volkswagen consolidated the tentative lead it had gained during the occupation period and established itself firmly as West Germany's dominant car producer by a considerable margin. Buoyed by burgeoning domestic and international demand, Volkswagen expanded its annual output from 46,154 automobiles in 1949 to roughly 960,000 in 1961, during which time the company's workforce in West Germany increased from 10,227 to 69,446. In the domestic market for passenger vehicles, VW's share oscillated between 34 and 40 percent between 1951 and 1961, while Opel, the Federal Republic's second-largest producer, trailed by at least 15 percentage points. The city of Wolfsburg showed impressive signs of growth as well: the number of inhabitants more than doubled, from roughly 30,000 to 63,000, between 1952 and 1960. By the end of the occupation period, with its housing still dominated by cramped barracks, Wolfsburg had begun to experience a construction boom that alleviated the town's dearth of housing and established an urban infrastructure in the course of the fifties. While the average number of occupants per residence stood at 6.9 in 1950, it fell to 3.6 over the ensuing decade. Much as

Americans were wont to label Detroit "Motor City," West Germans began to refer to Wolfsburg as "Volkswagen City."[8]

While journalists who wrote about Wolfsburg in the late forties had associated the town with the exceptional local electoral appeal of the radical Right, they treated the factory and town in the fifties and sixties as showcases of the country's economic dynamism. A visitor returning to the municipality after an absence of less than ten years hardly found his bearings in early 1958. "If you saw Wolfsburg yesterday, you will not recognize it today, and if you set out . . . to examine its present, be prepared, if you come back tomorrow, to start all over again: a new, yet again different, and previously unseen place will greet you." The town's vibrancy directly reflected the growth of VW, a company that, as a journalist from nearby Hanover pointed out, developed into "an exceptionally strong and powerful engine of . . . the West German economy." When a liberal reporter claimed in 1953 that the "German miracle" was above all "an industrial miracle, or, to be more precise, a production miracle," he found his main proof at VW, since this company had not only gained "European stature," but within just a few short years had achieved significance "even when measured by global standards." News coverage along these lines cast a car company in a novel economic role. If automakers had been minor players for the German economy before 1945, West German portrayals of Wolfsburg ascribed prime importance to Volkswagen within the postwar economic landscape. As VW increasingly appeared as a motor of growth in West German industrial society, the Federal Republic began to view itself as a country of car makers, a process also bolstered by rising production figures by Opel in Rüsselsheim, Ford in Cologne, and Mercedes-Benz in Stuttgart.[9]

VW aroused curiosity since the company outperformed the rest of the booming country. Observers seeking explanations for Volkswagen's exceptional strength often began with Heinrich

Nordhoff, who, while building on the foundations laid in the late forties, determined all important corporate strategies. Upon his arrival in Wolfsburg, Nordhoff had introduced a highly centralized corporate structure that gave him "very comprehensive powers." The general director consolidated his grip by promoting a string of blue- and white-collar workers, who had proven themselves capable organizers and coordinators, to leading managerial positions in which they served as loyal lieutenants. As a result of VW's exceptionally steep expansion curve throughout the fifties and early sixties, Nordhoff gained a reputation as an economic miracle worker second only to that of Ludwig Erhard himself, whose government policies he staunchly and publicly supported. In addition to business acumen, considerable (self-) promotional talents contributed to Nordhoff's fame. When he took to the public stage, Nordhoff displayed a charismatic persona, commanding and charming crowds with ease as he "celebrated" VW's sales and production figures. Some domestic observers found Nordhoff's public suaveness alien, likening him, despite his protestations, to an "American manager" rather than a typical German entrepreneur. Others found his public demeanor slick and "arrogant." Irrespective of these detractors, most West Germans thought of "King Nordhoff," as he was nicknamed in the press, as lord over an enviable empire of production with a specific set of norms and with a distinctive "corporate soul."[10]

Productivity ranked high among VW's values. Rather than merely push up production figures by placing additional demands on workers, Nordhoff explained in a speech to employees in 1956, the company focused on "creating an environment in which the most modern, labor-saving machines" could be installed. Although he repeatedly underestimated the scale of future demand, Nordhoff oversaw productivity gains at Volkswagen that allowed the company to increase its output per worker roughly

General Director Heinrich Nordhoff staging himself as the self-confident executive in front of thousands of workers and employees gathered outside the main factory in Wolfsburg. His penchant for (self-) promotion earned him the moniker "King of Wolfsburg" in the fifties. Courtesy of Volkswagen Aktiengesellschaft.

threefold during the fifties. Initially the company turned out more cars by using existing capacities more efficiently, but after 1955 further expansion required new methods. As part of the preparation for an investment drive, Nordhoff instructed a group of senior managers in 1954 to visit the machine tool show in Chicago and report back to him with "an outline of what our works are supposed to look like in ten years' time."[11] The most striking technological development the German visitors studied in Illinois was the automated machine tools that required neither manual loading nor removal of individual parts. Especially when combined with transfer lines that transported components between several manufacturing stations mechanically, automated machine tools offered significant potential for either a reduction of the workforce or for increasing output without further hirings. Relying on electronic control systems, automation presented an intensified version of Fordist mass production that preserved highly specialized and standardized forms of labor division as well as uniform work routines, while significantly increasing precision and reducing the amount of manual labor on the shop floor. General Motors and Ford were the American auto manufacturers that implemented automation most comprehensively during the fifties, in the process driving several smaller manufacturers out of the market who could not afford this costly investment. Nordhoff recognized automation's possibilities and authorized its gradual adoption. By the end of the fifties, the factory at Wolfsburg had introduced partial or full automation in its stamping, heat treatment, paint, and body shops. Roughly ten years after Nordhoff fired the starting shot for automation, Volkswagen's productivity levels in the mid-sixties were on a par with those in Detroit. In terms of its manufacturing methods, the company closely followed American models.[12]

Productivity increases, the general director insisted, did not entail harsher labor conditions. "People like to work in our bright, modern factories," Nordhoff declared on more than one occasion. Many employees would have disagreed with this rosy portrayal of their everyday lives, not least the commuters who made up half of Wolfsburg's workforce throughout the fifties. One expellee who joined the company as a line worker in 1953 recalled his daily routine without nostalgia. After cycling through the night to catch a Wolfsburg-bound train at ten minutes to four, he worked an eight-hour shift, holding down his job despite allergies to engine oil and smoke. While two colleagues who started with him soon gave up and went elsewhere, he persisted "because I was married and needed the money." In the early fifties in particular, the management was regularly confronted with workers who "were physically incapable . . . of keeping up with high work rates" because of lasting health problems "resulting from the war," another employee explained in an interview. Production areas with particularly taxing conditions included the body and the stamping shop, where workers lifting heavy components constantly "grumbled" and issued "threats of work stoppages, [which] did become a reality." These production areas in which harsh routines caused disciplinary problems were precisely those that were first targeted for automation. But even once the installation of new production technology reduced the amount of physical exertion in auto manufacturing, job satisfaction remained low. A reporter visiting the works in 1956 encountered workers who told of their "monotonous" occupations "with tired resignation." After eight hours of sanding down mudguards, "enough is enough," a semiskilled employee decisively put it. "The same thing all day! That got on my nerves! No, that's no job," a female worker recalled her years of employment at VW after she had retired. Just like at Ford's Highland Park plant decades earlier, toil and drudgery characterized life on the assembly line in Wolfsburg in the 1950s.[13]

Given the demanding nature of many jobs at VW, and Wolfsburg's location in the countryside less than ten miles from the border with East Germany, recruiting and retaining workers became a constant preoccupation for management. A West German labor market that, from the mid-fifties on, began to display shortages especially of skilled workers also left the automaker in a vulnerable position. The rationalization program Nordhoff launched in the middle of the decade went hand in hand with a rise in the proportion of skilled labor among the workforce from 32 percent to 37 percent from 1953 to 1961. At the same time, the training of semiskilled workers intensified. Quality concerns and an unwillingness to leave valuable equipment in the hands of unskilled operators motivated the hiring of a more highly qualified workforce, a trend that differed sharply from Henry Ford's employment policies at Highland Park. Unlike Ford, who had used mechanization to decrease the share of skilled workers from over 60 percent to below 30 percent in the

1910s, Nordhoff's management throughout the 1950s attracted a rapidly growing workforce with an improving qualification profile while keeping staff turnover at consistently low levels—a crucial precondition for efficient day-to-day operations.[14]

That "the scourge of the assembly line," as a journalist termed it in 1953, generated neither major public controversies nor mass defections in Wolfsburg despite workers' misgivings about their daily labor routines was related to Volkswagen's leadership culture, or so Nordhoff would have argued. The general director pursued a new approach to industrial relations, which he outlined in detail in a speech at the Swedish Chamber of Commerce in 1953. He unapologetically affirmed the need for firm leadership in business, since capable entrepreneurs counted among the "superior personalities who bring progress and change the shape of things." In light of wartime destruction, granting businessmen free initiative, he added, was of particular importance in "a country like Germany, which continues to feel the reverberations of the terrible catastrophe" and therefore possessed an economy that still lacked "stamina." As far as company strategy was concerned, Nordhoff tolerated no interference. He found ideal conditions to enact his vision of patriarchal entrepreneurial leadership at VW because the local trade union movement remained weak. Since the works dated only from the Nazi era, trade unionists in Wolfsburg could not reactivate organizational networks from the Weimar period, a strategy labor representatives used elsewhere in West Germany to strengthen their position. The composition of the workforce, with its disproportionately large share of expellees, many of whom had not worked in industry before 1945 and remained hostile to trade unions, hampered efforts at unionization.[15]

At the same time, Nordhoff emphasized that strong entrepreneurial authority did not necessarily entail tense industrial relations. "In my experience," Nordhoff explained to his Swedish audience, "the worker does not resist leadership, but he desires clear, meaningful directives. . . . He also desires senior executives whose superiority . . . he can acknowledge." To nurture the trust of the workforce in those at the corporate apex, the general director recommended an active information policy, pointing to his practice of holding quarterly, factory-wide assemblies in which he updated his employees on recent developments. Beyond stressing the need for effective communication within the company, Nordhoff also harbored the conviction "that the trade unions ought to be positive and desirable partners in social dialogue," especially if workers' representatives refrained from importing "party politics" into collective contract negotiations. He outlined his overall aim as working toward "social peace and social reconciliation in a manner diametrically opposed to the fruitless and completely antiquated idea of class antagonism."[16]

A bundle of motivations prompted Nordhoff's pledge in favor of collaborative industrial relations. As a former manager at the General Motors subsidiary Opel, he was particularly interested in the arrangements between GM's chief executive Alfred Sloan and the head of the United Auto Workers Walter P. Reuther, whose collective bargaining reduced ideological tensions while granting white autoworkers substantial improvements in wages and benefits in the immediate postwar years. Nordhoff was not the only West German industrialist to register with approval transatlantic examples of social reconciliation in the workplace. Otto A. Friedrich, who headed tire producer Continental in Hanover and with whom Nordhoff maintained cordial relations, also drew on American models to reorganize industrial relations. While German entrepreneurial circles of the Weimar Republic venerated Henry Ford's model of mass production and his hostile stance toward trade unions, American concepts emphasizing more collaborative industrial relations proved far more attractive in West Germany.[17]

Germany's recent past, the general director warned, also demonstrated the need to build bridges across social divisions in the workplace. "Especially after the terrible shocks of the last war . . . we simply can no longer afford the lunacy of infighting, if we wish to survive as a people and as free men," he insisted. Industrial relations, this stark formulation implied, were of existential individual and collective importance. At the same time, Nordhoff took great care to point out that his ideas were different from the notions of social harmony the Nazis had propounded under the label of the "people's community," declaring that in the context of industrial relations National Socialism had offered nothing but "politically colored cant." Nordhoff thus drew on both American models and lessons from the recent past to search for common ground between employers and workers.[18]

Nordhoff's commitment to a new approach to industrial relations went far beyond rhetoric. Hugo Bork, who headed the local branch of the metalworkers' IG Metall union between 1951 and 1971, became the general director's close and trusted negotiating partner. Nordhoff's overture met with favorable conditions during a decade in which West Germany's trade unions abandoned radical critiques of capitalism and instead adopted a more pragmatic strategy to secure tangible material improvements for the workforce. In Wolfsburg, the new collaborative style worked to mutual advantage. While the management benefited from the absence of disruptive strikes, the trade union seized opportunities to demonstrate its efficacy to the workforce by taking the lead in negotiations for beneficial contracts. By the mid-fifties, successful collective bargaining had helped IG Metall secure a majority in elections to the works council, the body that acted on behalf of workers' interests. Simultaneously, the candidates with right-wing extremist views who had attracted the largest number of votes in these elections in the late forties and early fifties lost their support, a clear indicator of diminishing social tensions and discontent. Nonetheless, union membership at Volkswagen remained comparatively low, amounting to only 50 percent of the workforce in 1967. Beyond the high number of expellees and the lack of a local organizational tradition, the continuing economic boom accounts for the union's recruitment problems that labor activists encountered in Wolfsburg and elsewhere in West Germany. In the economic climate of the fifties and early sixties, many workers were content to vote trade union representatives onto the works council but saw no need to join the organization itself.[19]

The collaborative approach toward industrial relations found its most important expression in the manifold material gains VW workers enjoyed. Similar to their American predecessors at Ford's Highland Park plant in the early twentieth century, Volkswagen workers received the most generous remuneration in all of West Germany from the 1950s on. At VW, average hourly wages rose from DM1.92 in late 1951 to DM4.78 in 1964, while for workers in the rest of the country typical hourly rates rose from DM1.48 to DM3.88. This positive wage curve constituted only part of Volkswagen's financial appeal. In the course of the fifties, the VW workforce gained a range of benefits including a company retirement plan, a life insurance scheme, loans to support the construction of private homes, and a Christmas bonus.[20] Copying a recent pay practice at GM, each worker and employee also received an annual 4 percent bonus as a share of the dividend. Other industrial employers in West Germany noted this initiative with alarm, since they feared VW had set a precedent that would fuel workers' expectations elsewhere. At the same time, the company reduced weekly working hours from forty-four to forty-two in early 1956, a time when the average West German working week stood at forty-eight hours. A year later, a Social Democratic newspaper reported that VW had "clandestinely introduced the 40-hour week"—a step that would have amounted to the fulfillment of one of the most prominent long-standing demands of the European labor movement. Fearing yet another precedent, businesses circles and the Conservative federal government were aghast and sought to extract an assurance from Nordhoff that he harbored no plans to implement a forty-hour week in West Germany's most prominent publicly owned company. Although it would take another four years for most of VW's labor force to work a forty-hour week, Nordhoff refused to stage a public climbdown on the issue in 1957, thereby indirectly demonstrating his company's tacit approval of a trade-union demand of prime symbolic importance.[21]

While the federal government and other employers regarded Volkswagen's corporate generosity with ambivalence, the West German press published numerous reports admiring the company's wage and benefits policy. The extent to which VW workers profited from rising incomes manifested itself most visibly in the large number of Volkswagens in Wolfsburg's streets during commuting hours. A visitor who arrived in the city at the end of a shift in early 1958 encountered "car after car" driven by those who had just finished work. A year earlier, a reporter for the conservative daily *Die Welt* counted no fewer than four thousand vehicles that awaited their owners' return in the "gigantic parking lot in front of the plant." The local proliferation of automobiles provided only one indicator of the comparative prosperity of VW's workers. When writer Erich Kuby took a stroll through Wolfsburg around 1957, he was amazed that the population of a factory town seemed to consist of "middle-class people, middle-class people, and nothing but middle-class people." Well-clad and busily shopping, the Volkswagen workers Kuby encountered contrasted sharply in appearance with the air of material deprivation Germans associated with manual laborers.[22] No longer marked by poverty, insecure employment, and meager retirement provisions, the workforce at VW displayed its discretionary spending power. What made social developments in Wolfsburg even more remarkable was the high proportion of expellees among the local population. While this group made up a disproportionately large share of the poor in many parts of West Germany during the fifties, the expellees formed an important core of the exceptionally well-paid industrial workforce in Volkswagen city. With its comparative wealth and social homogeneity, Wolfsburg appeared to have moved beyond class divisions and thus resembled the "leveled middle-class society" that, as influential sociologist Helmut Schelsky put it in a famous phrase in the early fifties, appeared to emerge as a defining hallmark of West Germany.[23]

Like the wider population, Wolfsburg's architecture radiated "solid prosperity," as a conservative journalist observed in the late fifties. Volkswagen's local taxes financed the rapid transformation of what had been a barrack settlement into a town with an up-to-date infrastructure, including residential and commercial buildings arranged along wide streets, new schools, a state-of-the-art hospital, large outdoor public swimming pools, and generous green open spaces. Drawing on British schemes for the so-called "New Towns," as well as Scandinavian urban planning models, the local government adopted a functionalist approach in the fifties. Dividing industrial, commercial, and residential zones, the city council built neighborhoods for four thousand to six thousand inhabitants with large lawns, parks, and easy access to the surrounding countryside.[24]

As a planned city, Wolfsburg differed fundamentally from West German urban centers like Berlin, Hamburg, Bremen, Nuremberg, and Munich. Up and down the country, local administrations faced the task of rebuilding cities with a substantial architectural heritage that lay in ruins after 1945, grappling with the question of which buildings and urban ensembles to preserve and which to replace. Since the Volkswagen works had been founded only in 1938, the space surrounding the factory had largely remained a tabula rasa. Containing neither wartime ruins nor many buildings predating 1945, the Wolfsburg that came into existence in the fifties struck many observers as the epitome of modernity and newness. West German journalists regularly drew attention to its "modern town hall," "modern apartments," and "modern houses." Admiration shone through in articles from the mid-fifties detailing local proposals "to create a new town, in which roughly 80,000 people can live in comfort and at ease." This coverage aroused curiosity far beyond journalistic circles and helped turn Wolfsburg into a "modern pilgrimage site" that attracted three thousand visitors each week by 1956. The city council sought to support the municipality's reputation through public-relations initiatives like a series of postcards issued on the occasion of the town's twenty-fifth anniversary in 1963 calling upon the West German population to visit "Wolfsburg, the modern industrial and residential city."[25]

That Wolfsburg celebrated twenty-five years of its existence in 1963 and thereby indirectly referred to its foundation in the Third Reich illustrates how little the city's origins in the Nazi era disturbed postwar tributes. Insensitivity to Wolfsburg's problematic beginnings partly resulted from the public silence that surrounded the crimes committed at the Volkswagen works during the war. Since the wartime foreign workers had been repatriated to their home countries, they were in no position to make their voices heard. Meanwhile, Volkswagen, led by a former Opel manager who had employed numerous forced workers at a truck factory in Brandenburg during the war, had no interest in drawing attention to wartime abuses, and thus fell in line with virtually all West German companies, which refused to accept liability for maltreatment in their plants. The public silence surrounding wartime offenses at VW reflected a broad consensus about National Socialism's criminal nature at the time. Although West German officials condemned the Holocaust and World War II as barbaric atrocities, they and the wider public only rarely acknowledged other criminal acts such as anti-Jewish initiatives before 1939 and the abuse of forced labor. Most West Germans argued that the murder of Europe's Jews had been committed by a small Nazi elite without the participation and knowledge of the general population. By cordoning off wider society from the majority of crimes during the Third Reich, this bundle of interpretations amounted to a collective self-exculpation and placed a strong emphasis on the regime's deceptiveness, by which it had allegedly concealed atrocities from the German people with great effectiveness.[26]

In addition to having misled the population, the Nazis, a prominent argument ran, had positively betrayed the nation by starting a hubristic war that eventually exposed the German people to aerial attacks, mass rape by the Red Army, and expulsion from their homes in eastern Europe. According to numerous postwar accounts, the German people viewed themselves as a collective victim of the Third Reich. Germany's Nazi past throughout the fifties took the shape of highly selective public memories that pushed German suffering into the foreground while remaining largely silent about issues of guilt. With its large number of impoverished expellees, Wolfsburg was predestined to be viewed as a site that revealed German hardship after World War II. This theme loomed large in both newspaper articles and Horst Mönnich's 1951 documentary novel *Die Autostadt* (The Car City), an instant best seller, which Heinrich Nordhoff praised for uncovering "truths that lie below the surface."[27]

In the course of the fifties, however, narratives of German victimhood were complemented by a different theme that had already shimmered through early accounts of postwar Wolfsburg.

Rather than as a site of helpless suffering, Volkswagen city struck visitors as emblematic of the wider reconstruction under way in the early Federal Republic. Composed with the support of the company and published in 1949, Heinz Todtmann's *Kleiner Wagen in großer Fahrt* (Small Car on a Roll) was among the first books to characterize Wolfsburg as a city where "stranded" Germans "from all strata and areas . . . prove through their current existence how serious they are about learning anew and rebuilding decently." Ignoring the local appeal of the radical Right that dominated several elections at the time, Todtmann detected "a new and roughly hewn form of democracy" born out of "hardship and existential necessity" in the factory town.[28]

Far more than new political values, an egalitarian ethos of hard work struck commentators as Wolfsburg's defining cultural feature. Todtmann, for instance, found that the citizens of Wolfsburg "made a virtue of necessity" and strove to create "a community in which nothing counts as much as achievement." A readiness for resolute self-exertion appeared to permeate all ranks. Nordhoff, who from the day he took up the reins in Wolfsburg exhorted his employees to dedicate themselves fully to their jobs, owed his high local standing in part to leadership by example. When the general director's Volkswagen broke down on his way to the plant on a hot summer morning, he turned this glitch into a demonstration of virtue by single-handedly "pushing [the car] to the factory gate" for over a mile. "Dripping with sweat, he arrive[d] there with his coffee-colored jalopy." Praise for this work ethic saturated a host of accounts about the town. While press reports consistently exaggerated the scale of destruction at the factory at the end of the war and downplayed the British contribution to Volkswagen's survival in the late forties, they drew attention to "proud" German workers who "saw their chance" in Wolfsburg and "rolled up their sleeves." In 1961, Nordhoff summarized the cause of VW's success in the following simple terms: "We did it under our own steam." Wolfsburg offered an early example of self-made German socioeconomic success. Born out of an exemplary work ethic and a cooperative approach between capital and labor, the city was seen as indicative of a culture of achievement that emerged as a central feature of West German identity in the fifties.[29]

Since the local radical Right collapsed amid the economic boom of the fifties, Wolfsburg no longer made negative political headlines as it had in the late forties. A prominent debate about Volkswagen that absorbed considerable political energies in the late fifties reinforced the dominance of economic motifs in Wolfsburg's public image. Since the factory's transfer into joint ownership between the Federal Republic and the state of Lower Saxony in 1949 was intended as a strictly provisional arrangement, the question of who actually owned the Volkswagen works essentially remained unresolved, leaving VW a booming "works that belongs to no one," as several journalists put it. Not until 1961 did West Germany's political establishment end Volkswagen's proprietary limbo by turning the company into a publicly listed corporation in which the Federal Republic and the state of Lower Saxony each held a 20 percent stake while the rest of the stock was issued to private investors as "people's shares." Designed to spread "co-ownership to the man in the street," as Ludwig Erhard explained when first announcing the idea for "people's shares" in 1957, this legal construction ensured that both prominent state actors and private shareholders continued to take an active interest in the corporation's economic fortunes.[30]

The "people's shares" came at the end of a protracted process that turned the Volkswagen works into "a symbol of how the German people manages to assert itself as a modern industrial nation after a terrible catastrophe," as a conservative weekly summarized in 1957. Indeed, some considered Volkswagen nothing less than an industrial "fairy tale."[31] With its highly productive, rapidly expanding, and technologically advanced plant, its collaborative industrial relations under a generous patriarch, and its prosperous workforce, Volkswagen struck numerous commentators as a paradigm of how their country moved beyond victimhood through postwar reconstruction based on hard work and a culture of achievement. Wolfsburg's up-to-date urban infrastructure and its calm local party political landscape further reinforced favorable impressions of orderly industrial, largely apolitical modernity. While Volkswagen offered one of the Federal Republic's earliest stories of self-made success, the social and economic results of the car maker's rapid development drew attention to the new role the auto sector had begun to play within West Germany's industrial landscape in the fifties. The countless accounts of the company's economic success formed an integral part of the iconography of the car itself, demonstrating its status as central to a larger boom that benefited both an expanding corporation and the national economy.

Of course, shining stories of an economic recovery that gradually led contemporaries to think of West Germany as a car-making country could also be told about Ford in Cologne, Opel in Rüsselsheim, and Daimler-Benz in Stuttgart. The fascination Volkswagen exerted in the early Federal Republic, however, far surpassed aspects of production. It was this corporation that manufactured the vehicle that advanced West Germany's mass motorization like no other, thereby turning the car itself into a symbol of the enticing prospects economic recovery held for West German society.

In the fifties and early sixties, the Volkswagen became omnipresent on West German roads, fulfilling the long-frustrated desire for private car ownership. The VW set the country's automotive standard and transformed into far more than an indicator of

West Germany's economic recovery. The car's symbolic eminence in the Federal Republic rested on the conviction that its technical characteristics, which the company consistently sought to improve, embodied the very values on which the country's resurgence rested. The Volkswagen's roots in the Third Reich did little to disturb its iconic postwar image because West German society retold the vehicle's history in highly selective ways that regarded its commercial success as proof of the Federal Republic's superiority over National Socialism. After all, it was the postwar order that delivered, en masse, the very vehicle that had remained nothing but an empty promise under the dictatorship. Frequently treated as the material manifestation of the qualities underpinning the postwar order, the Volkswagen served as the harbinger of an increasingly appealing normality that took shape in the fifties.

Although the Federal Republic remained far behind the United States, which led the world with a ratio of one passenger car per 2.8 persons in 1962, it rapidly caught up with its Western European neighbors. By 1962, the West German ratio of passenger vehicles per capita stood at 1:10, already close to French and British levels at 1:7.8 and 1:8.5 respectively. Translated into absolute figures, this trend amounted to a rise of registered passenger cars in West Germany from roughly 821,000 to over 6.6 million from 1952 to 1963. The majority of the vehicles responsible for this surge were small cars, a striking contrast to the lavish automobiles sold in the United States at the time. Since West German drivers could not yet afford more expensive models, over 950,000 of the 1.16 million cars purchased in the Federal Republic in 1963 featured an engine of less than 1.5 liters. While this pattern highlights the limits of the Federal Republic's material recovery, it nonetheless marked a fundamental shift with respect to individual transport, spelling the end of the motorcycle as the car's main alternative.[32]

An increase of average monthly pretax incomes from DM304 to DM950 between 1950 and 1965 provided the indispensable precondition for mass motorization, allowing West German society to translate the desire for the automobile that had already existed in the Weimar Republic and the Third Reich into actual economic demand. As incomes rose, the automobile industry's clientele underwent major changes. Initially demand for passenger cars primarily came from business circles, because the self-employed and entrepreneurs earned far higher incomes than the rest of the population and profited from tax breaks for their vehicles that were not available to private drivers. In 1952, a daily newspaper reported, merely 10 percent of passenger cars were acquired by private individuals. This figure had climbed to 40 percent four years later, but almost two-thirds of the roughly 2.3 million cars registered in the Federal Republic remained company vehicles. Only in 1960 did privately owned automobiles outnumber commercial vehicles in West Germany.[33]

The drivers best able to afford a new private automobile were overwhelmingly civil servants and white-collar employees, whose pay was roughly 15 to 20 percent above the national average. From 1957 to 1963, annual sales to this middle-class segment rose from 85,481 to 372,996, but most blue-collar workers, who made up half the working population and, despite pay rises, earned wages hovering approximately 20 to 25 percent below the national average, found a new car beyond their means well into the sixties. Although the number of manual workers who bought a new vehicle jumped from 74,774 in 1957 to 302,462 in 1963, they were far more likely to turn to the expanding market for used cars. In 1963, nearly 730,000 workers became automobile owners via this cheaper route. While one in four middle-class households owned an automobile in 1959, only one in eight working-class families possessed a private vehicle, mostly second-hand cars powered by an engine smaller than one liter. Following a fundamentally different social path of mass motorization from that of the United States with its high share of early rural drivers, the Federal Republic of the fifties and early sixties was nothing like Helmut Schelsky's model of the "leveled middle-class society" but bore the imprint of contemporary income inequalities.[34]

The cars that proved most popular throughout the fifties included microcars with engines smaller than half a liter, often developed by companies under threat from the collapsing motorcycle market. BMW reacted to this development by securing the production license for the Isetta bubble car from Italian motorcycle manufacturer Ivo. Introduced to the West German market in 1955 at a price of DM2,580, the three-wheeled, egg-shaped two-seater, which driver and passenger entered through a single front door, reached speeds of roughly fifty miles per hour thanks to a twelve-horsepower, four-stroke engine. By the time BMW ended production in 1962, it had sold over 160,000 models. Glas was another motorcycle producer in the microcar market, retailing around 280,000 of its Goggomobils at a cost of around DM3,000 between 1955 and 1961. Equipped with a two-stroke engine that produced between 13.5 and 15 horsepower, the "Goggo" had four wheels, a door on each side, and offered seating for two adults and two children. Before BMW and Glas, Borgward had brought out the Lloyd 300 in 1951 for DM3,330. This four-seater with a ten-horsepower, two-stroke engine consisted of a plywood body mounted on a hardwood frame. Selling about 132,000 vehicles, the Lloyd was mocked as a "bomber made of sticking plaster" (*Leukoplastbomber*) because artificial leather rather than metal formed its exterior skin.[35]

While these models provided basic forms of individual motor transport, they suffered from a lack of engine power as well as social prestige, offered little comfort, and gave insufficient protection in accidents. "Only those who do not shy away from death drive Lloyd," a characteristic German saying went at the time. These problems burdened cars with engines between 500 and 1,000 cubic centimeters to a far lesser extent. Auto Union achieved an early success with its DKW Meisterklasse with twenty-three horsepower and the thirty-four-horsepower DKW Sonderklasse, selling roughly 110,000 models between 1950 and 1955. Praised by owners as "the little miracle" (*das kleine Wunder,* taking up the brand's initials) for their streamlined bodies and sophisticated suspension systems, DKWs were fully fledged automobiles that retailed at prices from DM5,800. At similar cost, Ford offered its Taunus 12M, a thirty-eight-horsepower, 1.2-liter limousine that attracted around 250,000 buyers in the fifties.[36]

All these automobiles, however, existed in the shadow of one single competitor: the Volkswagen. Annual sales of the car made in Wolfsburg shot up from 61,522 in 1951 to 151,733 in 1956 and reached 369,746 by 1961. With over 2.1 million models on West Germany's roads in 1963, almost every third passenger car was a Volkswagen. The VW's strong market position reflected an enormous and seemingly insatiable demand for the small vehicle that the works struggled to satisfy despite "Nordhoff's single-minded pursuit of a single-model policy." Since the company could not turn out its star product fast enough, German customers faced average waiting times of four months in 1955, a figure that rose to over a year toward the decade's end. In small towns like Weißenburg in Franconia, the imminent delivery of a batch of cars from Wolfsburg became a local news item in the second half of the fifties because, as a dealer explained, "everybody wants a new Volkswagen."[37]

VW's model policy was a crucial element in this success story. While the company offered a highly standardized product that kept production costs under control, it did not go as far as Henry Ford, who stubbornly objected to most modifications of the Model T. Throughout the fifties and sixties, VW continued to deliver its main product as a pared-down "Standard" and a pricier "Export" limousine with more elaborate technical and cosmetic features (including chromium details). As the corporation kept the car's distinctive silhouette unchanged, in-house designers and technicians repeatedly modified the vehicle to maintain its appeal. Some alterations, including an expanded color range of body paints and interior fabrics, served to enhance the VW's aesthetic appeal, but other initiatives amounted to considerable technical changes. In the course of the fifties, engineers increased the engine power from thirty to forty horsepower, gradually introduced a fully synchronized gearbox to facilitate the operation of the stick shift, improved visibility through substantially bigger windows, modified the heating system, added more comfortable, adjustable seats, and fitted hydraulic brakes that allowed drivers better control. In most cases, Wolfsburg initially introduced technical alterations for the Export version before transferring them to the cheaper Standard a few years later.

This approach served a dual purpose. While technical changes kept the vehicle up to date by improving power, handling, and comfort, producing both an economy model and a more technically refined export version extended the company's market segment beyond those customers seeking a basic set of wheels. Volkswagen indirectly catered to the lower end of the West German luxury auto market, because, for DM2,000 more than the price of an Export, drivers could acquire a VW Cabriolet, which the Karmann body shop in Osnabrück produced in close cooperation with Wolfsburg. Volkswagen thus covered a broad range of a West German auto market characterized by demand for small vehicles. Sales also exploded between 1950 and 1960 because VW lowered the price of the Export model from DM5,450 to DM4,600, thanks to productivity increases and the expansion of production. At this rate, the Export was considerably cheaper than a vehicle in the class of a DKW or Ford 12M, but more expensive than a microcar. This was also true of the Standard, which retailed at around DM900 less than the price of an Export.[38]

While VW engaged in no systematic market research before the later sixties and thus did not possess precise information on the composition of its clientele, the Volkswagen's price made it a quintessentially middle-class vehicle. Many middle-class drivers were prepared to pay the difference between a VW and a microcar because, by common consent, a Volkswagen offered outstanding value in the small-car market. Test reports consistently praised it for combining "economy and power during operation, modernity in technical design, and meticulous finishing." It was, an auto journalist concluded in 1951, simply "the best car on the German market." Numerous owners agreed over the years, commending the stability of VW's frame, the car's road-handling characteristics, and its stamina. The four-cylinder boxer engine, which consumed around eight liters of fuel per 100 kilometers (or twenty-nine miles per gallon), emerged as a particularly noteworthy feature, winning a reputation for starting reliably under adverse conditions, "purring" steadily at top speeds of between sixty and sixty-five miles per hour on the autobahn, and navigating urban traffic swiftly.[39]

The Volkswagen was as rugged as the Model T had been. Requiring few repairs, comparatively little fuel, and limited maintenance, the car made in Wolfsburg may initially have cost more than some of its domestic competitors but proved cheaper in the long run—an important factor in a society still characterized by financial limitations. Its dependability also made the Volkswagen stand out from many German cars of the past as well as several of the microcars of the time. As an automobile perceived to be manufactured in accordance with the high standards of German "quality workmanship," the VW thus struck contemporaries as a simultaneously affordable and prestigious vehicle because its comparatively low price did not compromise its technical characteristics.[40]

Providing "a simple and reliable means of transport" remained Nordhoff's explicit priority, as he explained in an interview in 1957. Having overseen the introduction of stringent quality controls in the late forties, Nordhoff continued to pursue a policy that strove to eradicate faults and defects with an almost missionary zeal. Although sales were booming and the press was showering the company's product with compliments, the general director reminded top executives in no uncertain terms during a meeting in September 1954 that complacency was out of the question. "We must do everything in our powers to maintain quality and remove sources of complaints. This enterprise seems to make a specialty of recognizing deficiencies but taking far too long to deal with them," he stressed. To lend substance to his stern admonition, he enumerated a series of problems that demanded immediate attention: "At the moment, there are complaints about rattling doors and wheel caps, about door handles that work badly, about gear wheels [in the gear box] that are too loud, and then some. All of this is very annoying; we can't go on like this, or we will have a big setback one day."[41]

As Volkswagen made quality the cornerstone of its corporate strategy, the company vowed to keep its product's technical virtues intact. While competitors emphasized innovations when relaunching models, Wolfsburg took a far more conservative approach that sought to preserve the VW's widely acknowledged merits. This strategy resulted less from an antipathy to market forces than from the conviction that customers would turn their back on a product incorporating flawed modifications. Tried and tested features such as the air-cooled rear engine, the torsion-bar suspension, and the vehicle's distinctive rounded shape remained central characteristics over the years, providing tangible evidence of technical continuity. Although the material features of both the Standard and the Export models underwent significant alterations throughout the fifties and early sixties, the company stressed that it steered clear of what it deemed unnecessary changes. In the late fifties, for instance, Nordhoff delayed the introduction of an electronic fuel gauge for years because he deemed available devices unreliable and thus a danger to the car's reputation. Owners were required to note down their mileage conscientiously or risk running out of gas on the road, especially if they had forgotten to fill a small reserve tank the car featured for emergencies. Although drivers complained about the absence of the fuel gauge, Nordhoff insisted that VW would incorporate only equipment that was "cheap and reliable and completely, completely safe."[42]

Nordhoff's dedication to quality extended beyond immediate production issues. The expansion of Volkswagen's network of dealerships and service stations, which had already become a focus of attention in the immediate postwar years, continued unabated throughout the fifties. Only a car company, the general director's reasoning went, that assured drivers they would readily find help in case of a breakdown could hope to sustain growth in the long run. By the end of the decade, more than one thousand licensed workshops with qualified staff trained in accordance with guidelines from Wolfsburg featured the Volkswagen logo, offering maintenance and repairs at moderate prices almost everywhere in West Germany. A journalist visiting a VW service station in Hamburg in 1960 was impressed by the efficiency of the mechanics who, within two hours, changed a dented fender and fixed the bumper for a driver who had arrived without an appointment after a minor accident. "The bill came to DM53.20. Just try and have the same kind of repair carried out at the service station of another make," this observer wrote with considerable admiration.[43]

There were thus very sound material reasons for the Volkswagen's commercial success in the fifties and early sixties. As VW's hit product sold hundreds of thousands, the vehicle became ubiquitous on West German roads. "Every other car . . . seems to be a Volkswagen," an auto journalist summarized his impression as early as 1950, a visual presence that only increased over the years. In a country whose car ownership levels had long trailed those of other Western European industrial nations, the VW put West Germany's transformation on display for all to see. By the beginning of the sixties, a regional newspaper was only stating the obvious with a headline that read: "The VW—symbol of economic ascent."[44]

When contrasted with the Third Reich's history of betrayal, West Germany shone particularly brightly because the Federal Republic made good on a promise broken in the past by producing millions of Volkswagens. "Before the war," an auto journalist wrote in 1949, the car "existed only on paper, cheated people out of their savings, and yielded a lot of propaganda." As it turned the aspiration of car ownership into reality, the Federal Republic proved its superiority and gained legitimacy. In this respect, the Volkswagen's proliferation marked a clear and obvious break between the Third Reich and West Germany. At the same time, a fundamental cultural continuity underlay this process. Since the Federal Republic fulfilled a long-standing wish to own an automobile with the help of a model designed in the Third Reich, the proliferation of the VW was simultaneously bound to affirm consumer dreams of a "people's car" that had arisen under the Nazis. By retroactively meeting material ambitions from the twenties and thirties within a novel political and economic framework, then, the Volkswagen acted as an icon of the Federal Republic in which continuities and ruptures with the Nazi era were inextricably interwoven.[45]

The Volkswagen's ability to encapsulate both breaks and links between the Third Reich and the Federal Republic rested on a widespread postwar consensus that the car possessed a politically neutral past. West German commentators consistently played down the ideological dimensions of the Nazis' mass motorization scheme, at times reducing Hitler's involvement to little more than a "whim." Rather than Hitler, the argument ran, Ferdinand Porsche deserved credit for the vehicle. In 1959, Heinrich Nordhoff characterized Porsche as "a phenomenal man, a designer and an engineer not only by occupation but with all his heart. . . . He was a fanatical and exceptionally gifted engineer" whose creation, the VW, embodied "the creed of a whole generation of constructors." Nordhoff's appraisal was in keeping with the West German media, which uniformly eulogized Porsche as a "genius." As postwar accounts of the Volkswagen's Third Reich history shied away from critical scrutiny of Porsche's close relationship with Hitler, the car emerged after 1945 as a technical artifact that owed its existence to an outstanding individual engineer rather than the Nazi regime. In 1955, the weekly *Der Spiegel,* often regarded as a rare voice of critical journalism in the early Federal Republic, articulated the dominant view in an emphatically uncritical fashion: the Volkswagen amounted to a "timeless basic construction"—a phrase that, like so many contemporary assessments, dissociated it from its political past and turned it into an engineering classic, thereby lending it an unproblematic lineage.[46]

Not even the fact that the car's military version, the Kübelwagen, had been used in the Second World War left a significant stain on the VW's postwar reputation. On the contrary, the wartime services rendered by the Volkswagen's military incarnation enhanced its postwar fame as a quality product. At the beginning of the fifties, a *Spiegel* journalist anointed the Kübelwagen as the "most splendid war automobile" since its air-cooled engine had neither overheated in "Rommel's desert" nor frozen "during the Russian winter battles." Veterans who drove a VW in the early fifties concurred, openly attributing their automotive choice to positive wartime experiences. An aristocratic owner who took great pride in his pricey Volkswagen cabriolet explained in a car magazine in 1951 that his "predilection for the VW dates to the African campaign. There, starting in Tripoli, he became familiar with the Volkswagen in the desert sand."[47]

To a considerable extent, positive assessments of the VW's wartime service were supported by a powerful myth in the early Federal Republic that credited the Wehrmacht with having conducted an honorable campaign rather than one marked by numerous atrocities. Since the myth of the "clean" Wehrmacht held firm for decades, invoking the Kübelwagen's use in the war posed no danger of moral contamination for the Volkswagen during the fifties and sixties. But while references to the Kübelwagen did not compromise the postwar Volkswagen in moral terms, recollections of the VW's military version evoked war's lethal dangers. Horst Mönnich's best-selling novel *Die Autostadt* contains a dramatic passage in which two German corporals in a Kübel find themselves encircled and hopelessly outnumbered by British troops blocking all exits from an African wadi. As shells explode around them, they are convinced that "now, now it was over." In desperation, the driver steers the Kübelwagen up a steep slope. Unexpectedly, "the wheels gained traction. . . . Hits to the left and to the right, the sand dust threw up fountains. But they climbed. Inch by inch." Despite heavy fire, they reach the top of the sand dune. As the Kübel tips over the ridge, "the desert lay ahead of them. Freedom!" is the soldiers' overriding thought as they race away from their opponents. This fictional episode more than highlighted the car's versatile performance in inhospitable terrain; as it credited the Kübelwagen with a lucky escape, it cast the vehicle as an object that had faced the same kinds of mortal perils as countless soldiers. In the early postwar years in particular, many West German veterans may thus have seen the Volkswagen as an object whose history closely resembled their own: like them, and against the odds, this car was a survivor.[48]

Largely depoliticized portrayals of the car's origins as a quality product with roots in a time of broken promises and existential insecurity prepared the ground for the Volkswagen's immense cultural resonance in the fifties. As an economical, simple, and highly reliable commodity, the car was not simply indicative of West Germany's recovery; its technical characteristics also reflected a new normative framework that, observers hoped, lent stability to the postwar order. VW itself argued along these lines on more than one occasion. Similar to manufacturers of industrial design commodities for the home, the company went to great lengths to position its product as an incarnation of sobriety and solidity that exerted a strong appeal because its plain appearance contrasted starkly with National Socialism's overblown promises and the chaos of war.[49]

At the beginning of the 1950s, an article penned by Volkswagen's PR manager maintained that a new landscape of desire stood behind the car's rise to prominence. Forced to turn "necessity into a virtue" after the war, he declared, "people in Germany [had] become realistic. . . . Rather than the primitive which we came to despise the more the accursed war forced it upon us, we revere the useful and serviceable, the truly progressive. It is not external appearances that determine value, but what is intrinsic. . . . After hard experiences and in the midst of difficult times, one learns to appreciate . . . the true inner values." Detaching the automobile from the Nazi era, the company linked its product to values like moderation and utility, which dominated postwar culture. Nordhoff described his aim in similar terms when he spoke of his ambition to offer "people genuine value [through] a product of highest quality." The car's virtues, according to VW's publicity, stood for a new, stable form of normality that ordinary people could rely on.[50]

The company's efforts to identify the Volkswagen with a new peacetime normality struck a chord among the wider population. Through its ubiquity, the Volkswagen gave values like honesty and dependability a pervasive and visible presence in West German culture. Crucially, VW refrained from altering its external appearance by preserving its beetle-like shape. As an auto journalist pointed out, the car's improvement remained largely "invisible" and thus allowed the company to sell ease of mind. In the late 1950s, when the first journalists criticized the model as somewhat dated and began to lobby the company for a replacement model, many drivers defended it as a much-loved token of stability. One fan used the most conventional idiom of loyalty and obedient companionship available to middle-class man: "I like [my Volkswagen] like a dog on the street corner with its faithful eyes." Such affection was rooted, an auto journalist explained, in the car's absolute dependability. "How lightheartedly one drives it, how strongly one trusts it! This trustworthiness is a guarantee that VW drivers' nerves will be spared as much as possible—a very important consideration in our hectic times." Statements like these were more than mere expressions of private attachment; they illustrate the "need for compensatory stability" that ran through many spheres of life in the early Federal Republic. The Volkswagen could be viewed as material proof that the achievements of the fifties rested on foundations that were as solid as a VW. Put differently, the car's technical and optical characteristics came to reflect a collective desire for a lasting postwar normality. Rather than make false promises, here was an honest, resoundingly unflashy, reliable, and immediately recognizable product whose success signaled that the new postwar period with its incipient, appealing affluence had come to stay. Irrespective of its Third Reich origins, the Volkswagen turned into a symbol not only for the country's rapid postwar recovery but for the values that would turn it into a permanent feature of West Germany.[51]

As mass motorization became a defining element of West Germany's postwar normality, countless contemporaries integrated their Volkswagens into everyday life. The drivers of the Weimar Republic had already praised their small cars for enriching the private sphere, a motif subsequently taken up by the Nazi regime in its propaganda for the "people's car." Opening the door to new, pleasurable pastimes for millions in the fifties, the Volkswagen's arrival led numerous car owners to value their vehicles as profoundly intimate possessions. The treatment the VW received as a highly prized personal commodity often struck observers as bizarre and humorous. At the same time, its growing prominence in quotidian affairs also triggered acrimonious debates about the VW's appropriate place in the contemporary gender order as an increasing number of female drivers took the wheel. While these tensions suggest that the Volkswagen could give rise to social conflicts, its private appropriation by millions nonetheless anchored the vehicle deeply in West German everyday life.

Some West Germans could not wait to become car owners and jumped at the first opportunity to acquire a Volkswagen. In 1951, when Gerhard Kießling heard that a relative who ran a car rental agency offered a damaged VW at a discount, he did not have to think twice. While this trainee teacher could not afford to operate his Volkswagen day to day, he spent much of his limited disposable income on weekend outings and on extended holiday trips all over Europe, where he covered ninety-six thousand kilometers over the next five years. Piling a tent, sleeping bags, blankets, and other equipment onto the backseat, he traveled with friends and visited the Black Forest, Switzerland, France, Italy, England, and the Scottish Highlands. While numerous Wehrmacht veterans returned to places they had come to know as military occupiers to relive "youth, adventure, and domination," younger West Germans who had not served in the war sometimes saw international travel as a form of "public pilgrimage of atonement to distance [themselves] from the Third Reich [and] . . . build a bridge to a new Europe." Neither impulse appears to have motivated Kießling. Having survived the war as a private in the Wehrmacht, completed his studies, and secured a safe job, he simply "wanted to see something of the world." In his case, the VW transformed private life in a manner that allowed him to satisfy an apolitical curiosity about faraway places.[52]

Many contemporaries would have regarded this young teacher's long-distance excursions in the early fifties with considerable envy, because even at the decade's end fewer than one-third of all West German households could afford an extended annual holiday. Those who did overwhelmingly remained within the country, frequently staying with friends and relatives.[53] Nonetheless, throughout the fifties the automotive press brimmed with articles about international and domestic travel that could be undertaken with a Volkswagen, offering tips on routes as well as practical issues including passport regulations, currency quotas, fuel vouchers, foreign culinary customs, hotel categories, behavioral guidance, and, crucially, cost estimates. This flood of texts was of greater significance than providing practical advice for would-be tourists; as the economic boom and mass motorization progressed, it established trips both within and beyond West Germany's borders as a firm prospect for the wider population. Long before such outings actually became a reality for most contemporaries, travel turned from a distant dream into a viable expectation. Given its prominence in the Federal Republic, the

Volkswagen played a key part in the process whereby regular excursions became a normal aspiration in an increasingly affluent population.[54]

For most West Germans, private life centered on their immediate families, irrespective of the tensions that might exist between husbands, wives, and children. Often returning from the war with physical disabilities as well as emotional scars, men "found it difficult to reintegrate into the family," Hanna Schissler has noted, especially if their attempts to reclaim their conventional role as head of household met with resistance. Beyond gender tensions between spouses, a generational gap provided another fault line within the family when offspring proved unwilling to accept the authority of long-absent fathers. Finally, living conditions often remained cramped, even in those families enjoying the good fortune of moving into a newly built apartment, because planners allowed on average merely six hundred square feet (55 square meters) for four people in 1953.[55]

Although tremendous strains could permeate families in the fifties, most West Germans spent many hours off work in the company of immediate relatives, passing their time within their own four walls by listening to the radio, reading the daily paper, and simply relaxing. To some extent, this withdrawal into the family can be read as a belated reaction against the hyper-politicization that had characterized the Third Reich. After years of almost omnipresent political propaganda, as well as the chaos and humiliation of military defeat, most West Germans quite literally wanted to be left in peace. The domestic sphere's appeal also derived from a longing for an unspectacular and predictable existence. Viewed from this perspective, the family circle emerged as a prime site where West Germans hoped to find normality after the war. The long working hours of the fifties, which deprived the majority of the population of time and energy to pursue activities outside their home, further contributed to leisure patterns revolving around domesticity. Rather than devote a significant share of their rising incomes to hobbies and commercial entertainments that would have taken them out of their homes, many West Germans saved for big-ticket domestic purchases such as furniture, washing machines, fridges, and television sets.[56]

Short of building one's own house or buying an apartment, the most costly of these material objects of desire was, of course, an automobile that promised to extend West Germans' radius of activity beyond their home while offering new forms of family pleasure. Although half of the nine hundred drivers taking part in a 1959 survey stated that they owned their vehicle primarily for professional reasons, most contemporary commentators emphasized that an automobile, if used in a circumspect manner, was ideally suited to enrich family life through short trips. Unlike the small cars of the Weimar Republic whose diminutive size left them unsuitable for family outings, the VW offered sufficient space to accommodate parents and their offspring. Published in 1951, the *Buch vom Volkswagen* (Book of the Volkswagen) explicitly recommended the vehicle because it allowed couples and their children to undertake "weekend excursions." Such trips did not need to be expensive. "Within a thirty-mile radius, there is always a good opportunity," the author stressed, and a "clever housewife" could further limit expenses by bringing food and drink. Once the family had reached its destination, the children could play in the forest or on a lake while the parents relaxed: "Yes, that's it, one feels freer and happier. . . . Life is hard enough as it is, times are difficult, everybody should cast aside his inhibitions and give his body what is beneficial for it."[57]

As a family vehicle, this text suggested, the Volkswagen put drivers in a position to leave the mundane concerns and material restrictions of everyday life behind. A short journey thus offered an opportunity to enact a family ideal of peaceful happiness and undisturbed bliss in an idyllic landscape. Few images capture

Weekend idyll with Volkswagen. As West Germany's recovery gathered pace in the fifties, the VW emerged as a symbol of a new, family-centered postwar normality. Courtesy of J. H. Darchinger, Friedrich-Ebert-Stiftung, Bonn.

this vision of postwar normality better than the photograph of a young family picnicking in front of their Volkswagen parked on a forest strewn with various leisure items. Shot in 1957 by Josef Heinrich Darchinger, the pictorial chronicler of the economic miracle for several dailies and weeklies, it staged the Volkswagen as an integral part of an altogether pacific family ensemble that exuded self-contained contentment.[58]

The strong emotions that automobiles elicited in their owners counted among the curiosities that attracted constant attention.

An owner's bond with the car began with the act of acquisition, an event amounting to far more than a prosaic financial transaction. Buying a Volkswagen was a special occasion that testified to personal success and status, serving as an indicator that the new owner counted among the economic boom's beneficiaries. "My father's first car was a Volkswagen, a used 'Beetle,' which provided a decisive step forward in our family's motorization," a man recalled of his 1950s childhood. "An inevitable ritual of appropriation" in the form of "a small family drive" during which "father explain[ed] the advantages of the new vehicle" provided the first step in the vehicle's domestication, he continued. Apart from visibly demonstrating a family's good economic fortunes to friends and neighbors, the first drive launched a protracted process of appropriation during which the buyer transformed an automobile into an eminently personal possession.[59]

Many owners marked their vehicles as private territory, treating them like a canvas on which they expressed aesthetic predilections by adding accessories that helped either eradicate the traces a previous driver left on a used car or, in the case of a new vehicle, modified the highly standardized product VW offered. To be sure, extras such as tow ropes, first-aid kits, repair tools, and road maps, as well as metal staffs to measure gas levels in the tank before the arrival of the automatic fuel gauge, were primarily functional rather than expressions of personal taste. Many drivers, however, did not restrict themselves to such practical additions. A minority added supplementary lights, chromium strips, whitewall tires, stickers, tags, and more to their cars, abandoning themselves to decorative urges that invited ridicule for creating a "rolling Christmas tree."[60]

Highly conspicuous exterior alterations may have risked opprobrium, but hardly anyone raised an eyebrow about embellishments to the Volkswagen's spartan interior. Albeit a costly investment, fitting a radio proved a welcome diversion on long, potentially monotonous trips along the autobahn. Alongside cushions, protective seat covers guarded against wear and tear while lending a cozy feel to the VW.[61] In the fifties, a particularly popular aesthetic alteration of the car's interior were small vases, many of which were produced by high-end porcelain manufacturers including Rosenthal and Arzberg. While some couples attached a vase to their dashboard to preserve blossoms plucked en route before displaying them as a "visible memorial token" at home, others had little inclination to remove the flowers they found along the way from their car. "My husband and I most certainly do not want to do without fresh flowers in the vase of our VW," one Lilo Müller asserted in the mid-fifties. She regarded her floral bunches as a defining trait of the family vehicle. In keeping with their widely acknowledged role as homemakers, women left a pronounced aesthetic imprint on the inside of many a Volkswagen, thereby playing a crucial role in turning them into personalized possessions.[62]

According to contemporary auto manuals, a regular maintenance routine was a key activity through which male owners demonstrated their dedication to the automobile. Even a manual explicitly addressing readers with little mechanical expertise included detailed information on cleaning a carburetor, fixing a fuel pump, checking a distributor, and adjusting a generator. However, the best indicator of a man's attachment to his automobile was the increasingly common weekend ritual of washing the car. In some families this chore fell to wives and children, but numerous male drivers proved remarkably reluctant to delegate this activity. In and around the house, virtually all forms of dirt removal were exclusively female responsibilities, but when it came to the automobile, few men felt that a temporary suspension of the conventional sexual division of labor compromised their masculinity. Some writers encouraged owners to keep vehicles spick-and-span because a car's external appearance formed part of an individual's public persona. The leading manual for the VW driver portrayed washing the car as an act of appreciation: "When we decide to clean our car ourselves, we virtually never do this to save money . . . but rather out of a noble intention to reward our cars for offering us faithful service day in and day out."[63]

Beyond concern for their reputation and gratitude, deep affection prompted drivers to wash their cars themselves. Writing in the early fifties, an automotive journalist likened the impulses that prompted men to perform cleaning routines to those felt in the early stages of a romance. He commented that the sight of a middle-class man putting on "his oldest suit and a pair of even older shoes [before] getting a bucket of water, a sponge, [and] mother's best chamois leather" to "wash, polish, and clean his 'wheels'" with gusto revealed "a degree of love and care that [could] lead detached observers to believe he was flirting with a new lover." Cleaning a car thus illustrated the intimate bond that tied many owners to their automobiles. Another writer expressed the relationship between car and driver along similarly gendered lines as he pointed out that the automobile "is the only affair of which you can speak without censure and which you can take home without committing a sin." While both authors feminized the car by casting it as an object of male desire, they highlighted the exuberant emotions cars elicited.[64]

A VW advertisement from 1956 likened the relationship between a Volkswagen and its owner to that between husband and wife. Pictorially merging a man's smiling face with a frontal view of a Volkswagen body, the tagline pronounced the automobile "his better half." Instead of a stormy love affair, the stability of marriage provided the dominant motif in this ad about the unity between proprietor and car. Others described the Volkswagen as a "good friend" or a house pet, an apt analogy given the extensive grooming vehicles received. Small wonder, then, that numerous Volkswagens, like good friends and pets, received nicknames including "Fridolin," "Oscar," and, in the case of a red VW, "Tomato." West German citizens developed countless ways of expressing their automotive affection.[65]

As soaring sales in the fifties and early sixties ensured that the present overshadowed many unsavory details of the VW's Hitler-laden origins, the attachment owners developed to their cars turned the Volkswagen into a public symbol whose significance went beyond articulating the hope that the Federal Republic's normality rested on firm foundations. Rather, the car evolved into a national icon that millions of West Germans venerated as a personal possession. In this respect the Volkswagen differed fundamentally from the deutsche mark, another prominent symbol of postwar stability. Given the currency's character as a means of financial transaction, DM banknotes retained a far more prosaic everyday presence because they changed hands incessantly.[66]

While both women and men played important roles in turning Volkswagens into personal possessions, there could be little doubt that most contemporaries ultimately considered the automobile a male domain. Above all, this conviction derived from the assumption that operating a car competently required a considerable degree of technical expertise. To spread knowledge of the car's inner workings in a society in which automobiles had long remained the preserve of the wealthy few, the auto manuals of the early Federal Republic included lengthy passages devoted to mechanical aspects. Helmut Dillenburger's *Practical Auto Book,* a 450-page guide that sold more than two hundred thousand copies between 1957 and 1961, devoted roughly 150 densely printed pages to detailed explanations of, among other subjects, various engine types, brakes, clutches, transmissions, ignitions, and fuel injection systems. "The motor car of today,"

An early Volkswagen ad that pictorially celebrates the unity between car and driver. By referring to the car as the proprietor's "better half" (bessere Hälfte), the ad cast the VW as an automobile that "stood by its man." Courtesy of Volkswagen Aktiengesellschaft.

he pointed out, "is more powerful than its predecessors. . . . To recognize and apply its potential correctly, one needs to know more about it." Efforts to promote popular technical knowledge with an eye to familiarizing drivers with their vehicles, however, were largely lost on women, many male contemporaries argued. In keeping with long-standing beliefs about female technical incompetence, women were said to display only a limited interest in the functional aspects of automotive engineering. In 1957, an article in the middle-class women's magazine *Constanze* categorically declared that "a man who wants to buy a car . . . must know that for the vast majority of women an automobile is an item that consists of two parts only: the body and the interior decoration."[67]

West German women faced widespread suspicions about their ability to drive. Although few statements in which men explicitly charged women behind the wheel with gross ineptitude made it into print, a wealth of evidence testifies to the private prejudices female drivers had to contend with. The daughter of a VW dealer appealed to *Gute Fahrt* in the following manner in 1951: "Thank God I don't have a husband yet who forbids me to drive, but I do have a daddy who greets me with equal skepticism. . . . What shall I do?" Some husbands did erect insurmountable obstacles that prevented their wives from steering the family car. The daughters of Hildegard Eyermann, a university-trained pharmacist who had gained her permit in the Weimar Republic, recalled that "our mother never drove while father was alive." When Kurt Eyermann, a country doctor, took his family on outings in their dark-green VW in the fifties, Hildegard sat in the passenger seat, clutching the road map and warning her nervous husband against oncoming traffic on the rural lanes of upper Franconia. "Even in the winter, with all the snow and the hills in our town, she never got the car. She did all the shopping by bicycle," her daughters remembered, still shaking their heads decades later. As late as the early sixties, women who drove their VWs with self-confidence and panache felt it necessary to defend themselves against charges of being either "masculine" or "bisexual," since many contemporaries continued to regard the car as a male preserve. West German mass motorization thus bore the imprint of a deep-seated automotive misogyny that articulated itself in the private realm and severely restricted women's ability to take advantage of ownership.[68]

Reliable figures about the number of female drivers are difficult to find for the early Federal Republic, since national statistics did not count them as a separate category. A sociological survey taken around Frankfurt in 1959 shines a light on the sexual inequality that was in all likelihood part of a wider social pattern: fewer than 9 percent of those whom the social scientists encountered behind the wheel were women. Nonetheless, the fifties and the early sixties witnessed an increase in female drivers, as statistical records of women gaining a license to handle a passenger car, truck, motorcycle, or moped indicate. While 141,226 women passed a test to handle a motorized vehicle of some type or another in 1955, as many as 372,629 received a license in 1963, a rise from 16 to 23 percent. That women developed a more prominent road presence in the early sixties is also reflected by the decision of several publications to address female drivers' concerns directly. By 1961, *Constanze* ran a biweekly column for readers who regularly got behind the wheel. Two years later, *Gute Fahrt* introduced "Beate," a commentator who championed female drivers in debates on whether woman or man "was the better participant in road traffic." The emergence of columnists directed at female drivers in the sixties highlights that magazines ignoring gender issues in automotive matters risked appearing outdated.[69]

The rising number of women driving during the fifties and the early sixties had several causes. Growing levels of wealth and a labor market that saw the proportion of working married women below the age of forty increase from around 27 percent in 1950 to 40 percent in 1961 partly account for this trend, putting more women in a financial position to afford driving lessons or to buy and maintain a car. The way in which Marlies Schröder, a seamstress in her early twenties from the rural Eifel region, learned to drive in the early sixties offers a case in point. When a local driving school hired her for sewing work, she recalled using her earnings "to get a driver's license myself." Fearing parental opposition, "I registered in secret. Only after I had passed several lessons with success did I confess my project." A dose of subterfuge and an independent income put this young woman in a position to gain her permit.[70]

At the same time, public debate worked to isolate the camp that regarded the operation of passenger vehicles as a male prerogative. It is no coincidence that virtually no explicit statements about women's supposedly inferior abilities to handle cars appeared in the public sphere throughout the fifties. From its earliest days, West German public culture acknowledged women's contributions to the German war effort as well as to the country's reconstruction during the "rubble years" in the immediate aftermath of military defeat. In both contexts, German women had amply demonstrated that they could fill roles in arenas conventionally deemed male preserves. Against this backdrop, public assertions that categorically denied women's ability to drive were bound to prove deeply controversial. The Basic Law, which served as West Germany's political constitution after 1949, worked to similar effect. Although family policies of the fifties embraced the concept of separate gender spheres that defined women's roles as maternal homemakers while assigning to husbands the duty to act as providers, the Basic Law enshrined equality between the sexes as one of the Federal Republic's central constitutional tenets. In other words, no one could legitimately deny women the right to drive in public, regardless of conservative family policies.

Rising employment figures, public memories of women's contributions to the war and reconstruction, and constitutional stipulations on sexual equality provided the context that facilitated public challenges to private prejudices against female drivers. Some men overtly rejected suggestions that female drivers were incapable of handling their automobiles with assurance. Insecure female drivers lacked practice, one argument ran, because their husbands treated the family vehicle like a "toy" that they "wished to keep for themselves. . . . Above all, gentlemen, let your wives drive on their own, that's when they learn it." Women, meanwhile, advanced the public claim throughout the fifties that they—and not their male counterparts—ought to be considered the more reliable drivers, especially when it came to avoiding accidents. In 1950, *Constanze* printed a photograph of a secretary next to her VW whom American occupational authorities had officially crowned as the "safest driver" in their zone. "Women drive better," declared a headline in the same publication a year later, citing an expert of the German Automobile Association (ADAC) as well as accident statistics to back up the self-confident assertion. Beyond offering assurances of their automotive competence, women emphasized in the fifties that they regarded themselves as "passionate VW drivers."[71]

These interventions managed to change the tone of public debate on gender and the automobile, but only to a limited extent. Although some female writers in the early sixties still felt a need to remind contemporaries that "we are very well able to navigate road traffic drive as competently as our male competitors," by then others flatly refused to discuss stereotypical gender ascriptions and instead treated them with irony and sarcasm. "Every woman knows that men drive well—without exception," a female columnist ridiculed the prevalent male self-image in 1963.[72] For some women, charges of female automotive incompetence no longer merited serious consideration. At the same time, a number of male commentators began to argue in public that female drivers often displayed more talent than men. Several male interviewees in *Gute Fahrt* agreed that "women drive better than men, with a greater sense of responsibility and in a more relaxed fashion."[73]

Yet the very men cited who issued the most confident statements regarding women's driving abilities in *Gute Fahrt* ended their accounts with frank admissions that their own wives never drove in their presence for fear of attracting spousal criticism. Whatever partial success they had in challenging gender stereotypes in public contexts, female drivers continued to encounter skepticism in the private realm as increasing numbers of women took the wheel in the early sixties. Still, even those men who regarded female drivers with mixed feelings now felt compelled to acknowledge that, in principle, women could confidently handle a Volkswagen. By the early sixties, West Germans gradually came to accept that drivers of both sexes regarded the VW as a treasured possession. Despite the gender tensions that accompanied mass motorization, for countless West Germans the Volkswagen turned into an object of great affection that their owners washed, serviced, and embellished. The close bonds that car ownership created between man and machine—or woman and machine—were central to the Volkswagen's iconic status in the Federal Republic.

The automobile's everyday appeal to both sexes was closely related to its emancipatory potential. By the beginning of the 1950s, the notion that driving promoted individual liberty already had a long and, in the German case, checkered history. Henry Ford, back in the 1910s, had famously praised the automobile for its capacity to enhance individual freedom, as had owners of small vehicles during the Weimar Republic. Hitler, too, had invoked this motif when he envisioned the "people's car" as an exclusively "Aryan" instrument that would release the *Volksgemeinschaft* from the constraints of public transport. While none of the Third Reich's racial restrictions on car ownership survived into the Federal Republic, the notion of automotive freedom gained unprecedented prominence in West Germany, and not only because of the rising number of cars. The predominant Cold War climate, and the ways many citizens, including Volkswagen owners, narrated their driving experiences cast the automobile as an instrument of freedom in the fifties and early sixties. Nonetheless, a rising death toll due to car accidents soon cast a pall over the automobile's reputation as a liberating force in West German society. Public safety concerns triggered a protracted debate about the necessary limits of liberty on the road.

Given the country's membership in the American-led coalition of the "free world," "freedom" *(Freiheit)* provided a key political term in West Germany. Throughout the fifties, the Federal Republic's political and economic elites issued a barrage of public pronouncements in praise of individual liberty that underlined a sharp contrast with the collectivist ethos prevalent in the socialist bloc. The rhetoric of freedom permeated a plethora of political statements as well as tributes to "free enterprise," which, as Ludwig Erhard declared in a 1957 essay, had returned "life and liberty to a deprived and hungry people" within a few years. Beyond strictly political and economic liberties, the freedom of movement counted among the core values espoused in the West. In 1954, Wilhelm Röpke, one of Erhard's most influential economic advisers, went so far as to include the "freedom of traffic" among the Federal Republic's defining traits. He elevated the ability "to satisfy one's transport needs in accordance with one's wishes" to a "basic and inalienable right of the individual," an "elementary liberty" to which "a state on this side of the Iron Curtain needs to subscribe."[74]

Although political circles explicitly tied the automobile to the domestic agenda of the Cold War, the wider public displayed little overt interest in such politically charged discussions. This reluctance reflected not only a general lack of faith in political ideology in the wake of the Third Reich, but also the profound skepticism with which the new parliamentary order of the Federal Republic was greeted by a population that, as several opinion polls showed, retained strong "authoritarian, antidemocratic, and fascist inclinations" in the early fifties. As a result, journalists praising the automobile's ability to enhance individual freedom restricted themselves to succinct and general proclamations: "The car makes us free," a writer on automotive affairs declared laconically in the early fifties, repeating a point that had been made for half a century. One of his colleagues agreed, characterizing the automobile as the epitome of individual liberty: "Grasping the steering wheel is the essence of one's personal freedom." By employing a key term of West Germany's political language, praise for the automobile as a source of "individual liberty" indirectly associated the car with the Federal Republic's liberal political values. As it boosted individual mobility, the proliferation of the car—and the Volkswagen in particular—revealed that West Germany did indeed offer enticing new freedoms. Mass motorization played a crucial role in demonstrating to a politically skeptical population that West German Cold War rhetoric with its encomiums to freedom amounted to more than mere words.[75]

Yet it was not just individual mobility that lent the automobile the appeal of liberty. The car itself could serve as a zone of privacy that allowed people to circumvent restrictive moral codes and legal regulations. Unmarried couples seeking solitude, for instance, faced considerable difficulties in the early Federal Republic, given the cramped conditions in many households. Finding a hotel for a tryst was fraught with obstacles, because federal law threatened managers who rented rooms to couples without a marriage license with prosecution for abetting prostitution. Under these circumstances, even unmarried couples on vacations who booked themselves into separate rooms within the same establishment risked prosecution. Amorous activities inside the car, however, were beyond legal reproach, as a car guide pointed out in a section entitled "on the kiss in the car." Readers who were concerned about incurring a potential charge of being a "public nuisance" received the advice that auto-erotic shenanigans did not qualify as an indecency in the eye of the law, as long as a couple parked in a discreet location: "Only *public* nuisance exists [as a corpus delicti], but for this the public is necessary. The public begins with one person, but not with the policeman; one should inform him of this when he opens the door in case the windows are fogged up." The advice appears to have proved apposite, because no legal cases about amorous entanglements in cars made the headlines throughout the fifties. For young people in particular, the automobile served as a space of personal freedom, offering a safe site for erotic experimentation.[76]

Beyond enhancing individual mobility and providing a space of privacy, the automobile's liberating qualities became manifest in the act of driving. West Germans who handled a Volkswagen throughout the decade frequently described doing so in highly emotive, enthusiastic terms. "I so enjoy driving," wrote a young woman in 1951 about her outings in the VW. Even experienced professionals like an auto journalist found test-driving "a Volkswagen for three weeks . . . an express delight." The junior teacher introduced above still recalled driving as a "wonderful sensation" more than fifty years after he first took his VW for a spin. For a growing number of West Germans the Volkswagen added a new joy to life, an automotive happiness that stood in sharp contrast to the gloomy deprivations of the immediate post-war years.[77]

Mass motorization added to West Germans' personal identity a new dimension, which, for lack of a better word, may be termed the "driving self." This aspect of the self, which springs into action when an individual gets behind the wheel, took time to develop, however. Placing the operator inside an unfamiliar machine and engaging all senses, driving required the acquisition of a set of highly coordinated individual skills. Over time, repeated practice set in motion a process of habituation that gradually transformed what began as a series of discrete, initially tentative, self-conscious actions into a flowing, largely intuitive performance that could give women and men the sense of being in unison with a powerful technological device. As a sociologist observed in the fifties, a driver gained a degree of assurance that allowed him to move the focus of attention from consciously "pushing the stick shift [and] the complex operation of the vehicle [to] moving along the road. The driver himself [now] drives, he himself takes the turn, leaves another driver behind etc." "Being at one with the machine" was the turn of phrase a contemporary employed to describe the sense of mastery he developed while driving his Volkswagen.[78]

A heightened feeling of power formed part and parcel of the transformation that made the driver. An observer of West German road traffic in 1957 noticed that "at the steering wheel, we yield to the apparent increase of power, which inflates our 'ego.' Enormous forces obey at the touch of a finger; obedience is complete." Reflecting on the intimate unity between human and car, the author of an auto manual took conceptual refuge in ancient mythology, observing that "the good driver merges with his vehicle into a modern centaur, whose head rules the engine and the wheels without demanding the impossible." As such, drivers stood out as "people for whom the machine has become a friend, not a ruler or a slave, but an intermediary toward a higher sense of living." The Volkswagen was by no means the only car that turned citizens of the Federal Republic into drivers, but its proliferation on the country's roads did more than any other model to help West Germans experience the driving self and its liberating sense of power.[79]

Those who acquired an automobile during the fifties could enact their identity as drivers under remarkably car-friendly conditions. Although many roads were initially in poor repair, owners could drive with few interruptions, since traffic density outside built-up areas remained low during the early years of the decade. When car numbers increased significantly in the second half of the fifties, the federal government initiated an extensive road construction program to ensure smooth traffic flows. Beyond investments in infrastructure, the state promoted car ownership directly by granting a tax credit in 1955 that allowed private owners to write off the costs of their daily commute.[80] Most notably, the legislature afforded drivers considerable leeway while on the road after the parliament in Bonn followed a proposal by transport minister Hans-Christoph Seebohm to remove all speed restrictions for automobiles in 1953. Seebohm, a leading member of the ultra-conservative Deutsche Partei, celebrated the bill not only as an important step toward restoring national sovereignty because it reversed restrictions the Allied occupiers had introduced in 1945. Even more significantly, he explicitly described his initiative as a return to "our old, tried and tested regulations"—an open reference to the traffic laws of the Third Reich that had abolished all speed limits in 1934. Neither fellow parliamentarians nor the press took issue with Seebohm's historical line of argument, neglecting to point out that the transport minister had brought back a failed Nazi policy that had prompted the regime to reintroduce comprehensive speed restrictions in 1938 because of high accident rates. Instead, members of parliament and journalists were under the impression that the Nazi dictatorship had enacted speed limits only to prevent fuel shortages at the beginning of World War II. This convenient misperception led *Der Spiegel* to congratulate the government a few years later on having repealed a "Nazi law" passed in 1939.[81]

Seebohm's laissez-faire approach to speed limits rested on the assumption that drivers would handle their vehicles with a "sense of responsibility," moderating their speed so as to expose neither themselves nor other road users to unnecessary dangers. Yet not long afterward, when the country registered an increase in road accidents, serious doubts arose about whether the Federal Republic's car owners really could keep their driving selves in check. National statistics attributed 6,314 deaths and 150,416 injuries to road accidents in 1950, but these figures climbed to 11,025 and 298,231, respectively, in 1953, and reached 14,088 deaths and 412,036 injuries in 1962. Given a more than tenfold increase in passenger cars during this period, a less than threefold increase in accidents may look modest in retrospect, but this was not how contemporaries regarded the issue. They saw road accidents as a new and dramatic public risk whose causes and appropriate antidotes triggered controversies about the relationship between state, civil society, and the individual. At the core of this meandering public debate stood the question of how much freedom an individual ought to be granted while on the road.[82]

Although commentators identified as sources of accidents a lack of motoring skills, the bad state of the road network after the war, inattentive pedestrians who literally got in drivers' ways, and the presence of large trucks, many observers insisted that the lion's share of responsibility for road casualties had to rest with the drivers of passenger cars, since this group expanded at the fastest rate. Ultimately, those behind the wheel were suspected of lacking the character traits to handle a car in a manner that did not endanger fellow citizens. "Tell me how you drive, and I tell you whether you are a decent human being," one auto journalist asserted in a statement that elevated an individual's driving style to a reflection of personal temperament. West German automobilists, a popular argument went, displayed a far greater degree of aggression behind the wheel than their Western European counterparts, whose countries exhibited lower accident rates. "The French driver is polite and considerate by nature, just as the English driver is polite and considerate," proclaimed one source. Concern about West German drivers' mores was neither confined to the press, nor did it come to an end in the late fifties. In 1961, the Bonn Foreign Office asked embassies in France, Italy, Spain, and Switzerland whether aggressive road conduct by West German holiday drivers damaged the Federal Republic's international image. Although the diplomats wrote reassuring reports dispelling officials' fears in Bonn, the initiative itself illustrates that apprehension about West German drivers' behavior circulated in many spheres.[83]

When public commentators reflected on the causes of the aggressiveness that appeared to manifest itself on the country's roads, inexperience alone provided an unsatisfactory explanation. Instead, they diagnosed a civility deficit among West Germans. The Federal Republic's roads reminded some drivers of combat zones. "Whenever I take a journey, I feel as if I'm off to the front. You never know whether you will get back home safe and sound," a driver wrote in the early fifties. Military motifs continued to color West German assessments of driving conduct well into the late fifties and mid-sixties, when a prominent car journal described the autobahn as a "battlefield," while West Germany's leading news magazine impressed upon readers the scale of havoc through a reminder that the annual tally of victims stood at "more than one and a half times as many dead and fifteen times as many injured as in the Polish campaign" of 1939.[84]

Present-day social scientists attribute aggressive road conduct to a lack of communication among drivers, but the frequent military motifs in accident debates in the early Federal Republic indicate that numerous West Germans suspected an altogether different cause at the time. The rising crash curve, they feared, was an unwelcome inheritance left behind by the Third Reich and World War II. Of all people, it was transport minister Seebohm who made this point explicitly, crediting the supposed lack of "discipline" on the road to "the conditions of the war and the postwar years." Concerns about the moral legacy left by National Socialism emerged in numerous contexts in the early Federal Republic and triggered a host of calls to stamp out the remnants of National Socialism's militaristic ideals in everyday life. In a protracted effort to recivilize the young country comprehensively, numerous etiquette books, some of which became best sellers with six-figure print runs, offered extensive advice on how to remodel quotidian manners along the expressly civilian lines appropriate for a democratic polity. Similar motifs stood behind programs from the mid-fifties promoting more restrained driving styles on West German roads. Underwritten by the Federal Transport Ministry, these voluntary initiatives encompassed educational schemes, the dissemination of information in various media, and awards for exemplary behavior on the road.[85]

In light of steadily rising accident rates, however, many West German commentators argued that appeals to voluntary behavioral change were futile unless complemented by more extensive disciplinary sanctions that penalized reckless conduct. Controversies about road safety legislation, which gained urgency in the mid-fifties, revolved around the state's regulatory power over civil society and the extent to which authorities should curtail the freedom of the road in the interest of public safety. Since excessive speed invariably emerged as the main cause of car accidents, speed limits provided a particularly virulent issue in debates about traffic regulation. While politicians from all parties and diverse professionals including policemen, urban planners, and trauma surgeons pressed for compulsory speed restrictions, the auto industry and car associations established a powerful lobby that defended, with considerable success, the right to drive as fast as one wished as a crucial individual liberty.[86]

In its campaign, the latter camp could count on the vocal support of the expanding ranks of drivers, not least those behind the wheel of Volkswagens, who consistently reacted with indignation against regulatory initiatives on the part of the state. "In no other country does it rain as many bans" as in Germany, *Gute Fahrt* railed in 1952, adding a few years later that the Federal Republic's traffic regulations highlighted "a German love of prohibiting." While the federal government acted on public safety concerns in 1957 and limited passenger vehicles to thirty miles per hour in built-up areas, polemics against a German tradition of state authoritarianism proved remarkably effective. Despite increases in traffic-related deaths and injuries, Bonn refrained from imposing further restrictions until 1971, when it introduced an upper limit of fifty-five miles per hour on two-lane open roads but continued to exempt autobahns from similar checks to avoid a public outcry. By then, the Federal Republic's lenient approach to speed limits marked the country as an international regulatory exception whose roots, like the Volkswagen's basic design, dated back to the Third Reich.[87]

Given the largely successful campaign against speed limits, the VW's reputation did not suffer from protracted controversies about traffic safety. Moreover, the debate about speed limits themselves centered on individual conduct rather than on specific automobiles. Even those who identified car traffic as a public danger concentrated on the behavior of drivers rather than cars. As a result, safety features of automobiles escaped public scrutiny, and the technical flaws in vehicles that may have contributed to hazards on the road did not attract much attention in the Federal Republic during the fifties and early sixties. The increase of traffic-related deaths and injuries, therefore, left no scratch on the Volkswagen's iconic body. Instead, countless West Germans saw in the VW an instrument that turned the Federal Republic into a nation of drivers, allowed them to enjoy unprecedented individual mobility, and fostered a sense of liberty that found its most potent expression in the lasting right to drive as fast as one wished. Sidestepping the deep suspicions with which many citizens regarded ideologically charged political pronouncements at the height of the Cold War, the Volkswagen thus functioned as a West German icon that articulated a notion of freedom in a seemingly apolitical manner.

In the early fifties and sixties, the Volkswagen conveyed a host of flattering associations about the young Federal Republic as a socially and economically modern, albeit largely apolitical country. In the context of the "economic miracle," the car took up and amplified an early West German story of self-made success based on hard work and a culture of achievement. As sales soared, observers emphasized that VW's good fortunes benefited both corporate interests and employees. The Volkswagen, this story went, owed its existence to an exemplary and highly productive company that, under Nordhoff's patriarchal leadership, embraced cooperative industrial relations, rewarded its workers with the highest wages and best benefits in the land, and funded Wolfsburg's metamorphosis from a barrack settlement into a medium-size city with an up-to-date infrastructure. At the same time, the VW reassured contemporaries that the Federal Republic's seemingly miraculous transformation, in which automakers played an increasingly prominent role, was no fleeting episode. Against a backdrop of recent existential hardship and insecurity, the solid and reliable Volkswagen appeared as the epitome of trustworthiness, whose ubiquity indirectly expressed the hope that the enticing postwar order with its incipient affluence rested on foundations that were as solid as the little car itself. As an automobile embodying the virtues associated with the label "made in Germany," it conveyed a narrative of self-made success based on quality workmanship. Crucially, the VW's comparatively low price ensured that the vehicle remained anything but a distant icon. By becoming widely affordable, it demonstrated that broad sections of the West German population benefited from the economic miracle directly. As it helped West Germany become a nation of drivers, countless men and, from the early sixties, women began to treat the vehicle made in Wolfsburg as an object of affection that they integrated into their quotidian routines. The intimate bonds between West German citizens and their VWs owed their existence not only to new, car-based spare-time activities but to a liberating sense of enjoyment drivers encountered behind the wheel. The Volkswagen became a collective symbol of the economic miracle—a symbol that millions owned, and treated as a unique and private possession.

The car's Nazi origins presented virtually no obstacle to this ascent. Silence about the human rights abuses at the factory during the war supported the Volkswagen's postwar prominence, as did a host of highly selective public stories that presented the car's history in largely depoliticized terms. Rather than acknowledge the ideological motivations behind the Third Reich's plan for a "people's car," contemporary historical accounts focused on the vehicle's outstanding technical characteristics, crediting these to Ferdinand Porsche, who was cast in the role of a politically neutral engineering genius. Since most war stories of the car's involvement in military campaigns focused on the Kübelwagen's technical virtuosity, the VW's wartime record presented no major embarrassments for the West German media either. Above all, however, the Volkswagen's proliferation in the fifties threw into sharp relief National Socialism's failure. By delivering on the enticing promise of an affordable automobile for the wider population, the Federal Republic proved its superiority over the Third Reich while retroactively legitimating the consumer dream of individual car ownership that had arisen between 1933 and 1945. As it inextricably intermeshed the recent past and the historical present, the Volkswagen emerged as an unlikely survivor of the Third Reich that thrived in the Federal Republic, a trait that it shared with the numerous West Germans who viewed their own biographies in very similar historical terms.

The VW thus owed its outstanding prominence not only to burgeoning sales but to an uncanny ability to square several circles. As a much-loved presence in everyday affairs, it allowed contemporaries to translate their recent, deeply compromised history into a success story that bathed the postwar order in a thoroughly positive glow. As an affordable high-quality product, it cast West Germany as both a consumers' and a producers' republic in which the auto industry became a prime engine for prosperity. As an altogether unpretentious exponent of the Federal Republic's seemingly miraculous socioeconomic transformation, the Volkswagen managed to appear as simultaneously normal and exceptional, signaling that the postwar order with its long-awaited, incipient affluence rested on stable footings. And as an instrument that enhanced individual mobility, it drew attention to new personal liberties at the height of the Cold War without becoming entangled in overtly partisan political debates. Production and consumption, work and leisure, past and present, exceptionality and normality, liberal freedoms and seemingly apolitical privacy counted among the themes the Volkswagen invoked in West Germany in the fifties and early sixties. Integrating an unusually broad and at times contradictory set of associations, the VW functioned as a socially ubiquitous, culturally supercharged, and largely uncontroversial icon that shone very brightly indeed in a West German cultural firmament that was largely bereft of widely shared national symbols after the Third Reich.

The VW's rise to fame, however, was by no means an exclusively domestic affair. As the company from Wolfsburg established itself as one of the Federal Republic's leading industrial concerns, it owed its expansion increasingly to international sales, even more than to its dominant position in the home market. As such, the Volkswagen became not only a West German symbol of reconstruction at home but also an "export miracle" that quickly secured the vehicle a global presence. As sales abroad thrived, the VW developed an astonishing cultural valence on the international stage that ultimately aligned it, despite its origins in the Third Reich, with the counterculture of the late sixties.

5

An Export Hit

"The Beetle does float," declared the headline of an article in *Sports Illustrated* in 1963. As proof for its surprising claim, the magazine offered a two-page photograph of a Volkswagen swimming in a creek framed by plush Floridian swampland vegetation. The image, a caption explained, "was made in Homosassa Springs after a Volkswagen was lowered gingerly onto the water by a crane." While *Sports Illustrated* did "not recommend this experiment to others," it was deeply impressed by the car's amphibian properties because it only sank after remaining above water for almost half an hour. The writer freely admitted that this unorthodox test had been prompted by suspicions about the veracity of countless reports celebrating the Volkswagen as a commodity that, due to its solid engineering and quality, could function in the most improbable habitats. "Suspended in hyperbole, the Volkswagen cannot be all they say it is," the article set out its skeptical premise before conceding that "even the claim that the car is part water bug checks out." Buying more than two hundred thousand vehicles made in Wolfsburg in the previous year, Americans, *Sports Illustrated* continued, were conducting a "romance with a plain Jane. . . . In Florida not long ago, a bride stuck a miniature VW . . . atop her wedding cake [and] a Kansas couple sent out birth announcements when their VW was delivered." Boisterous demand and numerous declarations of affection showed that "the Volkswagen has found a home in America." In fact, strong global sales, the article concluded, had turned this vehicle "into the most easily recognized car on earth."[1]

The feature in *Sports Illustrated* was one of many articles to express astonishment at the enthusiastic welcome American drivers had accorded to the German import since the mid-fifties. The Volkswagen's success in the United States provided a deeply unusual business tale that puzzled numerous commentators. American observers readily acknowledged the new arrival's technical features as an important source of its appeal, but they nonetheless struggled to grasp the affection countless customers bestowed on their VWs. Some writers drew attention to the Beetle's unusual shape, declaring it an unlikely best seller. While *Sports Illustrated* was puzzled by the charms of this automotive "plain Jane," several business and auto journalists went a step further and dismissed the car as "inelegant," "lumpy," and "ugly." Clearly, the VW did not meet these male observers' aesthetic expectations. Beauty, however, lies in the eye of the beholder, and numerous drivers of both sexes begged to differ. Referring to its rotund shape, they immediately saw in the vehicle a "little ladybug" whose "beetle-back" silhouette lent it a likable appearance. In the United States, the Volkswagen's commercial ascent depended in equal measure on its striking shape and its solid engineering.[2]

VW's rise to international player profited immensely from a severe global shortage of vehicles in the first two decades after World War II. As a result of war-related destruction, military requisitioning, and the closure of civilian markets during the war, car ownership in all industrial nations had fallen to lower levels in 1945 than in the late thirties. Automakers stood to benefit from pent-up demand that, with the onset of the boom coinciding with the Korean War in the early fifties, led to a surge in orders far outstripping existing production capacities and creating a classical sellers' market. Throughout the fifties, the main challenge for the European motor industry was not so much to persuade consumers to buy new cars; rather it was to produce new cars in sufficient numbers. Boisterous demand remained an international phenomenon well into the sixties because the General Agreement on Tariffs and Trade, which established a new framework for postwar international trade, sustained the boom through negotiations lowering barriers for international commodity exchanges. As the postwar years provided ample opportunities for car exporters, Volkswagen was particularly well positioned to take advantage of the favorable conditions. Although it paid its employees generously by West German standards, its wage bill was far lower throughout the fifties than that of its Western European and, especially, American competitors. In conjunction with the productivity increases resulting from the company's investment in automation, VW's comparatively modest labor costs allowed it to offer a quality vehicle at a remarkably low price by international standards.

The strong international sales that established Volkswagen as one of the Federal Republic's premier global concerns left a profound imprint on the vehicle itself. Although the car that was on offer at home and abroad was largely identical in technical respects, it underwent a broad range of permutations in the wider world. As an export hit, the Volkswagen developed into one of the transnational commodities that "are constituted in the movement between places, sites, and regions."[3] In cultural terms, the Beetle's entry into the world market entailed numerous recontextualizations that lent the vehicle new meanings it did not possess in its homeland. When foreigners beheld a Volkswagen, they saw a car that frequently differed fundamentally from the one

West Germans beheld. As with the Volkswagen that *Sports Illustrated* plucked from the road and placed in aquatic surroundings, the car often displayed an uncanny ability to take to some national habitats, while in others it barely stayed afloat. Large sums were at stake in the games of cultural adaptation that determined whether this commodity managed to gain a foothold in export destinations. The VW owed its international appeal to corporate initiatives as much as to protracted processes of consumer reception and appropriation that reshaped its meanings in surprising and at times lasting ways.

While the company retailed its most famous product in dozens of countries, two select examples reveal the contrasting commercial and cultural fortunes of the Beetle. In the United Kingdom, which had played a seminal part in the plant's postwar survival, the VW remained a marginal commodity, a circumstance that highlights the car's limited success in European nations with strong domestic auto industries. Meanwhile, in the United States, the Volkswagen turned into a spectacular, long-running sales hit. This prestigious transatlantic success crucially contributed to the company's prosperity, attracted considerable attention at home, and played a major role in turning the Volkswagen into a global icon. What, then, happened when the "German miracle's favorite child" left home?

Nordhoff lost no time positioning Volkswagen as a company with an international customer base. Having already launched foreign sales on a small scale during the occupation period, VW rapidly expanded its international operations in the early fifties. Bearing in mind Germany's traditionally underdeveloped domestic car market as well as the Federal Republic's emasculated economy, Nordhoff was convinced that securing customers abroad was a vital affair. "A factory of this size," he explained to VW's advisory board in 1951, "cannot stand exclusively on a single leg alone that is as unstable and unreliable as the home market." The export business ranked highly among Nordhoff's personal priorities, prompting him to undertake several lengthy trips across the world with the aim of acquiring firsthand knowledge of international sales territories from 1949 on. Although foreign sales initially returned no profits, since the company had to introduce its car to markets abroad at a price barely covering production costs, Nordhoff insisted on taking a long-term view so as to gain terrain from competitors. While seeking to boost exports, the general director did not pursue a carefully planned strategy but pragmatically seized opportunities as they presented themselves, often relying on personal contacts with Germans who lived abroad.[4]

As a result, Volkswagen quickly developed into a car company that paid exceptionally strong attention to foreign markets. By 1952, Wolfsburg already shipped cars to forty-six countries in Africa, South and North America, and Europe, selling 47,000 automobiles, or one-third of its annual output, abroad. As early as 1955, the company was retailing more vehicles abroad (177,657) than at home (150,397). Over the next years, exports remained at the heart of VW's expansion, and it shipped over 620,000 cars, or more than 55 percent of its annual output, to destinations beyond West Germany in 1963. Volkswagen's export ratio reached far higher levels than those of other leading West German companies at the time. Krupp, the engineering conglomerate that abandoned arms manufacturing after the war to concentrate on civilian projects, sold 15 percent of its production abroad in 1954. In short, Volkswagen was at the apex of West Germany's export companies.[5]

With its orientation toward global markets, the company headquartered in Wolfsburg threw into sharp relief the central part exports played in the Federal Republic's wider dynamic recovery. Having posted its first trade surplus as early as 1951, the

Federal Republic witnessed a rise in the contribution of foreign trade to GDP from 9 percent in 1950 to 19 percent in 1960. Although the West German press did not call its homeland an "export world champion" before the 1980s, Volkswagen's expanding foreign operations highlighted impressively that the Federal Republic was not merely turning into a car-producing country, but a car-exporting country. Nonetheless, VW's foreign activities did not mirror West Germany's place in the international economy in a straightforward manner. In particular, Volkswagen differed in terms of its market structure. While most West German exporters concentrated their foreign trade on Western Europe as European integration gathered momentum in the second half of the fifties, Volkswagen expanded most rapidly in the United States.[6]

Initially, however, the lion's share of sales was concentrated in a small number of European countries. Before American sales took off in the mid-fifties, more than half of Volkswagen's exports went to Sweden, Switzerland, the Netherlands, Belgium, Denmark, and Austria, all countries that either lacked a car industry of their own or featured only a small domestic auto sector. Meanwhile, VW's efforts to gain a foothold in Western European nations with strong, established auto manufacturers hit a brick wall because governments in those countries shielded producers at home from foreign competition. In Great Britain, France, and Italy, protectionist policies largely prevented the import of Volkswagens. Like other Western European states, Great Britain imposed steep tariffs on import vehicles until the second half of the sixties. As a result, the VW did not achieve a prominent road presence in the country that had played a central role in securing its survival into the postwar world. Volkswagen began selling its main product in the United Kingdom in 1953, but the company's share still stood at less than 1 percent a decade later. Nonetheless, the Volkswagen attracted considerable media attention in Great

Britain during the fifties and early sixties because it appeared to encapsulate telling contrasts between West Germany's and Great Britain's postwar economic fortunes.[7]

After World War II, British car manufacturers counted among the first to take advantage of international demand for automobiles. Raising annual production from roughly 219,000 to 523,000 vehicles between 1946 and 1950, British firms sold no less than 75 percent of their output abroad and, with a 52 percent share of global automobile exports, briefly dominated the international trade in motor vehicles. This success, however, did not inspire confidence among the British public, since the car industry's exports unfolded amid alarming signals of the United Kingdom's eroding global economic and political stature. Britain had emerged from the war with heavy financial obligations to the United States, and servicing a massive international debt counted among the British government's priorities. In this context, the drive to export cars formed a central plank in London's international economic strategy to earn foreign currency, reflecting less the UK's strength than its straitened finances after the war. Beyond the diminished international economic position, the independence of India and Pakistan in 1947 shone a stark light on the increasingly shaky foundations of Britain's worldwide political role. At home, the sacrifices and deprivations of the war had lent urgency to calls to recast society along more equitable lines through state intervention in the form of welfare initiatives such as the foundation of the National Health Service in 1948. Nothing, however, threw the economic strictures of financial recovery and social reconstruction into sharper relief than the extensive rationing regime that remained in place for almost a decade after 1945. Food rationing, for example, ended in the United Kingdom only in 1954, long after Continental countries had abandoned similar schemes. In short, both domestic and international developments highlighted the fragility of Britain's status as a global power.[8]

Many Britons registered West Germany's economic resurgence with misgivings. Numerous British observers looked on with disbelief as the country that had started and lost World War II achieved an economic recovery that appeared to surpass their own. Economic development in the Federal Republic reinforced long-standing British concerns about German industry that dated back to the late nineteenth century when Germany had first challenged Britain's status as the "workshop of the world." In 1887, amid early fears of national decline, British legislators had passed a law that required German industrial products on sale in the United Kingdom to be labeled as "made in Germany." Motivated by hopes that this measure would stigmatize import commodities, the measure backfired as customers increasingly came to appreciate the workmanship of German manufactured goods, gradually turning the phrase into a marker of quality. In light of long-standing Anglo-German commercial rivalry, the fact that vehicles "Made in West Germany" emerged as serious competitors in export markets struck numerous British commentators as one of history's ironies.[9]

The British press followed with growing concern Volkswagen's efforts to boost exports. When, owing to the VW's increasing popularity across the Atlantic, the German car industry outsold British firms in the United States for the first time in 1954, British observers strove to identify the reasons for the German car's appeal among customers. Road tests revealed that, despite its prewar design, the VW offered a convincing choice for drivers seeking a small vehicle of high quality with low maintenance costs at an attractive price. Residents in the British empire were impressed by the VW's durability in heavy terrain. A Briton from colonial Tanganyika (present-day Tanzania) praised his used

Volkswagen for its ability "to stand up to our [bumpy] roads. . . . I'm so happy about the car I intend to buy a new one when I am on home leave this year." Beyond its undisputed mechanical qualities and its economy, the Beetle possessed an intangible charm. The author of a dispassionate test report found that it "grows on one. . . . There is something pleasantly deceptive about the manner in which the Volkswagen cruises along at whatever speed . . . the driver wishes." The Beetle, then, recommended itself to customers in search of an alluringly unpretentious product.[10]

Yet it was not the product alone that was responsible for Volkswagen's success. A journalist returning from a tour of "the European markets" in 1956 was struck by the German concern's dynamism: "The drive and enthusiasm at the top of the organization has been communicated in a striking way to the distributors and dealers," resulting in an aura "of invincibility" that reminded him of the nimbus "which surrounded the early successes of Rommel in the African desert." This energetic entrepreneurial culture manifested itself tangibly in VW's exemplary service organization, whose members were contractually required to maintain an extensive stock of spare parts for repairs performed at fixed prices. In British accounts of the company's success, the service organization featured prominently, since most VW owners were people of modest means for whom "the offer of swift, cheap and easily available service anywhere in any country comes as the revelation of a new motoring age."[11]

While Volkswagen's main product and its service network impressed them, British commentators also pointed to the supposed errors and mistakes by British politicians and entrepreneurs. VW's resurgence "is surely proof of our short-sighted policy" immediately after the war, a reader of the *Daily Telegraph* railed in 1953. Rather than declare the factory in Wolfsburg "war booty" or dismantle it, the incensed writer continued, "we installed British officials to reorganize the works, its production and finances," thereby creating a threat to one of Britain's most successful manufacturing sectors. Decisions of the British occupational authorities, as well as British companies' subsequent failure to counter VW's export offensive, allegedly revealed a "remarkable underestimation of Volkswagen's potentialities," another observer added. "British industry," the *Observer* quipped, "after a sniff at the 'beetle,' turned its nose up." Although it had done no such thing, British observers partly ascribed VW's success to British incompetence and negligence.[12]

In particular, critics charged the UK auto industry with a reluctance to develop car models that matched the VW's technical properties. One frustrated British resident of Nyasaland (present-day Malawi) claimed that cars made in the UK were too small and underpowered, lacked adequate suspensions, and could not cope with dusty conditions. The absence of an equivalent to the Volkswagen struck observers as symptomatic of the "arrogance" and "complacency" that supposedly characterized British manufacturers. An interview with a disconcerted dealer retailing British cars in continental Europe yielded an exasperated litany of complaints that began with delays in spare part deliveries, moved on to inflated prices, and concluded with a general disregard of the market: "You take so little trouble to study the Continental market that even the door locks are on the wrong side of the car, and you still print your catalogues and instruction books only in English."[13]

Mounting discontent led some British drivers to abandon domestic brands, opting for the small vehicle from Wolfsburg despite its compromised origins. When VW began to sell its main product in small numbers in the UK in 1953, the conservative *Daily Mail* announced the VW's arrival in Britain with a characteristic headline: "Hitler's People's Car Is Here." As a result of the car's roots in the Third Reich, Beetle owners in the United

Kingdom and the British empire felt under strong pressure to justify their automotive choice and defend their patriotic credentials in the fifties. "I would have preferred to buy a British car, but on a 'short list' of five, none of the other four had the good points of the VW," one convert assured his compatriots. Anti-German suspicions remained a staple in connection with the Beetle into the early sixties, as the following letter to the editor of a car magazine reveals. "I do not entirely support the contention that the only thing wrong with the Volkswagen is that it is not British, but this for many people must be its main drawback." The car gained customers in the British market despite its national and historical background, and some British owners lamented what one driver called "the constant denigration of the Volkswagen." They argued that, rather than the customers buying a Volkswagen, it was British car manufacturers incapable of developing a worthy competitor that failed their patriotic duty.[14]

Irrespective of its marginal share in British markets, commentators in the UK identified the Volkswagen as a serious competitor to their country's auto industry. As the fragility of Great Britain's global position gave rise to fears of decline, the German vehicle threw into sharp relief not only the Federal Republic's economic resurgence but drew public attention to the supposed inadequacies of British efforts to maintain a leading role in the postwar international economy. In fifties Britain, the VW remained very much a competitor from a foreign country, commanding respect for its technical accomplishments but also failing to shake off its stigmatizing birthmark.

American drivers, who first gained the chance to buy a Volkswagen in 1950, welcomed the car with far more enthusiasm. From the mid-fifties, the United States emerged as Volkswagen's most important and most lucrative export market by a wide margin. Commercial success in America played a key role in establishing Volkswagen as a global automotive player. American drivers' affection left a significant mark on the car. Transferring the automobile made in Wolfsburg to an environment dominated by far larger and more highly powered cars placed the German car in a new context, prompting a major and protracted metamorphosis in the VW. Although the vehicles sold in the United States and in West Germany possessed by and large identical technical characteristics, the VW acquired a new identity in the United States that it did not possess in its homeland. While the VW remained a "Volkswagen" in everyday West German speech until the mid-sixties, the vehicle immediately morphed into a "Bug" or a "Beetle" in the United States. Its technical features, its size, and its unusual shape set the German car apart from virtually all American vehicles. In West Germany, the VW set the automotive standards, but in the United States it functioned as a profoundly unorthodox product.

At the beginning of the fifties, the car's success in the United States was anything but a foregone conclusion. When Heinrich Nordhoff returned from a visit to the New York car show in 1949, he was under no illusions about the obstacles faced by any European auto company that wished to establish a sales presence in United States. In addition to the logistical complications of maintaining an effective network of service stations in a nation of continental proportions, breaking into the world's largest and most advanced car market dominated by the extravagantly wealthy "Big Three" seemed like an unrealistic prospect for a company that possessed no production record. "Exporting cars to America is like carrying beer to Bavaria," the general director feared. Nonetheless, the United States became VW's most important single foreign market, attracting 120,422 VW sales in 1959 and peaking at 563,522 VWs in 1968, when the company shipped no less than 40 percent of its production to the United States.[15]

Although the 28,907 VWs sold across the Atlantic in 1955 amounted to far less than a 1 percent share of the American car market that year, mainstream publications including the *New York Times, Business Week,* and the *Nation* devoted detailed articles full of praise for the West German vehicle that year. "The Volkswagen enjoys a glowing reputation for stamina," *Business Week* found. Time and again, owners praised this feature, noting the car's ability to traverse waterlogged fords "during unusually heavy rains," to climb "bunch-grass hillside[s] . . . with deep cow trails that . . . even stumped four-wheel drive vehicles," and to "grip the road like a determined ant." Its reliability and its commendable driving characteristics reflected the Volkswagen's high technical standards. "Everything about this car is top notch. Those Germans are real craftsmen," stated an early VW fan. That VW offered this vehicle at the comparatively modest price of $1,495 enhanced its appeal, as did its advantageous rates of depreciation and fuel consumption. In 1956, over two-thirds of drivers surveyed by *Popular Mechanics* named "cheapness of operation" as the VW's "best-liked feature." Volkswagen had entered the world's leading auto market with a reliable and affordable quality product.[16]

After decades of austerity caused first by the Depression and prolonged by the Second World War, the fifties turned into a "golden age of the American automobile" that saw car registrations jump from 25.8 million in 1945 to 52.1 million in 1955. Lifestyle changes underpinned demand for cars. As the postwar suburban population grew by 43 percent between 1947 and 1953, daily routines, not least those of women who wished to visit the new suburban shopping malls and plazas, increasingly relied on access to individual transport. The car and the housing booms were intimately linked, with average individual expenditure on these items tripling in real terms between 1941 and 1961. By the end of the fifties, 40 percent of American drivers were female—a far higher proportion than anywhere in Western Europe. As a dealer told a journalist investigating why Americans bought Volkswagens in 1955, the car "'was just the thing for the wife to run around town in.' Two-car suburbia is where Volkswagen has made its biggest inroads." Market surveys from the fifties and early sixties also revealed that, while over 60 percent of Beetles registered in the United States served as second cars, about two-thirds of the customers who signed a purchase contract in a dealership had attended college. In short, Volkswagen attracted a foothold among the materially secure, white middle class whose members made up the bulk of suburbanites in the fifties.[17]

Many drivers undoubtedly chose a Volkswagen for practical reasons, but from the outset the Beetle's appeal transcended its immediate use value. VW owners self-confidently displayed their vehicles, and not only as practical quality products. Many drivers established a deep emotional attachment to their new, small possession—behavior that took observers by surprise. To be sure, car ownership in fifties America frequently created highly personal bonds between man and machine, not least because many sizable vehicles served as prominent status symbols. Yet this small, modest, relatively inexpensive German car did not function as a conventional status symbol. Rather, this automobile possessed a charm that cast a uniquely enchanting spell over drivers, leaving them infatuated with an inanimate object whose shape contrasted sharply with America's automotive mainstream.

"Owning a VW Is Like Being in Love," *Popular Mechanics* titled an article in 1956 after readers had responded to a poll in "the most enthusiastic" manner "ever." Drivers' "replies are unbelievable. These owners have actually fallen in love with a car." The car conveyed none of the aggressiveness widely associated with the Nazi regime to which it owed its existence. On the contrary, people found it "real cute," comparing its form to a "bug" or a "ladybug." From the early fifties, the VW's American appeal derived from its unusual shape. Nicknames also testify to how much VW owners cherished their vehicles. Upon completing a lengthy trip in 1955, one female driver simply christened her car with the German words for "little love" *(kleine Liebe)*. Indeed, Americans welcomed the VW as a lovable material object with innocent, childlike traits. Women drivers took the lead in ascribing an endearing aura to the Volkswagen, casting the car as an object soliciting female impulses of care and devotion. Although they did not feminize the car in outright terms, the prominence of female drivers among early VW fans underlined the unthreatening and friendly air that surrounded the car. Women handled this car with enthusiasm and self-confidence. "I'll try anything in the 'bug.' It lets me be boss when I'm behind the wheel," explained one "Ohio housewife" in her declaration of love for the Volkswagen.[18]

The appeal of the "Bug" in America derived not only from its unique shape but was also directly linked to its size. While West German drivers praised the car for its reliability, Americans were far more likely to draw attention to its diminutive proportions, as the anecdotes of a woman who drove a VW from New York City to Florida for her winter break in 1955 bear out. The sound of the vehicle's air-cooled back engine was a technical trait that caught the driver's attention, reminding her not of a solid piece of automotive engineering but of a "hard-worked sewing machine." In addition, she reported how the Volkswagen's small dimensions elicited curiosity. When she checked into a hotel in Richmond, the "doorman asked: 'Shall we unload her here, or will you take her to your room?' " while attendants in Savannah, who had cleaned the vehicle, refused payment because "they couldn't charge for bathing a baby." In West Germany, small vehicles dominated the car market, but in the United States, the

VW's modest proportions struck drivers and observers as deeply unconventional. In fact, some Americans struggled to acknowledge it as a fully fledged automobile.[19]

The Volkswagen occupied a special place in the American automotive landscape from the very beginning. Beyond its size and distinctive silhouette, the car's air-cooled rear engine marked the VW as an unorthodox automobile in a country where Detroit's increasingly large and expensive vehicles provided an overpowering technical and aesthetic backdrop. Commanding a 95 percent share of the 7.9 million sales of new cars in 1955, Detroit owed its dominance in the decade after World War II to ever more spectacular vehicles, whose average price rose from $2,200 to $2,940 in the first half of the fifties. By the middle of the decade, V-8 engines generating over 150 horsepower, automatic transmissions, soft suspensions, and air conditioners counted among the standard features with which U.S. car manufacturers competed among themselves. In this context, the Volkswagen appeared as a positively harmless arrival that posed no significant threat to domestic producers. In fact, Detroit displayed supreme confidence in its own models. Unlike in Great Britain, where auto manufacturers called for protectionist measures, American car manufacturers counted among the most vocal industrial supporters of free trade, actively pushing for the reciprocal abolition of import tariffs.[20]

The new aesthetic that transformed the American automobile in a highly conspicuous manner during this decade indirectly added to the Volkswagen's harmless appearance. U.S. manufacturers followed the lead of General Motors' Cadillac range and introduced two-tone pastel paint jobs, ample chrome detailing, elongated bodies, and, most famously, tail fins. Detroit's increasingly baroque creations, then, provided the lavish context that marked the small "Volks" with its rounded body and thirty-six-horsepower rear engine as a "cute" car. The VW's charming and unthreatening image was further enhanced because it posed no overt danger to the American auto industry. Retailing at a price much below that of average American cars, the "Bug" had entered the U.S. auto market as a niche product, a fact underlined by the high number of Americans using it as a second family vehicle. Its highly visible yet marginal position in the world's leading auto market, as well as its size and shape, stamped it emphatically as a foreign, distinct, and unconventional object within America's commodity landscape.[21]

As a matter of fact, a significant minority of Americans did not feel comfortable in the "monsters" that emerged from Michigan's auto plants. A female Beetle owner claimed to be "scared to death to drive our full-size car." Outright annoyance that U.S. manufacturers chose to ignore the small-car market provided a further motivation to buy a VW. "Ah, Detroit. They're too complacent; they may be missing the boat," concluded a distributor of import vehicles. The author of an incandescent letter mailed to the *New York Times Magazine* in 1955 positively balked at Detroit's offerings and characterized his decision to acquire a VW as a form of consumer protest:

> There is much more to the Volkswagen than the engineering that results in the ease of parking, driving, fuel economy, safety, and downright comfort, not to forget the workmanship. Even my synchromesh floor shift is far more satisfying and safer than all the super-hydra-dyno-automatic power gismos. I've owned them and I know. Until Detroit finds out what the lowly Volkswagen is all about, it won't see another dollar of my money.[22]

The German company capitalized not only on demand for an affordable second car among the white, suburban middle class but also on growing disenchantment among American motorists about Detroit's product policy. The Volkswagen's success was connected to a nascent, middle-class consumer reform movement that began to articulate itself in the mid-fifties and castigated conspicuous consumption as wasteful. While exuberant hedonism and colorful excess undoubtedly left a striking and joyful mark on American consumer culture in the fifties, economizing had not disappeared. Thrift and prosperity were by no means mutually exclusive in American culture at the time, not least because notions of frugality survived from the Depression. Simultaneously, a new consumer ethos gained momentum in the second half of the decade, a growing scrutiny of the mechanisms through which big corporations allegedly manipulated consumers in what increasingly came to be known as the "affluent society." Beyond John Kenneth Galbraith's eponymous best seller from 1958, a host of writers and activists, including Vance Packard and Ralph Nader, made names for themselves in an ongoing critique of contemporary capitalism in the United States.[23]

From early on, these critics trained their sights on Detroit. Beyond singling out body and engine size as well as purchase price, they denounced the annual model changes that made U.S. vehicles depreciate rapidly by rendering them stylistically obsolete. Moreover, the widespread realization that many pricey consumer products, including American cars, had been designed intentionally to restrict their life span and were therefore "made to break" compounded accusations of calculated wastefulness. Next to the artificially ephemeral and costly automobiles from Detroit, the "Volks" looked more reliable not only because of its high quality but because of its unchanging appearance over the years.[24]

Above all, Volkswagen attracted none of the charges of consumer manipulation that were often directed at the automobile industry amid growing suspicions about market research and advertising. That the public relations initiatives of American automobile manufacturers became a target of consumer reformers was hardly a coincidence. After all, no other company splashed out as generously on Madison Avenue as General Motors with its $162,499,248 advertising budget in 1956. Detroit's promotional campaigns saturated the American media landscape, employing celebrity endorsements, tightly choreographed launch ceremonies of annual models, corporate sponsorship of TV and radio shows, and countless ads in newspapers and magazines. Nothing appeared to demonstrate more glaringly the disingenuousness of Detroit's hype than the $10 million advertising extravaganza that accompanied the introduction of the Edsel line in 1957, which Ford had to withdraw amid public ridicule within two years because of technical problems and an uncompetitive price policy.[25]

Next to such Hollywoodesque showmanship, the fact that Volkswagen expanded its sales while refraining from Detroit-like advertising for years underscored the car's budding mystique. "An uncanny word-of-mouth campaign has helped raise sales," found a journalist in 1956. "Each guy that buys one is good for another," confirmed a Volkswagen dealer. Rather than Madison Avenue's professionals, the consumers themselves spoke for this product, which appeared devoid of pretensions: "I love it because of what it is—a VW," explained one driver. According to *Fortune* magazine, the Volkswagen was that rarest of commodities: "an utterly honest car." Put differently, where other cars were deemed to hold out empty promises, the VW struck drivers as a trustworthy object because it delivered what it pledged.[26]

Volkswagen's early customers did not simply consist of white middle-class Americans in search of an affordable second car to complement their larger American-made vehicle. The German company capitalized on spreading disenchantment among middle-class motorists about Detroit's product policy. Unlike in West Germany, where its low price, quality, and durability stood for a new postwar normality, in the United States the Beetle's characteristics lent it a profoundly unconventional air in a car culture dominated by size and showmanship. In a culture of ostentatious display, acquiring a small, inexpensive automobile, whose distinctive shape neighbors were bound to notice in the driveway, amounted to an act of conspicuously inconspicuous consumption that demonstrated a driver's eye for a reasonable product. Although the expanding suburbs attracted much hand-wringing commentary for their supposed social and cultural conformity, they provided an ideal environment to cast the Beetle as a marker of individuality.

As much as it stood out, the Volkswagen did not strike American drivers as altogether alien. To many observers the VW radiated a diffuse sense of historical familiarity—and not just because its rotund shape harked back to the streamlined automotive aesthetics of the thirties. More important, commentators penned articles entitled "Herr Tin Lizzie" to praise the Volkswagen as "the postwar Model T" because its purchase price, technological simplicity, robustness, and economy reminded them of Ford's automotive legend. For some Americans, then, the arrival of the Beetle resembled the return of earlier, more constrained forms of consumerism.[27]

The Volkswagen's origins in the Third Reich, which were common knowledge in the United States, did little to disturb its budding reputation as an unconventional middle-class vehicle. That the VW constituted a failed Nazi prestige project facilitated its commercial success after the war because the car required far less denazification than would have been necessary had the Nazis actually mass-produced it. While memories of the war in Europe as well as of German atrocities persisted in the American public sphere, Volkswagen profited immeasurably from the recent incarnation of Germany as the Federal Republic that came to be seen as a Cold War ally. When the East-West confrontation gathered momentum with the war in Korea, the Federal Republic gradually emerged as a tightly supervised partner within the American-led alliance, not least because of West Germany's and West Berlin's strategic and symbolic location on the geopolitical fault line dividing Europe.[28]

The reassessment of West Germany as an ally was not restricted to American political and diplomatic circles; it took place in the media, too. The American press featured extensive coverage of the Federal Republic's domestic political and economic consolidation process that underpinned the new country's integration into the Western camp. In particular, American travelers were impressed by West Germany's fast economic recovery. A writer for the *New York Times Magazine* attributed the Federal Republic's economic resurgence to a work ethic that he related directly to West Germans' failure to address their recent past directly. In this "indefatigably industrial society," work offered "a nerve cure," because, as he explained, "work is the postwar opiate. Work is less a means to an end than an end in itself." A colleague writing for the same publication a few years later encountered the same phenomenon when he interviewed a trade unionist who confirmed that "we Germans live to work. Maybe we plunge into work as an escape from something. . . . I suppose it is a kind of escape, but we don't know what from."[29]

Whatever the doubts about West Germany's democratic credentials and its unwillingness to confront its recent past, American opinion makers openly admired the Federal Republic's economy and the policies propounded by Economics Minister Ludwig Erhard. *Time* magazine considered "Germany's rebirth . . . the kind of economic miracle Americans can understand." According to this publication, the Federal Republic shone economically in contrast to Great Britain and France, both of which had nationalized core industries after the war: "At a time when other European nations were leaning towards socialism, Germany plumped for free enterprise. . . . As a result, the free world is now blessed . . . by its strongest bulwark against Communism." The search for telling examples frequently took American journalists to the Volkswagen works, which struck them as "the pace-setter of West Germany's industrial comeback," as the *New York Times Magazine* put it in the mid-fifties. Rather than cast the company's Third Reich roots as a thorny moral legacy, American reports from Wolfsburg treated VW's past as a historical backdrop that accentuated the new nation's postwar transformation along Western lines.[30]

Beaming a smile from the cover of *Time* as early as 1954, General Director Heinrich Nordhoff embodied the "Volkswagen miracle" and personified the country's path to recovery. To begin with, his dedication to hard work struck American visitors as exemplary. When he took over at VW in 1948, *Time* recounted, "he moved a cot into one of the plant's drafty, rat-ridden offices and started a seven-day week with only a few hours off for sleep." Moreover, Nordhoff portrayed himself as a long-standing Americophile, who once had plans to move to the United States in the early thirties but instead joined Adam Opel Automobile Company, a General Motors affiliate. Nordhoff accepted the job of running the Volkswagen works in the British sector in 1948 only when it became clear GM would not rehire him. "Herr Nordhoff, using his American know-how, reorganized the administration, production, inspection system, sales and the actual development of the Volkswagen car," an interviewer summarized in the *New York Times* in 1958. Other reporters from the United States referred to Wolfsburg as "Dearborn-on-the-Luneburger-Heide," or "Klein Amerika."[31]

Press reports in the United States saw the Volkswagen corporation as an indicator of Germany's postwar recovery, whereby a former enemy became a Cold War ally that fit into a specifically American mold. It was not simply that Germany was turning into a car society with large auto manufacturing sites reminding observers from the other side of the Atlantic of Michigan's industrial landscape. Placing the Beetle at the center of tales stressing the value of hard work to escape from deprivation also aligned Volkswagen and Wolfsburg with the motif of self-made success so prominent in American culture. Finally, the U.S. media emphasized that the American business practices adopted at Volkswagen played a crucial part in making the plant in Wolfsburg into a resounding commercial success.[32]

These articles assured American drivers that a Volkswagen qualified as a commodity made by a company that was run in accordance with American business principles and symbolized the Federal Republic's economic transformation and integration into the Western Cold War camp. At the same time, the Volkswagen's likable aura enhanced increasingly positive coverage of West Germany in the American press. To be sure, no journalist went so far as to claim that the Volkswagen's cuteness and unconventionality reflected West German national characteristics. If Americans connected the VW with "typically" German national traits, they did so by praising the vehicle as a quality product that reflected a devotion to craftsmanship and hard work. It was thus a combination of press coverage of a country in transformation along themes eminently familiar to Americans and the characteristics American drivers attributed to the Volkswagen that paved the way for the car's rise in the United States. Since the American media ascribed both German and American qualities to the car, it gained prominence in the United States as an appealing foreign commodity with transnational traits embodying concerns of the early Cold War era.

The Volkswagen's appeal among white middle America in the first half of the fifties was indirectly enhanced by the circumstance that, despite deep private misgivings, most Jewish Americans refrained from public criticism of commodities "made in

Germany." Drawing public attention to the Third Reich and the Holocaust, many American Jews reasoned at the time, would portray them as victims rather than as successful members of American society then gaining a place within the white middle class. Many Jewish Americans were concerned that the domestic media would view open criticism of West Germany as an attack on a new Cold War ally and thus expose the Jewish community to charges of unpatriotic behavior. Then, too, sections of the American Jewish community read the Federal Republic's financial support for Israel after 1952 as an encouraging indicator that West Germany was turning its back on its anti-Semitic past. The public silence Jewish Americans maintained on West German issues in the national media was an important precondition for Volkswagen's commercial success in the fifties.[33]

The late fifties and early sixties, when new VW registrations rose from 120,442 (1959) to 240,143 (1963), provided a critical juncture for Volkswagen to expand its position in the American small-car market, raising its exports to the United States to one-fifth of Wolfsburg's annual output. A short recession in the late 1950s prompted more American drivers to switch to cheaper vehicles, increasing the market share of European import cars to over 12 percent in 1959. VW succeeded in profiting from this trend amid stiff competition. The "compact cars" retailing at around $2,250 that Detroit introduced in 1958 to win back customers posed no direct threat to the Beetle because of its lower price. Renault's elegant Dauphine model, however, did. Heavily marketed, this French sedan with a thirty-horsepower engine and a price tag of about $1,350 climbed to second rank among import cars in the United States in 1959 and caused great concern in VW's headquarters.[34]

VW defended its lead in part due to its superior dealership network that offered better service than its French competitor. Moreover, it relaunched its car with a stronger forty-horsepower engine, as well as an improved transmission, in 1960. Throughout the sixties, Wolfsburg engineers continued to modify the VW by further increasing the vehicle's horsepower to fifty-seven, expanding the size of windows to improve visibility, and introducing new paint colors in keeping with fashion trends. At the same time, they preserved the unique shape and the reliable air-cooled rear engine that lent the car a distinctive presence on America's roads. On the whole, the Volkswagen's proliferation did not trigger widespread animosity because, unlike Japanese imports in the seventies and eighties, it remained aimed at a market segment in which Detroit showed no significant interest.[35]

Volkswagen's service organization, its products' persistent quality, and the Beetle's comparatively low price explain the car's continuing sales. What they fail to account for is the Beetle's lasting reputation as an unconventional vehicle. It owed this durable aura partly to a strategic decision the management took at the end of the fifties. As other small-car makers, including Renault, invested heavily in PR campaigns to present their products as viable alternatives to the German vehicle, Volkswagen was forced to assert itself through its own advertising initiatives. Hiring an agency, however, bore significant risks for VW, given the vehicle's appeal among customers harboring suspicions about advertising in the first place. What Volkswagen needed was a promotional team that could square a circle by selling an unconventional product with a reputation for "honesty" to a clientele hostile to salesmanship and consumer manipulation.

The business partnership that emerged from the account competition forged a seemingly unlikely alliance. As a firm founded by Jewish Americans, Doyle Dane Bernbach (DDB) at first sight made for an improbable candidate to promote a product with roots in the Third Reich. DDB bid for the VW account as an expanding medium-size agency that was securing a reputation for inventive campaigns. Already commanding an expanding client list including Polaroid, Ohrbach's department store, El Al Airlines, and the Israeli Tourist Office, the agency was keen to add a car to its portfolio because automotive advertising stood at the apex of the PR industry's prestige scale.[36] If status considerations appear to have motivated DDB, Volkswagen opted for DDB because of a distinctive pitch that focused on the car's technical features and the service the dealerships offered. The emphasis DDB placed on product details is likely to have convinced VW's patriarchal boss Heinrich Nordhoff, who, like many among West Germany's business elite, initially opposed advertising as a wasteful expenditure rather than regarding it as a necessary investment.[37]

The prize-winning promotional material DDB designed for its German client after 1959 played an important role in perpetuating the Volkswagen's standing. William Bernbach explained that a visit to Wolfsburg made him realize that his agency needed to build explicitly on the vehicle's reputation as a trustworthy product: "Yes, this was an honest car. We had found our selling proposition," he recalled. Bernbach was aware of the challenge faced by an initiative that employed commercial publicity to craft a product image around notions of honesty at a time of widespread distrust of advertising. In fact, a 1961 poll conducted by the trade paper *Advertising Age* in three wealthy suburban neighborhoods revealed that not a single one among eighty respondents was prepared to describe advertising executives as "honest."[38]

To some extent, DDB succeeded in strengthening the Beetle's reputation as a trustworthy product because the agency fashioned a distinctive stylistic approach that broke with many customs on Madison Avenue. The agency's campaigns for Volkswagen, for instance, contrasted with the hyperbolic rhetoric of auto ads promising drivers "rocket engine action," "turbo thrust," and more. DDB instead addressed consumers in intentionally plain, folksy language. Moreover, it set itself apart from the ad industry's visual conventions. Typical automobile promotions employed color layouts placing cars in attractive settings such as suburban streets or family situations to suggest benefits consumers could expect to derive from buying a particular vehicle. Early VW magazine ads, meanwhile, stuck to a spartan black-and-white scheme that showed little more than the photograph of a Volkswagen in front of a neutral bright backdrop. This idiosyncratic, self-consciously unpretentious style helped DDB detach its campaign from the much-criticized advertising mainstream with its reputation for consumer manipulation. As the campaign gathered pace, product and advertising style seemed to be in perfect synch, creating a commercial synergy from which both Volkswagen and DDB profited.[39]

Entitled "Think small," an early and famous, award-winning DDB ad offers a good illustration of how the agency turned Madison Avenue's unwritten rules on their head to drape the Beetle in "honesty." While vehicles visually dominated the illustrations of most car promotions, setting a VW off-center at a medium distance downplayed the product's presence in the ad and thus underlined the theme of modesty implied in the tagline. The opening sentence succinctly set the tone by playing on the dual meaning of the claim that the "little" Volkswagen was no longer "a novelty," thereby characterizing the Beetle as a serious commodity that was not viewed as a supposedly bizarre gimmick but as an established quality product. To substantiate this assertion, the remaining text enumerated the advantages the VW owner reaped from driving "a little flivver." High gas mileage, low oil consumption, "small repair bill[s]," "small insurance," and "squeez[ing] into a small parking spot"—all counted among the small car's advantages. Size mattered, the ad asserted, albeit in a manner diametrically opposed to the rhetoric of technical superlatives that shaped most other auto promotions at the time.

This prize-winning advertisement recommends the Beetle as a vehicle for smart American consumers who had their own take on the phrase "size matters." Courtesy of Volkswagen Aktiengesellschaft.

The pitch concluded not with a direct purchase plea, but took up the ad's opening by encouraging drivers to "think it over." Allusions to the car's history as a quality product, its technical characteristics, its economy, and appeals to consumer rationality—all these were to remain principal motifs of VW advertising throughout the sixties.[40]

The distinctive visual and rhetorical styles of DDB's campaigns provide one explanation for their success. The ads' deft use of humor and irony were another reason they retained their appeal for more than a decade. Many promotions caught the attention of doubtful consumers through counterintuitive, surprising, and downright funny ads like one from 1961 that featured a blank space where readers expected a photo of a VW. "We don't have anything to show in our new models," proclaimed the tagline, an allusion to the Volkswagen's supposedly unchanging features. On the whole, DDB presented the "Bug" as an amusing, lovable, and curious automobile that signaled a driver's eye for a reasonable, quality product in a materialistic society abounding with false promises. The agency's strategy of taking consumers' "skepticism into account and [making] it part of their ads' discursive apparatus," as Thomas Frank has noted, enhanced the car's reputation as a trustworthy consumer durable.[41]

DDB's understated advertising campaigns not only aroused fewer suspicions than many other ad drives but also underlined the VW's exceptional place on the U.S. auto scene. Nonetheless, despite their effectiveness, DDB's ads did not singlehandedly transform the Beetle's image from Nazi car into a "cool" and "hip" consumer good. Rather, the agency bundled relatively loose preexisting associations and consolidated them into a coherent corporate iconography. The car's unconventionality, when compared with American vehicles, provided the starting point for DDB's unorthodox approach to advertising. By placing the car firmly in the historical present of the American sixties, DDB indirectly underlined the Volkswagen's detachment from its Third Reich past. In the world of advertising, the consolidation of the VW's reputation as a small, attractive, and unconventional quality product with an altogether unthreatening aura went hand in hand with the production of historical amnesia about the vehicle's origins.[42]

Beyond DDB's ad campaign, American drivers played crucial roles in accentuating the small car's unusual reputation during the sixties. Volkswagen launched a quarterly magazine under the title *Small World* that reached a circulation of half a million in the middle of the decade, but corporate initiatives to strengthen customer loyalty and exert control over the vehicle's image frequently struggled to keep up with many an owner's resourcefulness.[43] Because of its light weight, low purchase price, reliable engine, and sturdy frame, the VW came to be highly valued among Americans searching for an inexpensive automotive platform that lent itself to extensive reconfiguration in accordance with their wishes. The Volkswagen thus proved itself a flexible artifact that could be adapted to a range of lifestyles and subcultures. The products of these transformations, which often bore merely a faint resemblance to the vehicle that emerged from the factory in Wolfsburg, generated considerable surprise among contemporaries—including VW's managers. These automotive reincarnations appeared unconventional not so much because they contrasted with the American mainstream but because they differed so fundamentally from the unchanging, plain "Bug," accentuating instead the car's versatility. Consumer appropriation lent the Volkswagen a new reputation as a Protean object that could undergo baffling transformations.

The small car's appearance in American racing circles during the sixties illustrates this development. On the expanding "hot rod" scene, for instance, the successes of customized Volkswagens with high-powered engines attracted note. Moreover, a new variety of amateur circuit racing known as "Formula Vee," which featured vehicles with "a cigar-shaped body, roll bar and all . . . built around the components of the mass-produced Volkswagen," received national coverage from its inception in 1963 and gathered a following of fifteen hundred amateur drivers within five years. In these competitive arenas, the VW gained a sporty note.[44] Some youth circles also opted for completely transformed versions of the VW. From the mid-sixties, colorfully painted Volkswagens with broad tires and shortened, frequently open-top bodies became a ubiquitous sight on Southern California's beaches, admired for their ability to drive through deep sand and tear around steep dunes. Depending on their specific technical modifications, these vehicles came to be known and revered as beach bugs, dune buggies, and Baja bugs.[45]

These fun cars for the beach placed the Volkswagen squarely in California's surfer scene, which laid the foundation for the famous, heavily marketed strand of sixties youth culture celebrating the outdoorsy, sporty, and informal hedonism of beach life. The VW bus, which Wolfsburg had developed as a transporter in the early fifties and featured a boxlike body over a modified "Beetle" chassis, proved particularly popular in surfing circles. For those who wished to spend a weekend by the beach, a surfer from the sixties recalled, the bus served as a "home away from home. You just got your things—a few clothes, your surfboard—and off you went. Many took out the seats in the back and put a mattress in the back to sleep on." And the original Beetle developed strong appeal among young, mostly white middle-class drivers with expanding financial means. During the sixties, college students integrated the car into their pranks, for example, by organizing an "intercollegiate" sport called "Volks-tote" in which teams of contestants carried a VW one hundred feet before piling into the vehicle and racing back to the starting line.[46]

If the Beetle's involvement in fun and games integrated the car into hedonistic lifestyles and spare-time activities affirming contemporary cultures of affluence, young Americans embracing critical perspectives on material abundance and capitalism adopted the car, too. Members of the countercultural movements that sprang up during the sixties were drawn to the Volkswagen in part because they needed affordable rides. Irrespective of the high-minded seriousness and polemics of its political declarations, the counterculture contained numerous strands blending post-materialism and hedonism that manifested themselves vibrantly and noisily in a plethora of demonstrations, happenings, concerts, and festivals. The "Volks" and the VW microbus proved ideally suited to convey equipment and people to these events. These fans often came from the same white middle class whose members had flocked to the car during the fifties. By the late sixties, then, the Volkswagen had secured a following among those living comfortably in suburban affluence and the members of the counterculture who rebelled against suburbia as the epitome of conformity.[47]

Of course, the counterculture was predisposed to the Beetle because of its unconventional aura. In an environment that placed a high premium on inner self-cultivation, an automobile that, as an illustrator of underground cartoons pointed out, "stood out like a duck in a tiger's cage" provided an ideal projection surface for notions of individuality. Many youthful drivers played up this trait by painting their possessions colorfully with psychedelic swirls and daisies.[48] "Dropout" auto mechanic John Muir advanced the most widely disseminated countercultural interpretation in his manual entitled *How to Keep Your Volkswagen Alive* (1970). With sales in excess of two million copies over the next three decades, this book addressed the sense of insecurity plaguing most owners before and during car repairs through clear step-by-step instructions and lucid illustrations drawn in the style of underground cartoons. Most important, Muir's narrative chose to frame the Volkswagen not as an alien, complicated, and inanimate mechanical contraption, but as a living organism. To render the car approachable to his readership, he employed the rhetoric of Eastern spiritualism that enjoyed prominence in alternative circles, encouraging VW owners to tune into their car's soul. "Your car is constantly telling your senses where it's at: what it's doing and what it needs," he explained. Conceding that "the idea of feeling about your car is a little strange," he asserted that "herein lies the type of rapport which will bridge the communication gap between you and your transportation." Muir, therefore, urged his readers to develop a sentient understanding of their automobile through empathy. "Feel with your car; use all of your receptive senses and when you find out what it needs, seek the operation out and perform it with love. . . . [Your car's] Karma depends on your desire to make it and keep it—ALIVE."[49]

In keeping with the idea of reciprocity at the core of notions of karma, whether the car worked or failed supposedly reflected its owner's motivations and actions. Muir individualized the car by tying its performance to a proprietor's character. As he argued that maintaining a Volkswagen required persistent investment in a reciprocal relationship for the benefit of both car and driver, Muir also went far beyond older interpretations casting the Volkswagen as an object of love. Supported by its ironic, humorous tone, the manual construed the Beetle as an object that, at least metaphorically, ran on nurturing emotions. Regardless of its spiritualistic rhetoric, however, this quirky interpretation remained predicated on the Beetle's technical solidity. After all, only an artifact that could be trusted to function reliably held out the promise of repaying an owner's affection. Here, then, was a car that could be fixed again and again, heeding calls against the throwaway society. Put differently, the Beetle's technical characteristics directly supported its elevation into an icon of countercultural post-materialism.[50]

Although the counterculture elicited passionate public conflict in the late sixties and early seventies, the Beetle's prominence in alternative circles did not turn it into a contested object. Its air as a friendly and cute artifact did much to shield it from damaging polemics. Moreover, the counterculture never dominated the car's public appearance, since the Volkswagen gained prominence in many other settings, including the white suburbs, apolitical campus environments, the world of auto racing, and West Coast beach culture. Supported by sales that turned millions of American consumers into everyday Volkswagen drivers, the Beetle retained its reputation as an unconventional car into the 1960s as it came to be absorbed into vastly dissimilar cultural contexts.

Nothing illustrates more poignantly how deeply the heterogeneous process of appropriation rooted the Volkswagen as an icon of unconventionality in American culture than the 1969 hit movie *The Love Bug*. A 1969 Walt Disney production, it beat such classics as *Midnight Cowboy* and *Easy Rider* in box office receipts. *The Love Bug* assembled important elements that had shaped the car's cultural profile over the previous decade: set both in sunny California and the world of auto racing, it told the adventures of a down-and-out driver who found love and improbable competitive success thanks to an anthropomorphized Volkswagen named Herbie. As a movie critic put it, Herbie's clever ability to defeat mechanically superior opponents lent the car a "personality" with "specific human credentials." In fact, Herbie's "philosophy" ran along reciprocal lines reminiscent of John Muir's countercultural karmic motto. Rewarding friends with affection and punishing foes, Herbie enacted the maxim "Be nice to me and I'll be nice to you," as the press book explained. *The Love Bug* decidedly did not cast Herbie as helpless, but nonetheless staged the Beetle as a small, lovable outsider and underdog. As it took up and reinforced some of the oldest motifs in American assessments of the Beetle, *The Love Bug* illustrates that "this little foreign car" had traveled a long way in the United States by the late sixties. Unthreatening and much loved, the Volkswagen became the cute Beetle that conveyed notions of individuality and unconventionality.[51]

The Volkswagen's international progress quickly became a news item in West Germany. Given the car's eminence as a symbol of the Federal Republic's postwar recovery, its foreign good fortunes were bound to arouse attention at home. Beyond scrutinizing VW's sales abroad as signals of West Germany's return to the world market, the domestic media regarded the car as a symbolic messenger and an informal ambassador of their new country in the wider world. A profound desire for international acceptance, which was evident in diverse social and political arenas as the Bonn Republic strove to move beyond its status as an international pariah, fueled West German curiosity about the Volkswagen in foreign lands. Given the United States' undisputed leadership in the West and the car's soaring sales there, West German observers focused on the car in America, penning articles that interpreted its transatlantic popularity as evidence of a new niche the Federal Republic was carving out for itself in the postwar international order. At the same time, VW's American success reflected back to West Germany and reinforced the vehicle's iconic status at home.

West German writers were aware that Americans saw the Volkswagen as a very unusual automobile, registering with bemusement that, in the United States, the VW gave rise to various "cults and pranks" *(Kult und Ulk)*, including its involvement in Formula Vee. The Federal Republic's leading auto magazine directly attributed the car's popularity in the United States to its status as a "fun second car, a college car, a car for fans as well as a much sought-after platform for tinkerers." American owners, German readers learned, loved their vehicles regardless of numerous mild taunts about its size and shape. As such, the Volkswagen's increasing presence on American roads and highways, visiting journalists noted, resulted from its ability to attract discerning, self-confident customers. "These people know who they are. They do not need a big car to look like more than they are," a German provincial paper found. This reading implied that the car appealed to self-assured customers supposedly devoid of status anxiety, the American middle class, or "the technical and cultural intelligentsia."[52]

As German observers found, the VW's reliability and quality, which had turned it into an emblem of national recovery at home, commanded respect among American customers, too. They repeatedly noted that Americans praised the car for its dependability and its low running and repair costs. These reports cast the car's appeal in the United States in categories that were eminently familiar to German readers, implying that VW's dedication to hard work and high production standards offered a promising model for returning Germany to the international scene. A conservative journalist expressed this conviction succinctly in the early sixties when he stated that the Beetle "revives the somewhat faded shine of the label 'made in Germany.' " By arguing that that the Federal Republic owed its revival in world markets to a dedication to "quality work," this line of reasoning took up a long-standing conviction about the foundations of Germany's international economic strength. In the eyes of West German observers, the VW and its six-figure sales abroad illustrated like few other products how an "export miracle" based on solid quality underpinned the domestic "economic miracle" and helped restore Germany's ruined reputation abroad.[53]

While West German observers registered the Volkswagen's American success with pride, a consciously sober, restrained tone permeated German coverage. The reserved reactions to the VW's fame in the United States corresponded with the "style of

Allen voran . . .

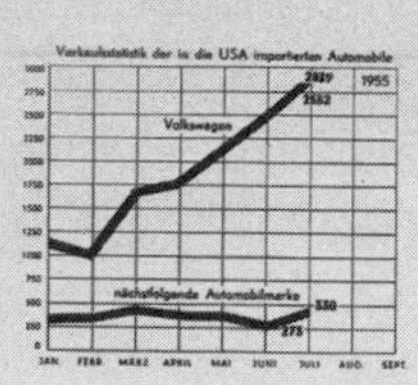

In Deutschland, in Österreich, in der Schweiz, in Schweden und Dänemark, in Belgien und in den Niederlanden ist man es gewöhnt, daß der Volkswagen die Spitze hält.

Auch in den Vereinigten Staaten steht der Volkswagen ganz oben auf der Zulassungs-Statistik (der importierten Wagen natürlich). Das ist besonders bemerkenswert. Im Land der Automobile ist man anspruchsvoll und kritisch.

der Volkswagen

"Ahead of all others—the Volkswagen," reads the tagline of this ad from 1956. Placing the small Beetle in front of an impressive downtown skyline, VW celebrated the fact that the car was turning into a successful export commodity not only in Western Europe but in the United States. Directed at a German audience, the ad highlights the pride the company took in gaining a foothold in the world's most prestigious auto market. Courtesy of Volkswagen Aktiengesellschaft.

modesty" that Bonn's diplomats had adopted since the fifties to leave the international world under no doubt that the Federal Republic rejected the violent power fantasies that fueled the Nazis. To some extent, the VW's image as a fun car counteracted Germany's tainted international image. In fact, West Germany, contemporaries thought, could hardly have found a better informal ambassador to the United States than the Volkswagen to illustrate the country's transformation—a deeply ironic stance, given the car's origins. "The Beetle is a good German," a glossy weekly stated in 1967 as it pondered the sympathies that greeted the car across the Atlantic. The VW's commercial success set an example of a civil route to international recognition that differed fundamentally from the confrontational approaches that Germany's leaders had pursued to disastrous effect in the first half of the twentieth century.[54]

While cherishing the Volkswagen as an altogether unthreatening international ambassador, West German commentators readily accepted the Federal Republic's subordinate position in the Western alliance when they pondered the car's presence in the United States. Volkswagen's advance as an export concern did not generate public statements of German industrial superiority or revanchist resurgence. Nordhoff personally intervened when he gained the impression that triumphalism reared its head abroad, because he feared that overt national assertiveness would damage the reputation of the Federal Republic and its companies, thereby hurting sales. In the early sixties, some German visitors may have found that it was "no longer possible to imagine America's roads without the Beetle," but these voices need to be placed alongside the measured assessments of the Volkswagen's international profile that dominated West German coverage. Marveling at the sight of several VWs speeding across Times Square on a summer evening in 1965, a German visitor reminded himself that VW had a market share of only 3 percent in the United States, although the company exported one-third of its production across the Atlantic. Despite the United States' prime economic importance for VW, the article emphasized the Beetle's status as a niche product across the Atlantic. Irrespective of domestic self-congratulation that the Federal Republic was establishing itself as an exporter and enjoyed a growing trade surplus, the German media acknowledged the undisputed Western leadership of the United States, and, in analogy to the VW's secondary market position there, assigned a secondary role to the Federal Republic within the world economy.[55]

As the Volkswagen's international success demonstrated to West German society the lucrative advantages of a circumscribed role in the wider world, the car's American reputation as an unconventional automobile left only minor traces on the car itself back home in Germany—despite a change in colloquial language that appears to suggest the opposite. From the mid-sixties, West Germans no longer referred to the car primarily as the "Volkswagen" or "VW" but increasingly called it *Käfer,* the German word for "beetle." In fact, Volkswagen adopted "Käfer" as the car's official name in its German publicity material in 1968. Taken in the year when the international counterculture stood at its peak, this decision, however, did not mean that the VW began to be viewed as an unconventional vehicle in its home country. To be sure, many a West German Käfer possessed a youthful appeal as growing numbers of young West Germans, who could afford their own small set of wheels in the second half of the sixties, opted for a Volkswagen. Numerous members of the Federal Republic's counterculture drove VWs, adorning them, like their American counterparts, with colorful pictorial compositions as well as anti–Vietnam War peace symbols. Nonetheless, these appropriations by rebellious youth did not transform the VW's dominant domestic reputation from symbol of postwar normality into marker of unconventionality.[56]

In the late sixties and early seventies, owning a "Käfer" allowed young Germans to venture near and far during their holidays. This photo is a memento of a trip by a young engineer and his wife—a recently qualified medical doctor—that took them from northern Bavaria to Morocco. Photo courtesy of Gisela and Reinhold Löhberg, Weiher.

The effect achieved by VW advertising campaigns in the Federal Republic bears this out. Doyle Dane Bernbach, buoyed by its growing reputation in the United States, quickly expanded internationally and opened a West German branch in Düsseldorf in 1963. Rather than conceive its West German campaigns from scratch, DDB recycled much of its material designed on Madison Avenue, in many cases translating American ads literally into German. From 1963, most of the promotions for the Beetle that appeared in West German magazines and newspapers were direct imports from New York. The West German media immediately singled out these Madison Avenue products for their exceptional look as well as their ironic, humorous tone. German advertising professionals also praised their unusual approach to marketing, awarding numerous prizes to DDB.[57]

Whatever their stylistic distinctiveness in a German context, however, these ads did little to lend the car an unconventional air. The characterization of the Volkswagen as an unpretentious, sturdy, economical, and reliable vehicle, which ran consistently through DDB's material on both sides of the Atlantic, could not strike West Germans as surprising or unusual, because these very features had turned the Beetle and its mystique into a West German symbol of postwar normality. In the Federal Republic, DDB's activities effectively enhanced the Volkswagen's existing reputation as a dependable and normal automobile. First conceived in the head office in New York City in 1963, DDB's most popular ad in Germany depicts a series of photographs of a Volkswagen disappearing toward the horizon beneath a headline asking why customers bought "so many Volkswagens." Among the "many reasons" the ad gave, "the most important" one was the car's reliability, asserting that "it runs and runs and runs . . ." Within a few years, this phrase had become a standing expression and entered everyday speech in West Germany. Regardless of its international sales and its reputation as an unconventional automobile in the United States, in its home country the Käfer remained a token of postwar normality and stability throughout the sixties.[58]

In the fifties and the sixties, the VW's stellar international sales hinged on the same qualities that turned this vehicle into West Germany's most popular automobile by a wide margin. In addition to the company's service organization, the Volkswagen's comparatively low price and running costs, as well as its quality and dependability, counted among its main draws in burgeoning auto markets across the world. The VW developed a prominent international profile not just in countries in which it attracted

Warum werden so viele Volkswagen gekauft?

(Bis heute über 7 Millionen. 3,5 Millionen in Deutschland. 1,2 Millionen in Amerika. 2,5 Millionen in der übrigen Welt.)

Dafür gibt es viele Gründe. Das ist der wichtigste:

Der VW läuft

und läuft

und läuft

und läuft

und läuft

und läuft

und läuft

und läuft

und läuft

und läuft

und läuft

und läuft

und läuft

First designed on Madison Avenue and subsequently translated into German, this ad claims that "so many" drivers buy a VW because "it runs and runs and runs . . ." The phrase quickly turned into a standing expression in everyday German. Courtesy of Volkswagen Aktiengesellschaft.

large numbers of customers but also in Western European nations like Great Britain, which protected their domestic auto industries through high import tariffs. As the Volkswagen developed into a global commodity, it gained international fame. At the same time, the car's proliferation across the globe went hand in hand with major semantic transformations.

Despite the Volkswagen's negligible standing in the UK market, the British media regarded the VW as anything but a harmless, normal car. Against the background of Great Britain's increasingly pressured position on the global political and economic scene and West Germany's industrial resurgence, British commentators cast the Volkswagen as a serious challenger to the international prominence the UK auto industry enjoyed immediately after World War II. As Volkswagen gained market shares, the West German company's dynamism, British journalists argued, contrasted sharply with complacency among British car manufacturers, who, after initially underestimating the commercial potential of the former "people's car," failed to develop models with comparable economic clout. Although individual British drivers praised the Volkswagen's charming aura, the British press first and foremost viewed the car as a substantial competitor with a Nazi pedigree.

While the VW's origins in the Third Reich were common knowledge in the United States in the fifties, they did not diminish its appeal across the Atlantic because Volkswagen and its product struck American observers as prime examples of West Germany's successful postwar reconstruction under American auspices. The car recommended itself as an eminently reasonable commodity to white middle-class drivers of both sexes either in need of a second car or disenchanted with Detroit's lavish and expensive offerings. Most important, in an automotive culture shaped by extravagantly large vehicles, the VW stood out as an appealing economical alternative that, with its sturdy engineering, diminutive proportions, and, crucially, its round shape, lacked all menacing connotations and struck numerous observers as both "cute" and thoroughly unconventional. The Beetle extended its American reputation as an unusual automobile throughout the sixties, displaying a rare ability to fit into a bewildering range of social environments. By the late sixties, the Beetle had sunk deep roots into American popular culture. While proving itself an exceptionally versatile commodity in the United States, the Volkswagen nonetheless retained its solid notional core as an idiosyncratic commodity that allowed owners and drivers to project a sense of individualism. Commercial success in the United States allowed the Volkswagen to embark on an unusually long cultural journey in the course of which it gained lasting fame as an irreverent, uncontroversial, and much-loved icon of unconventionality.

Focusing on the car's American good fortunes, the West German press followed Volkswagen's international success with interest. VW sales abroad did more than highlight the advantages of international economic success that gradually allowed the Federal Republic to view itself as a car exporter. Given the VW's domestic symbolic salience, West Germans viewed the vehicle as their informal ambassador that restored Germany's battered reputation abroad. At home, contemporaries took pride in the Volkswagen's warm reception by the West's leading power but studiously avoided triumphant, nationalist overtones in celebrations of the car's prominence in America. Observers from the Federal Republic noted the Volkswagen's American appeal as an unusual car, but emphasized far more strongly that its foreign sales hinged on its quality. While growing numbers of West Germans embraced the vehicle's American nickname and relabeled the Volkswagen as the Käfer in the sixties, the car itself remained a synonym for dependability and solid engineering in the Federal Republic.

Irrespective of its technical nature as a standardized, mass-produced object, the Volkswagen led numerous lives as an international commodity—and not only because individual owners regarded it as a personal possession. As a result of burgeoning global sales, it entered into different national cultures that left a deep semantic imprint on it. In Great Britain, the Volkswagen appeared as a competitor of infamous ancestry; in the United States, it emerged as the cute and unconventional Beetle; and in the Federal Republic, it continued as the quality vehicle signaling reliability. That West Germans viewed the Volkswagen as an automobile that runs and runs and runs reflected its strong international sales throughout the sixties. Nonetheless, toward the decade's end, more and more commentators at home and abroad noted problems in Wolfsburg and wondered whether the Beetle was beginning to run off the road.

6

"The Beetle Is Dead—Long Live the Beetle"

In December 1971, Carl H. Hahn, VW's head of sales, presented plans for major festivities marking the rapidly approaching day when the Beetle was scheduled to become the first car to replace Ford's Model T as the world's best-selling automobile. The manager, however, had misjudged the mood among his fellow board members. Citing cost considerations and fears of negative publicity, the executive board rejected Hahn's proposals, and, in sharp contrast to the lavish celebrations that had highlighted production jubilees in the fifties, authorized nothing more than "press releases, a small press conference, and advertising initiatives." After the record-breaking vehicle had rolled off the assembly line in Wolfsburg in February 1972, Volkswagen, therefore, treated the West German public not to an extravagant party but to an advertising spot whose upbeat message betrayed none of the executives' reservations. Surrounded by an ecstatic crowd cheering as if it had just witnessed the triumphant finale of a nail-biting bout, the viewer saw a light-blue Beetle in a boxing ring as the voiceover set in and informed the audience that "never in the history of the automobile has one car been produced in such numbers." Mimicking an excited sports commentator, the spot went on to applaud the Beetle for having knocked out countless "tricky opponents," thanks to its "refined technique, indestructible conditioning, and unpretentious lifestyle." As the camera glided over groups of dancing people, it briefly rested on the silhouette of a young African American strongly resembling Muhammad Ali before declaring the Beetle the undisputed "champion of the world."[1]

As it turned out, Volkswagen's show of comparative modesty on the occasion of the Beetle's remarkable jubilee in 1972 proved well advised. Two years after celebrating its longtime star product, the company headquartered in Wolfsburg resembled a punch-drunk giant, staggering about the commercial arena after posting unprecedented losses of hundreds of millions of deutsche marks. Making headlines not for profits and expansion but for deficits and layoffs, Volkswagen was suffering a dramatic reversal of fortune that many West Germans found deeply unsettling. Since VW's troubles of the early seventies coincided with the onset of wider economic and social difficulties in the Federal Republic, news emanating from Wolfsburg signaled to anxious West German commentators that what Eric Hobsbawm has called the "golden age of capitalism" was coming to an end.

West Germany's novel economic problems derived from several overlapping causes. The 1971 collapse of the Bretton Woods system of fixed exchange rates hit the country's exporters, including Volkswagen, hard, while the oil crisis of 1973 fueled inflation. Exposing problems that had been building in numerous companies for several years, these external shocks triggered the return of substantial unemployment, which was to remain a persistent feature of West German society. While the highly subsidized mining and steel sectors proved particularly susceptible during the downturn of the mid-seventies, the workforce at car makers and electronics firms by no means escaped unscathed. Growing job insecurity in manufacturing provided a potent indicator that West Germany was gradually transforming from the industrial society of the "economic miracle" into a service economy, albeit one with a far more pronounced and successful manufacturing sector than other Western "postindustrial" nations, including the United States and Great Britain.[2]

Numerous industrial enterprises failed to extract themselves from problems that became virulent in the course of the seventies. Persistent difficulties could stem from a neglect of new managerial approaches, underinvestment in research and development, a reliance on dated production modes, fractious industrial relations, and an inability to take into account novel consumer wishes. Although Volkswagen managed to cope with the challenges of the middle of the decade, it owed its survival and future success to a painful, dramatic, and drawn-out process of adjustment. The managers in Wolfsburg had recognized before the seventies that they could no longer primarily rely on the Beetle. Devising a convincing entrepreneurial strategy for a new automotive landscape, however, required time, because VW's search for an updated product range occurred in an inauspicious economic climate. At VW, a complicated period of corporate transition was exacerbated by the effects of the global recession of the early seventies.[3]

By that time, it was common knowledge that the Beetle had turned from Volkswagen's source of strength into its Achilles heel. The firm's troubles, numerous observers agreed, resulted primarily from depending for too long on the model that Ferdinand Porsche had designed in the thirties. As commentators worried that Volkswagen was running the risk of becoming a victim of its erstwhile success, the Beetle struck contemporaries as increasingly obsolescent. Nonetheless, the company's difficulties did not undermine the affection with which contemporaries regarded the small car both at home and abroad. Although the Beetle was quickly evolving into a highly marginal commodity in markets in which it had previously sold en masse, both West Germans and Americans continued to hold it in high regard.

"Is the Volkswagen dated?" the weekly *Stern,* a West German magazine loosely modeled on *Life,* asked in 1957 in an article cataloging numerous deficiencies that allegedly plagued the car made in Wolfsburg. With its thirty-horsepower engine, the Volkswagen, *Stern* argued, was woefully underpowered, leaving it with disappointing acceleration as well as an unsatisfactory top speed of seventy miles per hour. The VW, the publication continued, offered little comfort for drivers and passengers, owing to a cramped interior, a loud engine, and a smelly, erratic heating system. *Stern*'s litany culminated in the charge that the Volkswagen was difficult to handle and hence dangerous: its shape exposed it to crosswinds, while the engine's location in the back accentuated a tendency to oversteer in corners. Small, slow, stinking, erratic, and unsafe—these were the epithets *Stern* attached to the car that shaped the Federal Republic's auto culture like no other.[4]

Penned at the very moment the West German auto boom gathered momentum, the report in *Stern* was a calculated provocation and produced a predictable result. In letters to the editor, VW owners dismissed the magazine's criticism, instead drawing attention to its technical and commercial virtues, including its low price and running costs, its high quality, and its exceptional reliability. Although *Stern* ran into a solid wall of customer loyalty in 1957, the issues it had broached would accompany the Volkswagen henceforth. Only two years later *Der Spiegel* employed *Stern*'s headline as the title of an interview with Heinrich Nordhoff. In addition to reiterating the earlier technical charges, including the car's "treacherous" tendency to oversteer, *Der Spiegel* maintained that the Volkswagen suffered from restricted luggage capacity, poor visibility, and higher fuel consumption than the company admitted. On this occasion, the criticism of the VW produced a far more mixed mailbag. While no one faulted the car on overall economy and quality, several readers vented their frustrations at uncomfortable back seats, small windows, a malodorous heating system, and difficulties in crosswinds. By the late fifties, then, some consumers and journalists laid their finger on limitations in terms of size, speed, safety, and comfort that inhered in Ferdinand Porsche's design. The Volkswagen dominated the domestic market and provided the undisputed symbol of West Germany's postwar recovery, but it gradually attracted critical technical scrutiny.[5]

Although Nordhoff waved away suggestions of the Volkswagen's imminent obsolescence with demonstrative impatience, he took charges along these lines seriously. He was, he assured *Der Spiegel,* well aware of the grave problems that Ford had encountered in the 1920s when drivers had tired of the Model T, thereby triggering a major crisis at Highland Park that cost the company its leadership in the United States. Ford, Nordhoff argued, had refused to modify the Model T despite customers' complaints. To avoid replicating a similar slump in Wolfsburg, Nordhoff frequently ordered in-house engineers to redesign those aspects of the Beetle that attracted censure. Throughout the sixties, Volkswagen, for instance, enlarged the windows to improve visibility, increased the volume of the baggage compartment, strove to enhance the performance of the heating system, and equipped the vehicle with more powerful brakes. In the mid-sixties, the company also introduced larger engines of 1.3 as well as 1.5 liters, thereby raising capacity to forty-four horsepower and boosting acceleration as well as top speed. These technical modifications were accompanied by countless cosmetic changes to render the car visually attractive to an increasingly demanding customer base. While these changes undoubtedly helped secure clients, they could not overcome complaints about a lack of space, the car's tendency to oversteer, and its vulnerability in crosswinds.[6]

Since domestic demand remained robust and exports developed dynamically until 1966, early complaints about the car's limitations did not translate into falling sales. Into the mid-sixties, Nordhoff's prime task consisted in managing the corporation's growth. From 1962 to 1966, the number of VW workers and employees in West Germany rose from around 78,000 to over 91,500, while annual output of the Beetle increased from 819,326 to 988,533. When the company reported a net profit of DM300 million in 1966, it produced almost 1.5 million automobiles in factories in Wolfsburg, Hanover, Braunschweig, Kassel, and Emden, the latter a plant solely dedicated to the assembly of export vehicles for the United States. Since VW's continued expansion occurred in a labor market characterized by a shortage of manual workers due to West Germany's booming economy, the corporation followed the example set by other companies and began to hire so-called "guest workers" in 1962. Drawing on contacts Nordhoff had established with the Vatican during a private audience with Pope Pius XII, VW concentrated its recruitment efforts on Italy. In 1966, around six thousand Italians on temporary contracts staffed lines in Wolfsburg. Expansion appeared to provide Wolfsburg's omnipresent motif, not least after Volkswagen had acquired Audi in 1965, already an established manufacturer of high-end vehicles, to secure a foothold in a lucrative market segment.[7]

These impressive growth indicators hid the fact that Volkswagen was entering into an economic environment for which it was ill equipped. Wolfsburg's problems, which first manifested themselves in the second half of the sixties before reaching dramatic proportions in the mid-seventies, did not stem from a single cause. Rather, gradual changes in West Germany's automotive and consumer landscapes, as well as increasingly volatile economic conditions, provided the context in which the company's longtime strength—the Fordist mass production of a basic, highly standardized, and reliable automobile—turned into a liability. In response to the new conditions, Volkswagen was forced to go through a painful process of reinvention between the mid-sixties and the mid-seventies that relegated the Beetle to the corporate margins.

In the early sixties, Volkswagen's West German competitors set out to challenge Wolfsburg's lead by launching small cars directly targeting VW's core sales territory. First introduced in 1962, new versions of Ford's Taunus 12M and Opel's Kadett presented the most important newcomers. While these automobiles cost only marginally more than the Beetle with its price tag of around DM5,000, tests revealed that they compared favorably in several respects. With water-cooled front engines and slightly more horsepower than the Beetle, they offered better acceleration and higher top speeds while avoiding the susceptibility to crosswinds that plagued the vehicle made in Wolfsburg. In fact, the Opel received particular praise for offering a smooth ride. The Kadett and the 12M also boasted more-effective heating, better visibility, more-comfortable interiors, and larger trunks. In terms of fuel consumption, all three automobiles resembled each other. Only when it came to the workmanship that characterized the final product did the Volkswagen have an edge over the small Ford and Opel. Auto journalists came to the conclusion that, while by no means "sensational," the newcomers presented "serious competition for Volkswagen" because "everything about them makes sense."[8]

As many drivers took to these and other new automotive arrivals, Volkswagen's overall domestic market share fell from 45 percent in 1960 to 33 percent in 1968 to 26 percent in 1972. To some extent, this downward curve reflects an adjustment from a position of exceptional dominance that VW had owed to the absence of convincing competitors in its main market segment. With the appearance of the Ford Taunus 12M and the Opel Kadett, this unusual period came to an end. Nonetheless, competition in the small-car sector alone does not fully explain VW's falling share. Changes in the national car market reflected wider trends in West German consumer society. Demand for new automobiles underwent a major structural shift as a result of rising disposable incomes. While cars costing between DM4,100 and DM6,200 had attracted the largest number of sales in 1960, vehicles priced between DM6,200 and DM9,300 formed the center of commercial gravity in 1969. In the sixties, West German society made the transition from a small-car to a midsize car market.[9]

VW's failure to develop a convincing model range that took full advantage of the new lucrative sales territory lay at the core of its problems. In 1961, Wolfsburg responded to calls for a midsize model by presenting the Volkswagen 1500, but it proved of limited appeal. Costing around DM6,400, it had a chassis that owed much to the Beetle, as did its drive train with a 1.5-liter, forty-five-horsepower air-cooled rear engine. The 1500 avoided some of the Beetle's weak points, running at higher top speeds, offering better visibility, and featuring a roomier interior, as well as a bigger luggage capacity. At the same time, its air-cooled rear engine was responsible for replicating some of the Beetle's shortcomings, including an erratic heating system, noisiness, and difficulties in crosswinds. Despite respectable sales, the 1500's share of domestic production did not rise above 16 percent in 1968. As a result, almost 70 percent of the passenger vehicles manufactured by VW in West Germany that year were still Beetles. The corporation thus remained heavily dependent on a small car designed in the 1930s, which, despite repeated alterations, was incapable of penetrating the new, more profitable markets that came into existence in the sixties.[10]

Marketing studies showed that the company's narrow model range rendered it progressively difficult for VW to attract drivers in search of their first vehicle. Even more worrying in the long term, however, was the increasing number of West German car owners who wished to "trade up" to more powerful and comfortable vehicles. Beetle owners shopping for more up-market automobiles found that VW offered only one option. As a result, Volkswagen faced significant problems in retaining the loyalty of established Beetle drivers. Compared with Opel and Ford, brand loyalty rates for Volkswagen displayed unfavorable—and falling—levels in the second half of the sixties.[11]

As VW struggled with its customer base, the Beetle's profit margins came under pressure. In part, VW's bottom line eroded because the firm continued to pay workers West Germany's highest wages and benefits at a time when all sectors of society enjoyed substantial income rises. Many technical and cosmetic modifications intended to keep the Beetle attractive also affected profitability. In 1968, for instance, the company wished to increase the trunk, but, as a manager recalled, "this undertaking required redesigning the front axle," which in turn entailed investments in excess of DM100 million for new manufacturing technology. Rather than give VW an edge over Opel and Ford, the manager added, the company incurred this additional expense merely to compensate for a long-standing disadvantage vis-à-vis competitors. Since VW had to price its main product with an eye to the 12M and the Kadett, it could not pass on to customers the full costs of modifications.[12]

Another commercial problem developed as a result of booming exports to the United States. As Beetle sales jumped from 232,550 to 423,008 in the United States from 1963 to 1968, sustaining this momentum generated problems of its own. Although the Beetle experienced little effective competition in its niche, a more stringent regulatory environment, which emerged in the second half of the sixties, placed new burdens on Volkswagen of America. In response to effective lobbying efforts by environmental and consumer activists, Congress passed a series of laws designed to reduce emissions, curb price-inflating business practices, and enhance safety standards. To fulfill the new regulations, the Beetle had to be fitted with, among other things, a collapsible steering column to lower the risk of impalement injuries, sturdier locks to prevent doors from opening in crashes, larger indicator lights, and an engine with an emission control system. All of this increased production costs. In 1967, for instance, VW was required to implement technical modifications costing around $200 per vehicle but could only raise the car's retail price from $1,645 to $1,700. In short, in the second half of the sixties, Volkswagen of America's profits eroded.[13]

Beyond generating financial pressures, VW's American operations exposed the company to significant reputational risks. American law required manufacturers to call back and fix free of charge entire production runs of vehicles affected by specific design faults that potentially endangered drivers. The reputational damage of comprehensive product recalls far outweighed immediate financial burdens because the American media regularly covered these announcements in great detail. While American car makers were the first to fall foul of the new rules, Volkswagen's turn came in July 1966 when it had to call back almost two hundred thousand vehicles with problems in the front axle ball joints that could result in loss of steering control. Eighteen months later, forty-two thousand VWs needed modifications in the fastenings of the spare wheel. Media reports of these deficiencies not only weakened Volkswagen's reputation as a producer of high-quality automobiles; they also played into the hands of consumer advocates who deemed the small German import a dangerous automobile that offered drivers insufficient protection in crashes. First articulated by consumer activist

Ralph Nader during a congressional hearing on auto safety in 1966, this charge continued to hang over the American Beetle like a Damocles sword in the early seventies.[14]

As falling profits and safety concerns left Volkswagen in a risky position in the United States in the late sixties, the company became dependent on the U.S. market to an unprecedented degree. Selling almost 40 percent of its Beetles in the United States in 1968, the company was largely sustained by its success in a single niche market for small automobiles. By then, VW's potentially precarious position had become an open secret in the Federal Republic. In fact, West German Finance Minister Franz Josef Strauß summarized the situation in Wolfsburg in a rhetorical question in 1967: "What happens if the Americans stop being amused by the Beetle?"[15]

Strauß's polemic resonated in the boardroom and beyond because Volkswagen had experienced its first significant slump in 1966. After more than one and a half decades of uninterrupted dynamic growth, the West German economy suffered a setback that year that led to a GDP contraction of 2.1 percent as well as a mild spike in inflation. This recession triggered deep anxiety in a population that took pride in the "economic miracle" but, as opinion polls showed, had not developed lasting trust in the new economic and political order. The public anxiety caused by the contraction resulted not only in a tortured debate about West Germany's economic future but also in electoral gains of a nationalistic party on the far Right, which entered into the Hessian and Bavarian state legislatures in 1967. The Right's political success, in turn, played a part in energizing the youthful, student-led Left, which took its activities to unparalleled levels that year. Amid a public climate of insecurity, economic and political conflicts dominated the national agenda in 1967.[16]

At Volkswagen, the downturn laid bare difficulties that had been building up over the previous years. Domestic sales took a hard hit, falling from almost 600,000 in 1965 to 370,000 two years later because apprehensive consumers decided against car purchases. This dramatic decline indicated that customers delayed replacing older vehicles during an economically insecure period, a demand pattern typical of highly volatile auto markets that have moved beyond the phase of immediate mass motorization. In this new situation, executives at VW conducted discussions in an atmosphere crackling with tensions. Challenged by about two hundred thousand unsold cars in March 1967, the head of sales tried to defend himself by pointing out that General Motors also struggled with unsold stocks in similar circumstances. "In that case, why don't you join General Motors?" Nordhoff snapped. As an internal review showed Volkswagen to be overstaffed by up to 30 percent, the executive board, dreading the blow mass layoffs would deliver to the company's public image, put workers on short hours and took the unprecedented step of closing its German factories on forty-two workdays in the first half of 1967. "Where is the Volkswagen works headed?" *Süddeutsche Zeitung* asked anxiously.[17]

VW's immediate difficulties proved short-lived. After a Keynesian boost from the federal government had restored economic growth, West German customers returned and bought Beetles in their thousands. By November 1967, the executive board struggled to cope with enhanced rather than reduced demand, discussing the need for additional hires to expand production capacities. The recovery, however, did not lead to a return to the status quo at Volkswagen. The dramatic collapse of sales in the preceding year had demonstrated to executives, including Nordhoff, that as the market demanded more pricey, better-equipped, and more-comfortable automobiles, Volkswagen had tied its fortunes too closely to the Beetle.[18]

Were it to survive, VW needed to alter its business model fundamentally. Already beyond the official retirement age and suffering from health problems, Nordhoff died from a stroke in the spring of 1968. The baton passed to Kurt Lotz, a former general staff officer who had risen through the ranks at the electrical engineering firm Brown Boveri. To modernize the company, the new CEO initiated a host of measures, including the professionalization of Volkswagen's business administration through electronic data processing, the introduction of management routines adapted from recent American models, novel training schemes, and a significant expansion of research and development.

Above all, Lotz ordered a new, more lucrative product range capable of anchoring VW in the expanding market for midsize automobiles. Since this strategic aim required more highly powered cars with less noisy interiors and better handling characteristics (not least in windy conditions) than VW currently produced, Lotz's initiative hinged on abandoning the air-cooled rear engine, which had so far provided the core of Volkswagen's engineering identity. Assuming that VW would require at least five years to transform itself, the CEO authorized experiments on no fewer than three different drive-train configurations—a midengine, a front-engine with front-wheel drive, and a front engine with rear-wheel drive. At the same time, technical staff started to design four different future car models with three different engine sizes to secure a broad commercial reach. While this ambitious development program was an engineer's dream, it was an accountant's nightmare. In conjunction with stuttering Beetle sales, rising costs for research and development played a central role in lowering the firm's profits from DM330 million in 1969 to 12 million in 1971. Since work on the new model range had to progress in secrecy, the media focused on the concern's disappointing financial news. Although Lotz pushed through a host of measures behind the scenes, the company's worsening results made VW appear rudderless. Amid pressure from political figures on the supervisory board, Lotz resigned in the fall of 1971.[19]

He was replaced by Rudolf Leiding, who had spent his entire career at VW. Leiding's tenure at the helm occurred under the most challenging conditions of the company's history since the immediate postwar years. As Leiding pushed Volkswagen toward a new model range, he had to steer a steady course in a dramatically deteriorating business climate. In the early seventies, the boom that had sustained Western economies since the early fifties came to an end. An early indicator of brewing global economic trouble came in 1971 when the United States allowed the dollar to float in response to rising inflation and budget deficits, thereby precipitating the collapse of the system of fixed exchange rates that had underpinned the Western economic order since the late forties. In conjunction with a 10 percent duty imposed on all imports by the Nixon government, the new currency regime caused problems for Volkswagen. As the deutsche mark rose by 40 percent against the dollar in the following two years, the company faced plummeting revenue from American sales. In 1973, the situation went from bad to worse. After OPEC's proclamation of an oil embargo to retaliate for Western support of Israel in the Yom Kippur War, exploding energy costs triggered a global recession. The oil shock severely intensified Volkswagen's problems as demand for automobiles collapsed. In 1974, the company posted an eye-watering loss of DM807 million, in part because Volkswagen of America had generated a deficit of DM200 million.[20]

VW's deepening difficulties coincided with the beginning of a protracted crisis in Western industrial societies that quickly engulfed the Federal Republic. At the end of 1973, West German inflation stood at 7 percent as a result of rising oil prices. To prevent fuel and energy shortages, federal and local authorities took drastic, previously unimaginable emergency measures that intensified a widespread sense of gloom in West German society. In addition to issuing a ban on the use of private cars on four

e stance failed not least because the Bonn
Social Democrats paid scrupulous attention
ues at Volkswagen. During a time of rising
ssues at a partially state-owned and symboli-
pany provided a vital test case for the govern-
t to social welfare. Bonn, therefore, was bound
on that the trade unions could support. At the
vernment unequivocally acknowledged VW's
ing, including mass layoffs. Chancellor Helmut
irectly involved in Toni Schmücker's installa-
n Wolfsburg in February 1975.

was an apt choice. After a career at Ford,
uccessfully rescued the struggling steelworks
early seventies. This trajectory marked him out
he West German industrial landscape: a restruc-
n proven expertise in the auto sector. Beyond
edecessor's organizational reforms and model
confirmed the need to curb worker numbers
il 1975, he broke the dreaded news that Volks-
e to shed 25,000 members of its 133,000-strong
st Germany. Schmücker's draconian announce-
media storm and a fractious debate in parlia-
across the entire political spectrum found the
of Volkswagen deeply unsettling. In light of the
ing deficits, as well as its revolving-door policy at
gel's title page asked: "Mass lay-offs—millions
ement crisis: What will become of VW?" Public
e fact that no one offered an authoritative an-
he nation worries about a symbol of its economic
conservative daily commented in the spring of
, the liberal weekly *Die Zeit* added, no longer
ity of the economic miracle" but had morphed
ter."[22]

consecutive
temporarily
miles per hc
Numerous l
measures by
lic places—a
nity during t
in 1974 and
nomic difficu
joblessness. In
the one millic
society that l
growth for tw

In this harsl
as one of Leid
decessor's lavi
els with wate
a result of this
Passat, Golf,
1975. At the sa
routines as a r
vealed, reachec
its workers cor
in West Germa
hours. The trad
a 15 percent wa
confrontational
tensions by lau
these dismissals
contentious issu
low managers, t
supervisory boa
end of 1974, Lei

Leiding's abrasi
government led by
to employment iss
joblessness, labor i
cally eminent com
ment's commitmen
to push for a solut
same time, the go
need for restructur
Schmidt became c
tion as chairman i

The new man
Schmücker had s
Rheinstahl in the
as a rare breed in t
turing expert wit
continuing his pr
policy, Schmücker
drastically. In Apr
wagen would hav
workforce in Wes
ment triggered a
ment. Observers
changed fortunes
company's mount
the top, *Der Spie*
of losses—manag
concern about t
swer ran high. "T
power," an arch
1975. Wolfsburg
figured as the "c
into a "crisis cen

Up and down the country a gnawing sense that VW's problems signaled a temporal watershed fueled apprehensions. Sebastian Haffner, one of the Federal Republic's most respected journalists, made this point lucidly in an editorial in *Stern,* as he summed up the current state of affairs: "What upsets people," he explained, "is the mutilation, nay desecration, of a national symbol. After all, VW isn't just any company. . . . Until yesterday, it was the embodiment of the German economic miracle. . . . Those who had found a job at VW were set for life, it was widely thought," because the firm seemed to possess a "guarantee for success. . . . And now, all of a sudden, VW is in the red . . . , its future insecure." As in the fifties and for much of the sixties, the Wolfsburg-based company highlighted national economic trends, albeit now under the fundamentally altered conditions of a recession.

That the company's difficulties sprang from its long-term reliance on the Beetle only intensified public concerns. Haffner reminded his readers that the car that "ran and ran" had long stood for "German industriousness and indestructibility," as well as for the Federal Republic's "unexpected rise from rubble to unknown mass affluence." Moreover, its largely unchanging appearance and basic technical design had turned it into a symbol of modesty as well as of economic and social stability. In Haffner's words, the Beetle represented a "rejection of the automobile as a fashion object and as a status symbol. . . . It was classless, and it was timeless." As such, "every Volkswagen was . . . an avowal of constancy in the midst of change." The most worrying aspect of the Volkswagen crisis, he found, was "that none of this is supposed to be valid anymore." As Haffner put it succinctly, the "Volkswagen epoch" embodied by the Beetle was drawing to a close.[23]

The Beetle threw into sharp relief nothing less than the end of an era characterized by social security and economic growth.

While the car retained its iconic potency, it was fast turning into a symbol of the fading economic miracle. In the mid-seventies, the Volkswagen owed its cultural resonance to a pronounced contrast: as it continued to convey its long-standing associations with reliability, stability, and prosperity, it highlighted a historical present marked by an amorphous sense of insecurity. The first Volkswagen model increasingly struck the citizens of the Federal Republic as a thing of the past—and that past was seen as emphatically West German. It is indicative that neither Haffner nor other commentators writing about Volkswagen in the mid-seventies paid much attention to the car's Nazi origins, which struck contemporaries as largely irrelevant. An erstwhile political émigré from Nazi Germany, who at the time was putting the finishing touches on a best seller on Hitler, Haffner of course knew of the Beetle's brown past—as did his colleagues. Yet this aspect of its history appeared to have no bearing on the historical present of the mid-seventies. At the very moment the socioeconomic conditions that underpinned the Beetle's commercial success were in a state of dissolution, the crisis at VW indirectly consolidated the car's status as a genuinely and exclusively West German icon.[24]

Despite the Beetle's close commercial connection to Volkswagen's difficulties, the company's most famous product did not turn into a symbol of failure. On the contrary, West Germany's largest car producer extricated itself from its troubles with remarkable speed. Toni Schmücker managed to establish a climate of trust between the executive board and the trade unions, and industrial relations remained largely peaceful as the concern reduced its workforce. In part, the comparatively orderly atmosphere at VW during this tense period derived from generous severance packages the company offered to those workers willing to resign voluntarily. This arrangement proved attractive in particular to the overwhelmingly Italian, nonunionized "guest workers" filling low-level positions on temporary contracts, as well as to female, unskilled laborers. These groups provided by far the biggest contingents among the thirty-two thousand workers and employees who left the company voluntarily between April and July 1975. As management bought its way out of potential social friction, it proved crucial to the company's future that the new model range immediately performed beyond all expectation. Launched in 1974, the Golf, with its front-wheel drive and fifty-horsepower water-cooled engine, which sold as the Rabbit in the United States, established itself as an instant best seller in Europe. Within two years, the company had already produced more than one million Golfs, a model that would become VW's commercial backbone in Europe for decades. Thanks to burgeoning sales, Volkswagen posted a staggering profit of DM1 billion in 1976—a result no one would have dared to dream of only two years earlier. While the press marveled at the new "miracle of Wolfsburg," Toni Schmücker gasped that VW's turnaround was "almost incredible." Nonetheless, he immediately sounded a note of caution. "We cannot rest on our laurels." A conservative editorial also warned against complacency: "The crisis has demonstrated far too clearly how easily the company can reach the edge of the precipice, how unreliable the business cycle of the car sector can be." Although Volkswagen prospered again, the sense of insecurity that had first manifested itself in the mid-seventies persisted.[25]

Aware of the fragility of commercial success after the crisis of 1975, the management in Wolfsburg continued to keep a close eye on costs. As sales of the Golf soared, demand for the Beetle shrank dramatically. In 1977, fewer than fifteen thousand West Germans decided to buy VW's first model. The following year, the executive board reacted to the collapsing market for the Beetle by suspending its manufacture in the Federal Republic. To cushion the blow for the West German public, the company combined this announcement with an assurance that dealerships would henceforth sell Beetles imported from Mexico, where Volkswagen had maintained a comprehensive manufacturing plant since 1967. This compromise allowed writers to console themselves that the Beetle was not "dead" because it would "live on" for years as a "simple economical car" as well as a "cherished, memory-laden second car."[26]

The press accepted Wolfsburg's decision as an inevitable commercial measure that reflected an irreversible trend in the West German automotive landscape toward more comfortable vehicles. Changes in consumer preferences left the company no choice. Nonetheless, commentators struggled to make sense of the announcement, because it remained unclear what the future held in store for the little car. The end of domestic Beetle production, writers of different political stripes agreed, marked a "caesura"—and not only because "one of the biggest symbols of the economic miracle" would no longer carry the label "made in Germany." Observers were equally struck by the new international flow of commodities Volkswagen's decision initiated. Previously a West German export article par excellence, the Beetle now became an import. One journalist surmised "history" was "turning around." A colleague watching the unloading of the first Mexican Beetles at the very dock VW had long used for its export activities expressed a similar sentiment: "It was as if a movie was running backward." No commentator felt in a position to offer an authoritative interpretation of how the car's new international role related to the Beetle's past and future. "The Beetle is dead—long live the Beetle," declared one article, without expanding on the implications of this opaque statement. In short, the vehicle, until recently a prime marker of continuity, now threw into relief an uncertain, tentative future in West Germany.[27]

The insecurity surrounding the car's future in West Germany in 1977 reflected the widespread conviction that the Beetle was now living on borrowed time. The vehicle's fate in the United States bore out this assessment. In contrast to the parent company in the Federal Republic, Volkswagen of America continued to struggle financially. As the management searched for ways to revitalize its American operations, it soon became clear that the Beetle could play no further role. A persistently adverse dollar exchange rate hampered sales, while the introduction of additional safety and emission regulations stretched designers beyond the model's technical limitations. Above all, Volkswagen encountered sharp and unprecedented competition from the Japanese auto manufacturers, who flooded the American market with competitively priced and technically superior models in the small-car segment. Although forced to offer the vehicle at a considerable loss, VW retained the Beetle in its product range to maintain sales volume and prevent a dissolution of its extensive dealership network throughout the mid-seventies. In 1977, the management abandoned this strategy. For the Beetle's many American fans, it was a sad announcement. After the company issued the press release about the car's imminent demise, the media wistfully reminded readers of the car's unconventional appeal, its ad campaigns, and the profound love many Americans had developed for their "Volks." The *CBS Evening News* ran a report full of anticipatory nostalgia at the prospect of the "Bug becoming extinct. . . . As the Beetle chugs its way out of our lives," the reporter noted ruefully, "you've got to say a part of our lives goes with it."[28]

Eight years later, management proclaimed the car's end in West Germany. With sales hovering around ten thousand vehicles in the Federal Republic in 1984, import operations became commercially unsustainable when the federal government made catalytic converters obligatory for all vehicles. At the time, commentators paid far more attention to the consequences of this new ecological regulation for the national auto scene than to the Beetle's demise. After all, the "hard, loud, cramped Beetle," as *Süddeutsche Zeitung* wrote, had gradually become a marginal automotive product since the mid-seventies, thus preparing the West German public for its eventual disappearance from a car market increasingly shaped by demand for more powerful, more comfortable, and more ecologically sound vehicles.[29]

In the late seventies and early eighties, the Volkswagen Beetle was halted in its two main markets. In response to consumer criticism and the launch of viable competitors, the company enhanced the Beetle's appeal through design modifications. Nonetheless, in a West German car culture increasingly favoring midsize vehicles, Volkswagen's automotive star lost its shine as the cumulative effects of technical alterations, a rising wage bill, and regulatory reform in the United States eroded profit margins. When the early seventies heralded the end of the postwar economic boom, Volkswagen, which was only then introducing its new model range centering on the Golf, experienced a sharp crisis that required layoffs of tens of thousands of workers. Amid persistent economic turmoil and recalcitrant mass unemployment in the Federal Republic, the Volkswagen Beetle mutated into a symbol of the fading "economic miracle" that highlighted the contrast between a historical present characterized by insecurity and a past shaped by full employment, rising incomes, and stability. Since hardly anyone reflected on the car's Third Reich origins as VW fought for survival, the crisis of the seventies inadvertently strengthened the car's status as a West German icon, albeit one referring to a historical period that was fast disappearing. At the same time, however, the Beetle did not turn into a symbol of failure, because VW quickly returned to profits, thanks to the Golf.

VW withdrew the vehicle from the United States in the late seventies, much to the regret of many American drivers who continued to cherish the small rounded car. In the Federal Republic, meanwhile, the first Volkswagen gradually vanished from West German showrooms after domestic production ended in 1978. The termination of sales in the car's two most important markets, however, did not "kill" the Beetle, as some observers had feared. On the contrary, it survived for decades because of demand in societies that had not yet experienced mass motorization. That Volkswagen was in a position to import the vehicle from Mexico while phasing out sales in West Germany in the late seventies and early eighties was no coincidence. As part of its wider global operations, the company had maintained a prominent presence in this Latin American nation since the sixties, speculating on business opportunities that would replicate the West German car boom. Although the Beetle followed a distinctive national path in Mexico, it again proved exceptionally attractive and, just as in West Germany and the United States, emerged as a much-loved, prominent icon.

7

"I Have a *Vochito* in My Heart"

"It was an impressive sight when all the taxis had taken their stations on the city's great square," wrote Volkswagen's senior manager in Mexico to his superiors in Wolfsburg on November 26, 1971. Helmut Barschkis had every reason to be proud: the taxis that had lined up in neat columns on Mexico City's vast Zócalo before driving off in search of passengers to a "concert of honking horns" consisted of no fewer than one thousand VW Beetles. The German manager had ensured that the delivery of the first substantial fleet of Beetle taxis to Mexico City's local government was more than a commercial success quietly enjoyed. Gaining permission for a mass display of VWs on the Zócalo was a first-rate public relations coup for the company, an achievement Barschkis's German bosses probably failed to appreciate. Framed by the National Cathedral and the Presidential Palace, the small Volkswagens briefly took over an architectural ensemble at the center of Mexico's national iconography. Like the demonstrators that routinely flock to the square to launch themselves and their political causes onto the national stage, the Beetle self-confidently announced its growing presence in the country as 1971 drew to a close.[1]

This public ceremony for the first Beetle taxis foreshadowed the car's subsequent exceptional fortunes in Mexico. Over the years, the vehicle not only developed into a best seller, but came to rank highly among the country's venerated objects. In fact, numerous Mexicans have gone so far to as to declare the small, rounded Volkswagen, locally known as *el vocho* and *el vochito,* a typically Mexican automobile. While the literal meaning and etymology of these endearing nicknames remain in the dark, the car's ascent into Mexico's national pantheon is a remarkable development. Commodities are rarely adopted as national icons in countries that played no role in their initial design and production; but irrespective of its German roots, the Beetle counts among the artifacts that have achieved this feat in several nations. Many in Brazil, where VW has maintained production facilities since the late fifties and manufactured the car intermittently for almost four decades, regard the Volkswagen as their very own *fusca*—another untranslatable moniker. The Beetle has thus evolved into an icon with multiple nationalities.[2]

Even in countries where the vehicle does not count among the core national symbols, it still conveys distinctive meanings that are widely recognized across the social spectrum. When Ethiopian military rebels deposed Emperor Haile Selassie in September 1974, they sealed their coup with an automotive ritual whose implications were not lost on the ruler. A keen collector of glamorous cars who had accumulated a treasure trove of twenty-seven Rolls Royces, Mercedes-Benzes, Lincoln Continentals, and other luxurious automobiles, Haile Selassie found himself led to a green Volkswagen Beetle as he was banned from returning to his palace. "You can't be serious!" the emperor bridled. "I'm supposed to go like this?" For a longtime ruler who traced his ancestry back to biblical King Solomon, being forced to ride in the back of the modest Beetle was the epitome of humiliation.[3]

The Volkswagen's rise to cultural prominence in countries as diverse as Mexico, Brazil, and Ethiopia shines a stark light on its character as a global commodity. The vehicle's widening international profile testifies to the determination of VW's executives to position their company as a global concern whose operations reached beyond Western Europe and the affluent nations of North America. Backed by the federal government's export-friendly policies, Volkswagen persistently searched for new international opportunities and became a dynamic agent of globalization, at times resorting to forms of corporate behavior abroad that would have aroused storms of disapproval at home. The pursuit of global markets left a deep mark on the corporation itself, reshaping it time and again. At the same time, numerous actors in the countries in which the company established a commercial presence lent the car global prominence, ascribing distinctive national meanings to it in the process. In the fifties and sixties, Latin American and African governments strove to attract Volkswagen as part of development policies that achieved varying degrees of success. Beyond those holding high political office, crucial roles in domesticating the German vehicle were played by manual laborers, employees in VW's sales organization, and of course owners and drivers. Rather than treat the car as a foreign invader, they often went so far as to incorporate the Beetle into the stock of national symbols in protracted processes of appropriation. An interplay between Volkswagen's corporate strategies and an array of actors in the countries to which the company extended its activities accounts for the Beetle's global iconic stature.

In postwar Germany as well as the United States, the Beetle's history frequently reads like a success story. By contrast, it proves impossible to construct similarly upbeat narratives for the Latin American and African regions that have repeatedly experienced deep economic crises. Nonetheless, these parts of the world played a crucial role in the Beetle's history because they prolonged its commodity life at the very moment when it its attractiveness evaporated in Western Europe and affluent North America. Historians of Germany readily acknowledge the contribution of exports to the wealth of the Federal Republic, a country that proudly laid claim to the label "export world champion" in the 1980s when its foreign trade surpassed all competitors, including Japan. At the same time, scholars have yet to consider in detail the international effects arising from German companies' global activities. The proliferation of the first Volkswagen provides numerous opportunities to examine a West German company's impact abroad in a focused manner. Our examination here, rather than spreading across several countries, singles out the Beetle in Mexico. The *vochito* brings into view Volkswagen's changing global strategies, local workers' involvement in its manufacture, and the reasons why consumers came to regard the car as a typically Mexican commodity despite its German origins. Because of the *vochito*'s loyal local following, the company continued to manufacture it into the new millennium. Indeed, its uninterrupted production at Volkswagen's factory in Puebla between 1967 and 2003 meant that the car enjoyed a longer manufacturing run in Mexico than in West Germany. The long, twisting history of the Volkswagen in Mexico provides a case study that uncovers why, as part of wider processes of globalization, this small car extended its iconic status beyond the affluent Western societies and into Latin America.[4]

The aim of securing a foothold in what contemporary observers deemed a lucrative future market prompted Volkswagen's chairman Heinrich Nordhoff to instruct his managers to pursue a "farsighted policy" in Mexico in 1965. VW had already entered the country through an import subsidiary in 1954, selling Beetles assembled in Mexico from kits shipped from the factory in Wolfsburg. When the Mexican government issued a decree in 1962 that required all cars sold in the country to contain a 60 percent share of components produced in Mexico itself, Wolfsburg responded by erecting a new, large auto plant at a cost of 500 million pesos in Puebla, where production began in 1967. While this factory initially produced only the Beetle, starting in the early seventies it also turned out other models, as well as components for plants that VW operated in other countries.[5]

Wolfsburg's decision in favor of this substantial investment amounted to a step toward Volkswagen's gradual transformation from an exporter with a global distribution network into a hierarchically organized German multinational with subsidiary production centers as well as distribution networks on several continents. By the early sixties, Volkswagen had already opened a comprehensive manufacturing plant with almost ten thousand employees in Brazil, produced more than half of the parts for its Australian vehicles on the outskirts of Melbourne, and was in the process of constructing sizable factories in South Africa and Mexico. By 1967, the overall share of Volkswagens manufactured outside West Germany had risen to 14 percent.[6]

Volkswagen's move toward a global manufacturing base was less the result of a calculated corporate strategy than a pragmatic reaction to economic policies through which developing countries sought to stimulate domestic growth in the fifties and sixties. Countries as diverse as Brazil, South Africa, and Australia aimed to strengthen their manufacturing base by adhering to variants of the "import-substituting industrialization" strategies that assigned the state an active, interventionist role. In the Mexican case, the authorities pursued a multidimensional approach. Beyond enhancing the agrarian sector through land reform, the governing Partido Revolucionario Institucional (PRI) aimed at establishing a dynamic manufacturing sector behind high tariff walls to reduce the country's dependence on foreign imports. In tandem with protectionist measures to shield domestic producers from international competition, the state hoped to stimulate the country's industrial base by stipulating that manufacturers operating in Mexico use primarily nationally made components in their products. The decree about the "national integration" of the car industry that President Adolfo López Mateos issued in 1962 reflected the widely held view that the auto industry counted as a key sector in a developing country's drive to industrialization. By erecting the factory in Puebla, VW established a firm presence in Mexico next to a limited number of competitors including Ford, General Motors, Chrysler, and Nissan.[7]

In the sixties, Mexico's recent economic performance appeared to justify the massive investment. Although per capita gross domestic product stood at a mere 4,573 pesos (or roughly US$366) in 1960, as 55 percent of the population still worked in low-paid agricultural jobs, the country's annual economic growth averaged 7 percent during the ensuing decade, while inflation remained at a modest 3 percent. To some extent, Mexico's economic growth was driven by a rapid increase of the population from thirty-five million in 1960 to forty-eight million people a decade later, as well as by dynamic urbanization that resulted in Mexico City's expansion from 5.2 million to 8.9 inhabitants over the decade. Moreover, Mexico distinguished itself from other Latin American countries like Brazil or Argentina through its political stability. While the PRI monopolized state power between 1940 and 2000, the country undoubtedly suffered from clientelism and corruption but never descended into a dictatorship, not least due to the military's limited involvement in national politics. News of Mexico as a politically steady nation with excellent economic prospects circulated widely in Europe at the time. "'Mexican economic miracle'—that is the label with which to characterize the last years," West Germany's leading business daily informed its readers in 1969; another paper christened Mexico the "Japan of Latin America."[8]

Whatever the country's promise, Volkswagen found Mexico a difficult environment. Beyond overseeing policies of import substitution, the Mexican state was involved in numerous aspects of the national economy that directly affected Volkswagen's business. Struggles with government officials proved a recurring problem for managers in Puebla. While taking minor administrative obstacles in its stride, VW management displayed an acute awareness that a good standing with state officials was a precondition for commercial success in Mexico. As the company was preparing for the plant's opening in Puebla in 1967, Helmut Barschkis emphasized the importance of staying in the local elites' good graces. "Especially now that we are erecting our plant . . . it appears particularly necessary to cultivate our relations with all government departments," the head of VW de México wrote to his German superiors and singled out Mexican president Gustavo Díaz Ordaz as a person to be showered with attention. Wolfsburg clearly took Barschkis's advice about the "cultivation of contacts" to heart. In 1969, VW chairman Kurt Lotz received a letter from Díaz in which the politician thanked his "dear and good friend" for "the Karmann Ghia car that you had the kindness to obtain for me. It is a wonderful automobile and . . . very beautiful." Other leading lights of Mexican politics could also count on favorable treatment from Wolfsburg. After Luis Echeverría had assumed the Mexican presidency in 1970, Lotz found time to welcome the politician's son in Wolfsburg during a private visit.[9]

While these gestures served to secure direct access to Mexico's political elite in an economic environment prone to considerable state regulation and intervention, they were by no means a guarantee that Volkswagen would be granted its commercial wishes on all occasions. Although the company consistently protested against the price controls the government imposed on all vehicles, demands for "free price setting" fell on deaf ears for decades because Mexican officials feared that such a step would lead manufacturers to inflate their revenue.[10] The administration's active role in structuring the Mexican car market also provoked discontent because the government regulated the number of vehicles each corporation could sell annually in Mexico, initially granting Volkswagen quotas the company deemed insufficient. In 1968, the works in Puebla delivered 23,709 automobiles to Mexican car dealers, a production volume that led Volkswagen de México to post a loss of 80 million pesos, or DM25.5 million. With a break-even point at thirty-two thousand cars and an annual production capacity of fifty thousand, the plant in Puebla had been conceived for much larger operations than initially materialized. An exasperated top manager vented his frustration at Mexico's "totally planned economy" in 1967. The management was granted permission to produce almost sixty-three thousand Beetles in Puebla in 1973, but complaints persisted. While VW's leaders were quick to criticize government regulation, they failed to admit that they also reaped significant benefits from the state's involvement. After all, schemes like the quota program operated across the entire auto sector and therefore curbed competition. Government interventionism thus created a captive, if limited, market.[11]

The company may have blamed the state for its losses, but the fundamental problem lay in Mexico's economic underdevelopment and unequal wealth distribution. It was an open secret that the car market of the sixties and early seventies remained tight because only few people could contemplate purchasing a vehicle. The Mexican car dealer association pointed out in 1966 that despite rising average incomes over the past decade, "we are a poor country." Moreover, there existed "a high concentration of individual incomes in a relatively small layer of the population."

Although the country's narrow elite could afford to buy and maintain one or more vehicles, a mass market for automobiles remained a prospect for the future because of the vast income disparities. To be sure, the economic growth of the second half of the sixties resulted in a 26 percent expansion of the market for *populares* or small cars, swelling the overall number of new registrations to 130,000 in 1970. VW profited from this trend because the *vochito* came to dominate the segment of the *populares* and achieved more than thirty-five thousand sales that year. Nonetheless, a skeptical mood took root in Wolfsburg in the early seventies that contrasted sharply with the earlier optimism about the Mexican market. In 1973, VW's CEO Rudolf Leiding declared bluntly that in the medium term, sales prospects were "exceptionally bad" in Mexico.[12]

When the "Mexican economic miracle" began to lose steam in the mid-seventies, VW had to cope with a car market whose volatility amplified the ups and downs of the wider national economy. While annual sales stood at 88,158 in 1974, they descended to 54,511 three years later as a delayed consequence of the oil crisis. A subsequent energy boom, from which the country profited by expanding its oil production, put VW de México in a position to sell over 110,000 cars, including 42,330 *vochitos,* in 1982. The economic turmoil that characterized Mexico throughout the eighties and nineties exposed VW to persistently unpredictable demand patterns. Beetle sales languished at a mere 17,532 in 1987, but rose sharply to 97,539 in 1993, only to drop back to 14,830 in 1995. Not for nothing did numerous financial crises cause the eighties to be termed Latin America's lost decade. Despite widespread underemployment, rapid urbanization, and growing social inequality and insecurity, however, car ownership slowly proliferated in Mexico. While the ratio of car per persons stood at roughly 1:17 in 1980, it rose to 1:12 ten years later.[13]

In an atmosphere initially characterized by state intervention and subsequently rendered highly combustible by unpredictable business cycles, establishing and directing a comprehensive auto plant immersed Volkswagen in numerous conflicts. The acquisition of the factory site measuring 2.5 million square meters (almost 1 square mile) on the outskirts of Puebla, which formed part of the Mexican government's strategy to spread industrialization beyond an increasingly overcrowded and polluted Mexico City, caused considerable strife. In 1967 a West German embassy official reported that *campesinos* on whose former *ejido* (communal land) VW erected its works threatened open revolt because of the price at which they had been forced to sell their farms. While the land was worth around twenty pesos per square meter, he wrote, Volkswagen had purchased it from the regional government at a price of three pesos per square meter. The peasants were incensed not only because the state had pressured them into giving up their holdings, but because they had received merely twenty centavos per square meter—1 percent of the land's market value. There is no doubt that the farmers' rage at this "virtual expropriation" went beyond state officialdom and extended to the German company. "Like the Nazis," screamed the headline of a left-wing magazine charging VW with robbery. Concern about the farmers' protests ran high at the time, not least since they were accompanied by student riots that led to pitched street battles in Puebla's city center. President Díaz Ordaz, who had agreed to officially open the factory in May 1967, canceled all scheduled appearances in Puebla because of "demonstrations local farmers associations planned against the government's expropriation policies," as Helmut Barschkis cabled to a disappointed Heinrich Nordhoff. While on this occasion VW's cooperation with state authorities may have worked to the company's material advantage, it also embroiled it in a local conflict that complicated its relationship with the political elite.[14]

Once the construction of the factory with a foundry, stamping shop, paint shop, and facilities for the assembly of engines, axles, and other mechanical parts had been completed, the requirement to produce cars with components made in Mexico gave rise to numerous apprehensions. Collaborating with local suppliers drove up prices, a report from Puebla complained in 1968. A few years later, laments were directed at the quality of sheet metal available in Mexico. Concern about the company's international reputation persisted among board members in Wolfsburg into the seventies amid complaints about quality problems.[15]

To increase efficiency and quality in Puebla, VW took several measures. In 1966, the company opened its Escuela de Capacitación, a training center that has since served as a vocational school. To finance this institution, Volkswagen received help from the development ministry in Bonn, which acknowledged "the lack of skilled workers [as] . . . one of the biggest bottlenecks of Mexican industry." While most unskilled or semiskilled personnel spent only brief periods at the Escuela de Capacitación, the skilled workers who acted as an important hinge between those staffing the lines and the mostly German upper management underwent lengthy, intensive training. Applicants for the comprehensive apprenticeship scheme faced a protracted selection process over three months that consisted of theoretical tests in numeracy and general knowledge as well a medical and a psychological exam. In the late seventies, VW accepted one hundred trainees per year out of a pool of more than three thousand young men. Rather than develop a new instructional regime, the company imported West Germany's much-celebrated training format for skilled industrial labor, which combined theoretical and practical aspects over a three-year period. Trainees spent one day per week in the classroom, while the rest of the time they received practical instruction in a wide range of technical subjects, specializing in areas such as industrial design or combustion engines.[16]

Beyond training skilled laborers, the firm consolidated a workforce that expanded from 2,619 employees in 1967 to 11,067 in 1974. To restrict disruptive turnover, VW offered generous remuneration from the early seventies, thereby replicating an entrepreneurial principle that had proven successful in West Germany. In the early years the company rewarded its employees by complementing wages with material goods and gestures of goodwill. One line worker who had joined the company as a sixteen-year-old from impoverished Oaxaca in 1971 recalled his surprise one December evening when the managers distributed free turkeys, irons for workers' mothers and wives, blankets, and toys for their children.[17]

Most important, however, Volkswagen de México replicated the high-wage policies that underpinned its reputation as an exceptionally benevolent employer in West Germany. Between the early seventies and the eighties, VW paid wages at around twice the average local rate, a factor that lured unskilled workers to move from textile factories surrounding Puebla to the car manufacturer. Volkswagen also offered its employees attractive benefits such as membership in the national social security scheme, health insurance, paid vacation time, and from the eighties, help with school fees. Skilled workers received particularly good benefits that included a "gold fund" consisting of a 10 percent income share that VW invested on favorable terms for one year before paying it out. A thirteenth monthly payment at the year's end also formed part of the remuneration for skilled employees, a measure that directly copied West German pay practice.[18]

In addition to attracting and retaining an expanding workforce, Volkswagen's comparative generosity served to reinforce self-portrayals of the company as a Christian and charitable enterprise. When the archbishop of Puebla officially blessed the plant soon after its opening in December 1967, he assured the German leaders that their factory was "Puebla's pride and glory" because of the "social work" it performed in "giving employment to many people, who provide the livelihoods for many families." Helmut Barschkis played along and replied: "We believe the factory is important, but the people who have made it a reality and keep it going are even more important." The German managers realized the need to adopt Catholic customs that had no parallels in Germany. In particular, they recognized the central importance of the Virgen de Guadalupe as the national patron saint who offered support and protection from physical harm. As in many other factories in Mexico, altars to the Virgen were erected along each line; each year on December 12, moreover, the company threw open its gates to celebrate the saint's holiday with workers and their families. After a religious service, the management treated employees to food and beer, staged a lottery as well as games that, in light of the varying states of inebriation some revelers succeeded in reaching, could turn into rambunctious contests. To ensure that matters did not get out of hand, in the seventies the company stopped serving alcohol after a few workers had sought to impress visiting relatives by operating heavy machinery after possibly one beer too many.[19]

Irrespective of its leadership's paternalistic rhetoric of social responsibility, VW confronted its employees with high expectations, urging Puebla to emulate production standards set in Wolfsburg. Although line workers virtually never had direct dealings with the German managers who headed the plant, VW's Mexican workers experienced the factory as an alien environment that contrasted starkly with the world outside the factory gates. The management's insistence on "disciplina, orden y limpieza" (discipline, order, and cleanliness)—internally known as "DOL"—left employees under no doubt that they worked for a German company. Yet it was not foreign ownership alone that made the factory a difficult place for many employees. Within the highly regulated surroundings of the shop floor, most line workers performed repetitive work, often assigned the same specialized task for years.[20] Employees complained about noise, stench, and boredom, all results of Fordist production processes that Volkswagen had developed during the fifties and early sixties in Germany before exporting them to Mexico along with machinery no longer considered sufficiently up-to-date for operations in Wolfsburg. Many workers found their jobs dangerous, not least those who operated the powerful presses stamping out large metal parts. "One feels very lonely, it is very cold . . . it is dangerous and the work is hard, running at a high pace," explained one operator of a press. A colleague who was fired from his job after committing some errors on the line attributed his problems to the physical toll that factory labor increasingly took as he reached his late forties. Most line workers experienced the factory as a highly pressured world. To be sure, some employees were impressed by the size of the company's operations and took pride in working in a big, powerful factory rather than a small workshop, but they had precious little to say in praise of the day-to-day activities they performed for forty-six hours each week in the mid-seventies.[21]

In the late sixties and early seventies, the lack of effective trade union representation exacerbated the social tensions that were bound to arise in this working environment. In keeping with widespread national practice, the PRI-affiliated trade union organization Confederación de Trabajadores Mexicanos (CTM) appointed representatives who had never worked in the factory and thus possessed little knowledge of laborers' concerns. In April 1972 discontent about the existing union exploded in a wave of protests. Wages and benefits, the enraged laborers pointed out, compared badly with other car factories in the country; the union representatives had never worked at Volkswagen; they had failed to negotiate forcefully; and, to add insult to injury, they had embezzled membership dues. To pursue their interests more powerfully, the workers elected a new "Executive Committee of Real Workers," which henceforth led an independent trade union. Its aim, a public statement summed up, consisted in defending "our constitutional right to strike . . . and our absolute right to a dignified life." When it came to negotiations about wages and working conditions, VW's manual workers insisted on representing themselves. Securing the right for the union to be consulted on issues of work organization and to intervene in everyday matters on the shop floor were was a key demand put forward by the laborers.[22]

The workers succeeded in breaking the link with CTM and set up one of Mexico's rare independent trade unions. To some extent, the expansion of the workforce in Puebla from roughly twenty-six hundred in 1967 to almost six thousand in 1972 rendered both VW's social paternalism and CTM's detached approach to industrial relations incapable of addressing the organizational complexities that accompanied the growth of the plant. The attempt to establish an independent union benefited from the absence of a viable organizational model with regard to workers' representation. Beyond the need for organizational change, the laborers profited from a climate of political reform in Mexico in the early seventies. Upon assuming the presidency in 1970, Luis Echeverría sought to repair some of the political damage resulting from the authoritarianism of his predecessor Díaz Ordaz. The latter's heavy-handedness culminated in the notorious "Night of Tlatelolco" of October 2, 1968, when government troops opened fire on peaceful student demonstrators and killed an estimated two hundred people. A more lenient stance on labor issues counted among Echeverría's concessions to popular pressure, thereby aiding the emergence of independent unions at several car manufacturers at the time. The workers in Puebla played their cards well by choosing not to advance a fundamental critique of the institutional framework of Mexican labor relations. As a result, the authorities could regard the dispute at VW as an internal problem. At the same time, the workforce indirectly benefited from problems VW faced elsewhere. Confronted with falling American sales and in the midst of preparations for the launch of the Golf, the management in Wolfsburg paid little attention to the labor dispute in Puebla in 1972.[23]

During the seventies and eighties, the independent union played an important role in securing and defending the Mexican workforce's favorable material conditions at VW. Many people in and around Puebla sought employment in the factory, because a job at VW was widely deemed "a stroke of good luck." The workers scrutinized union activities closely, reserving the right to oust corrupt officials, as they did in 1981. As the workforce defended material gains with great determination, it acquired a local reputation for rebelliousness. Volkswagen's management for its part found self-confident workers' representatives a burden, complaining in its celebratory company history that by the late 1980s the independent union had created an "unacceptable situation" in terms of wage levels and quality control. In light of the material advantages they had secured for themselves, it comes as no surprise that the workers fought back against their employer's attempts to reduce the wage bill as well as union influence in work matters.[24]

When the union sought to reopen salary negotiations in 1987 to compensate for an annual inflation rate of more than 100 percent, the company responded with a meager offer of 30 percent. The workers rejected this and voted to strike for a higher raise. "We will not give in," the company declared, since wage levels in Puebla stood at twice the national average at a time when the factory was allegedly operating with heavy losses. The union retorted by pointing to the profit of US$110 million that VW de México had posted the previous year and added that incomes in Puebla were low by international comparison. While car manufacturers in the United States paid $10 per hour on average, 98 percent of Mexican car laborers made less than $1.45 per hour. Indeed, VW's Mexican line workers earned far less in a day than their colleagues in Wolfsburg made in a single hour. Faced with persistently high inflation, the Mexican workforce reacted with anger to the management's recalcitrant stance. One female assembly line worker summed up her view: "The Germans allegedly come to teach us something, but in reality they only take away the money we have earned."[25]

Although starting a strike entailed forgoing all pay for its duration, the workers held out for sixty days in the summer of 1987. A month and a half into the dispute, they were forced to approach the regional government for food donations because many union members could no longer afford basic staples. To raise the strike's public profile, the union did not shy away from militant tactics. In the first half of August it requisitioned dozens of buses to block a national highway as well as major roads in and around Puebla, lighting tires in the street and sending huge plumes of black smoke into the sky. The company's intransigent stance as well as hostile coverage in the local media only further inflamed the situation. Since Puebla manufactured select components for VW factories in other countries, the stoppage affected the company's global supply chains. Moreover, news trickling back from Puebla to West Germany made the company, which was keen to retain its reputation at home as an exceptionally generous employer, appear in a distinctly unfavorable light. In mid-August, VW's headquarters publicly criticized the management in Mexico in the West German press for its "headstrong" tactics. When leaders from both sides returned to the negotiating table in late August, they quickly reached an agreement granting workers a 78 percent retroactive pay rise.[26]

In Mexico, this outcome lent VW's workers high visibility among an embattled Left that passionately resisted the free-market reforms launched by the government in 1982. None other than renowned Marxist historian Adolfo Gilly, an erstwhile "sixty-eighter" who had spent years as a political prisoner after the "Night of Tlatelolco," celebrated the union in the leading left-leaning daily for "turning the battle of Volkswagen into the Battle of Puebla to break the siege of indifference and fatigue the multinationals and their allies have tried to lay on [the workers]."[27] As he drew on the muscular rhetoric of class struggle, Gilly cast the laborers as heroic resistance fighters against foreign aggressors, likening them to the soldiers who had participated in the battle of Puebla of 1862, when Mexican forces defeated a superior French invading army in a military encounter that subsequently ranked highly in national mythology.

Gilly's historical analogy proved uncannily apt—albeit in a manner contrary to his intentions. Just as the Mexican army could not prevent Mexico's eventual defeat in 1863 at the hands of the French, the trade union had won the battle in 1987 but not the war. In response to major losses in its global business in the early nineties, VW set out to shed its character as a centralized multinational whose peripheral plants secured emerging markets through models and production technologies no longer deemed cutting-edge in Western Europe. Instead, the company turned itself into a transnational car producer with more uniform manufacturing standards whose global model range paid close attention to differences in regional demand patterns. This strategy assigned new importance to Volkswagen's Mexican production site. Puebla's appeal now derived much more from Mexico's membership in NAFTA than from its status as a gateway to the domestic and Central American markets. To ready Puebla for the North American free trade area, Volkswagen initiated a host of changes, including major technological investments to prepare the plant to make models like the Jetta. To control costs and bring quality in line with international standards, VW introduced new forms of labor organization on the basis of lean-production methods first developed in Japan that prioritized group-work approaches on the assembly line.[28]

While advancing its agenda of transformation, the management came to an agreement with Mexican union leaders in July 1992 to modify the work contract in the plant in return for wage increases. The laborers, who had not been consulted about the contractual changes by the union negotiators, immediately suspected union corruption. Apart from the fact that members of the workforce were to lose seniority rights and receive pay in accordance with new performance criteria, workers were incensed that the union had effectively relinquished its influence over work organization. A stipulation allowing subcontractors to perform maintenance work in the factory attracted particular ire because their arrival threw up the specter of wage depreciation and future job losses.[29]

After the Mexican courts had declared a strike against the new labor contract illegal, Volkswagen was able to implement a measure that would have been unthinkable in Germany: it dismissed all fourteen thousand Mexican workers. Although most were rehired, VW not only refused to reemploy the protest leaders, but insisted on new regulations that curbed union rights drastically. Unlike in 1987, the company emerged victorious from the conflict of 1992, paving the way for subsequent increases in annual auto production in Puebla from roughly 188,500 in 1992 to 425,700 cars in 2000—during which time wages and benefits as part of total costs fell from 8 percent to around 4.5 percent.[30]

In the Federal Republic of Germany, meanwhile, tense industrial relations in Puebla received only limited coverage. To be sure, the antagonistic dynamics between management and workers in Puebla attracted the attention of the *tageszeitung*, a left-leaning daily sympathetic to economic critiques of neocolonialism, as well as of the *Frankfurter Rundschau,* a national newspaper with close ties to the trade unions. But many other mainstream dailies and weeklies, including *Frankfurter Allgemeine Zeitung, Süddeutsche Zeitung,* and *Der Spiegel,* either displayed little curiosity about conflict-ridden industrial relations in Puebla or covered these events in brief articles sympathetic to the company. When President Richard von Weizsäcker toured Volkswagen's Mexican production site as part of a state visit in November 1992, the recent mass firings received no mention whatsoever. Instead, German reporters praised the works as an example of Mexican-German economic cooperation and stereotypically proclaimed relations between both countries to be as cloudless as the sky over Mexico. That German criticism of Volkswagen's business practices in the late 1980s and early 1990s emerged only in papers harboring suspicions about global capitalism as a matter of principle proved fortuitous for Volkswagen at home. As long as a large part of the German public remained unaware that VW's Latin American operations relied on methods that were within Mexican law but would have been illegal in the Federal Republic, the company ran no danger of losing its domestic reputation as one of Germany's model employers. Silence about developments in Mexico shielded the corporation from the stigma of worker exploitation and, by extension, helped the Beetle retain its iconic shine at home.[31]

Fractious labor relations in Puebla did little to diminish the appeal of Volkswagen's most famous car in Mexico. More than 1.4 million *vochos* left the factory in Puebla during its period of production between 1967 and 2003. The *vochito*'s appeal across the social spectrum counts among its most striking features. People from many walks of life have owned and driven this vehicle, making it an integral part of Mexican society. As a result of its ability to straddle deep social divisions, the Beetle has gained a firm reputation as Mexico's first "carro del pueblo" or "car of the people"—a literal translation of the German word "Volkswagen."

Rather than as a consumer item for the country's narrow social elite, the *vochito* emerged as a car for much more ordinary Mexicans. The vehicle's standing as an item for non-elite drivers dates to the sixties, when VW began to produce it in Puebla. At the time, General Motors and Ford dominated the Mexican market with so-called "standards" and *compactos,* larger automobiles targeting an exclusive band of wealthy drivers. As the Mexican industry minister stated in 1967, the lower-priced Beetle filled a "gap" in Mexico's automotive landscape because it targeted Mexico's small urban middle class, which expanded only slowly.[32] Composed of independent as well as salaried professionals, managers, teachers, technicians, bureaucrats, and merchants, this heterogeneous social group grew from 9.4 percent of the population in 1960 to only 13.4 percent in 1980. In 1970, the country's leading car magazine *Automundo* still identified the "unequal distribution of wealth and income" as the main obstacle facing the proliferation of the automobile in Mexico. At the time, a mere two hundred thousand Mexicans—or 0.4 percent of the overall population—commanded monthly incomes of between five thousand and ten thousand pesos required to afford a small car. As a result, the Beetle attracted "professionals and managers in mid-level positions," the magazine concluded. In the strictly limited auto market of the Mexican "miracle years," the *vochito* became the automobile that brought car ownership to the slowly expanding middle class. As such, its proliferation on the streets not only reflected the social status of successful middle-class Mexicans but the country's gradual economic progress at large.[33]

The Beetle remained a socially exclusive commodity into the eighties. Sales contracted substantially with the onset of the country's economic problems in 1982 that saw middle-class incomes erode dramatically. Only toward the end of the decade did the car come within reach of social circles beyond the professional middle class. In August 1989, the Mexican government decreed that vehicles selling below 14 million pesos (roughly US$5,000) be eligible for substantial tax breaks. Volkswagen was the only manufacturer to react to this announcement, implementing a price cut of 20 percent to 13,750,000 pesos to "make cars more affordable in Mexico," as the company declared in full-page newspaper ads. At a press conference, the president of VW de México emphasized that the discount served to facilitate "the acquisition of an automobile for the wider population. Together, we are building a new era in Mexico." Hoping that recent economic difficulties would not recur, VW affirmed its commitment to "a fairer Mexico and a more prosperous future for our children." Even critical observers of Volkswagen regarded the fact that "the most popular vehicle has fallen in price" as a significant event. An economist who regularly advised the independent union at the Puebla plant acknowledged that "yes, here it is, the car for the poor." Tax breaks and a short boom ensured that *vocho* sales peaked at almost one hundred thousand in 1993. Its comparatively low price and tax incentives secured Volkswagen's classic a new market in the late eighties, extending its social appeal and literally prolonging its lease on life at the very time that the company was phasing it out in Western Europe. Persisting amid Mexico's political and economic ups and downs, the *vochito* became a material fixture in everyday national life.[34]

While the Beetle's long-standing presence may have turned it into an instantly familiar commodity for most Mexicans, it owed its adoption as a national automobile to more than its status as a quotidian item. Volkswagen indirectly supported the process of national appropriation through professional marketing campaigns. VW de México delegated the creation of suitable advertising material to Doyle Dane Bernbach, the agency also in charge of campaigns in the United States and West Germany. In 1965, DDB opened a branch in Mexico, which soon established itself among the country's top advertising businesses. At first, DDB mainly adapted campaigns from the United States to local requirements, a modification that required "detailed knowledge of the Mexican driver's psychology," as its Mexican director Teresa Struck explained in 1970. But from the mid-seventies, the agency also developed *vochito* ads exclusively for the Mexican market. One such ad from the 1980s featured a photograph of a loaf of bread in the shape of a *vocho,* declaring the vehicle to be as much part of everyday life as "daily bread" *(el pan de cada día)*. While adopting a clean, functionalist aesthetic that lent the Beetle an air of modernity, the ad associated the vehicle with a dietary staple that possessed foreign origins but had featured on daily menus since the 1940s when Mexican bakeries began the mass production of bread. DDB thus framed the car as a familiar, domesticated, and indispensable commodity.[35]

Rooted firmly beyond the social elite, the car's reputation also profited from widespread awareness that Volkswagen maintained a large production site in the country. Over the years, the German company gained a high public profile on Mexico's industrial landscape as its directors employed press contacts to emphasize how their Mexican initiatives benefited the corporation and the country alike. The celebrations in September 1980 marking the production of the one millionth car in Puebla provide a good case in point. After a gleaming red *vocho* had rolled off the assembly line to mark the jubilee, VW's Mexican director Helmut Barschkis addressed an audience including a representative of President López Portillo, the governor of the state of Puebla, several European ambassadors, VW's chairman Toni Schmücker, and hundreds of workers. Although the company had faced difficult times, Barschkis assured his audience that he had never "lost confidence because we have never been cheated." He also praised "the strength of the hardworking Mexican. It is to his hands that we owe this wonderful triumph: the one millionth VW produced in Mexico." As he paid tribute to Mexico, Barschkis peppered his speech with reminders of VW's contribution to the Mexican economy through thousands of jobs and sizable investments, as well as production of over ninety thousand cars for export during the previous decade. According to official statistics, VW de México ranked seventh among the country's exporters. Barschkis thus painted his branch of Volkswagen as an enterprise with a leading role in Mexico.[36]

Neither the strikes in Puebla nor the economic turmoil of the eighties undermined VW's prominence in Mexico, as festivities staged in Mexico City in October 1990 illustrate. On this occasion, it was the simultaneous production of the one millionth Mexican Beetle and the two millionth car overall that prompted Mexican president Carlos Salinas de Gortari to host a reception for a party from Puebla including the state governor and VW de México's new director Martin Josephi, as well as VW dealers and workers. Josephi painted his enterprise in the brightest colors, mentioning its growing exports to North America and Europe, as well as its 28 percent share in the Mexican car market. Salinas for his part thanked the company for "dedicating itself to the nation's development." Neither forgot to pay tribute to the workers for their "daily extraordinary efforts" that ensured the company's competitiveness.[37]

While festive speeches at company jubilees glossed over the social and political tensions in Puebla, they did more than place a foreign company in the national limelight and cast it as an important element in Mexico's industrial fabric. Speakers repeatedly drew attention to the contribution of the workforce, thereby emphasizing that Mexicans left a profound imprint on VW's products. An ad Volkswagen ran in 1977 provides the most striking statement to this effect, as it aligned the *vochito* with one of the country's best-known gastronomic dishes. Entitled "Don't accept any *mole,*" the advertisement referred to the sweet and spicy, chocolate-based sauce for which Puebla is famous all over Mexico and whose preparation demands great skill and patience. Below the tagline, the reader beheld a color photograph of a clay pot filled with a rich brown liquid simmering over charcoals. The text beneath the photo explained how the clay pot had ended up in an auto ad:

> When one wants to make an authentic *mole poblano,* one cannot accept any substitutes since, as everyone knows, it is an incomparable dish, which is made with very special ingredients and requires a laborious preparation process that allows for no improvisation.
>
> In the same manner, when one wants a good car, made conscientiously and with the best parts, the necessary choice is a VW Beetle, which, by the way, is also produced in Puebla. . . .
>
> And the truth is, in Puebla, we are particularly proud of our *mole* . . . and of our VW Beetle.[38]

By pushing into the foreground the expertise of the Puebla workforce, the company underlined the car's credentials as a properly Mexican product. Within the company, VW emphasized "discipline, order, and cleanliness" as core values in day-to-day routines on the shop floor, thereby engendering a work culture that struck many employees in Puebla as German. In its public relations material, however, VW de México did not explicitly paint the *vochito* as a commodity produced to specifications imposed from the outside. Instead, it stressed how the expertise of Mexican workers underpinned the vehicle's appeal. If not in its original design, the *vochito* had become a Mexican automobile via Mexican quality workmanship.

While the extent to which VW succeeded in actively shaping public opinion through promotional drives and well-publicized jubilees remains unclear, some Mexicans disagreed with efforts to ascribe a Mexican national identity to the *vochito*. Skeptics readily acknowledged VW's production site in the country as well as the car's ubiquity as reasons for its cultural prominence, but argued that employing local workers is not enough to turn the vehicle into a product with Mexican qualities. The *vochito*'s initial German design and the German technology used in its production eclipsed the contribution of Mexican labor, one worker thought. Another argument against the Beetle's inclusion among "typically Mexican" objects hinged on claims that, because of its foreign origins, it "has nothing to do with our culture," as a skilled worker pointed out. This observer maintained that the car essentially remained a foreign object, no matter how widespread it may have become.[39]

Yet others begged to differ: "For me, the car is more Mexican than German because it was made by Mexicans," explained a VW worker with considerable pride. It is not only those who had a hand in auto production who share this view. A vendor of street food in Mexico City volubly replied in the affirmative to the question whether the small rounded VW was Mexican. "The *vochito* is me!" he exclaimed agitatedly before providing the main reason for his assessment: "It is made in Mexico." Mexicans also concurred that the car's broad social appeal and wide proliferation had given it a special place in Mexican society. "Everybody had one," another man exaggerated in an attempt to convey an impression of the car's omnipresence in the early nineties. And even if one never owned a *vochito,* most people are likely to have ridden in it, not least since it was the main taxi vehicle all over the country until the mid-nineties.[40]

For many Mexicans, the *vochito*'s technical features rest at the core of its supposed *mexicanidad.* As the company modified the

In Mexico, the VW became the ubiquitous *vochito* from the seventies to the nineties. Beetle taxis could still be seen on every street corner in Mexico City in 2008. Photograph by the author.

vehicle by equipping it with a stronger engine, a new fuel injection system that worked reliably at high altitudes, new brakes, a more effective suspension, and a catalytic converter, the car struck Mexicans as exceptionally sturdy and dependable. As early as 1971, *Automundo* ran an article that celebrated "the already mythical VW" for its "quality" and its "solid construction." Mexican drivers singled out the car's reliability and its sturdiness as important factors for its success. Here was a car a driver could trust because, as another fan explained, it was as tough as "a small tank."[41]

In many a Mexican's eyes, these technical properties lent the vehicle a cultural affinity with central features of everyday life in Mexico. One driver, who recalled his VW as an "excellent auto," deemed it a Mexican car because it did "not need much." Owing to its comparatively simple yet reliable technology, the Volkswagen, he explained, was an ideal vehicle for a country with a rough infrastructure. The Beetle, in other words, proved a worthy match for Mexico's hard conditions. A Mexican car joke exploits this theme for a punch line by relating the characteristics of drivers from different countries to their vehicles of choice.

> An American is driving along a lonely road in a Chrysler when he hears a funny noise. What does he do? He immediately stops and calls for road assistance on his cell phone. After 15 minutes, a tow truck shows up as well as a rental vehicle, in which the driver continues his journey safely.
>
> A Japanese man is driving along a lonely road in his Toyota when he hears a funny noise. What does he do? He immediately stops, opens the hood and examines the engine carefully, discovers the fault, invents a solution, implements it, and continues his journey safely.
>
> A Mexican is driving along a lonely road in his *vochito* when he hears a funny noise. What does he do? He takes a large sip from his beer bottle, turns his car radio to maximum volume, steps on the pedal, and continues his journey safely.

While casting American cars as unreliable and Japanese vehicles as requiring qualified expert attention, the joke self-ironically portrays the *vochito* as a hardy, robust, unpretentious Mexican form of transport whose minor ailments were no cause for concern. The VW's comparative plainness and robustness, then, struck Mexicans as attributes that highlighted a cultural similarity between life in their country and the car. Viewed from this angle, the Beetle was a tough, thick-skinned Mexican capable of handling both actual and metaphorical bumps in the road.[42]

For those affirming the car's inherent *mexicanidad,* the *vocho* filled a vacuum in Mexican auto culture that resulted from the absence of a popular vehicle made *and* initially designed in Mexico. While many in the United States regard the Ford Model T or the Cadillac as expressions of national technological creativity, and Britons revere the Morris Minor, Mexico lacks a direct automotive equivalent to these models. The cars retailed in the country first materialized on drawing boards beyond its borders. To some extent, the *vocho* came to be embraced as a car with a Mexican national identity for lack of a better domestic alternative. Moreover, the Beetle's German roots possibly favored its cultural adoption as a Mexican commodity in a country in which most other cars possessed U.S. pedigrees. Compared with vehicles originating in the United States, the Beetle lacked the imperialist associations that many Mexicans attribute to their northern neighbor as a result of long-standing unequal power relations. While VW's business practices were indeed occasionally denounced with references to Germany's Nazi past, such invectives remained isolated and never developed anything remotely resembling the virulence of *anti-yanqui* rhetoric. Next to its production site in Puebla, its social proliferation over an extended period, and its technical characteristics, the *vocho*'s German rather than U.S. origins may well have assisted its acceptance as a national product.[43]

Whether it is cast as an authentically national product or as a successful import, the stories Mexicans tell about the Beetle attest to its prominence in everyday life and its notable place in the recent past. Over the years the *vochito* has entered the stock of recollections many Mexicans share and become one of the sites of memory that affords a sense of national connectedness. People from many walks of life readily make and remake stories that incorporate the Beetle in a collective body of experience and that perpetuate the car's prominence in the national imagination.[44]

A central theme in such stories is the car's capacity to undergo simple yet imaginative repairs. Being able to fix the car themselves without complicated tools and elaborate components provided a crucial prerequisite for success in a country where many drivers had little money, roads were frequently rough, and small workshops could not always be trusted to offer good service. Mexican owners relished stories about how, after breakdowns, they returned their cars to working order through the simplest aids imaginable, including pieces of string, rubber bands, and bits of wire. One aficionado ended a spirited exchange on the subject with a neighbor by declaring: "When it ran out of gas, you could run it on water!" As these tales credit Mexican drivers with the talent to solve automotive difficulties resourcefully despite material constraints, they highlight common experiences and cast the Volkswagen as a simple car ideally suited for a challenging automotive environment.[45]

That many memories of the *vochito* are essentially private and lack an overt political dimension does not undermine their collective potency. The absence of political themes may well allow Beetle anecdotes to overarch the deep ideological chasms separating the Left and Right in Mexico. The car and stories about it rank among the common cultural reference points that fail to arouse political discord and are thus a precondition for a shared sense of national identity. Nor should one assume that all personal anecdotes need to be altogether truthful to fuel the collective and private imagination. Often, their effectiveness depends primarily on the verve with which they are told to sympathetic listeners. "The *vochito* was a great car for sex," one former owner recalled in a conversation about the vehicle as a commodity with Mexican traits. When asked how this statement related to the topic at hand, he elaborated: "It is not just that many *vochitos* were made in Mexico: many Mexicans were also made in

The shell of a *vocho* parked next to a private home in the Mexican coastal town of Puerto Ángel. The car's comparatively simple engineering characteristics allowed drivers to fix and modify their vehicles themselves. Having reached the end of its life, this *vocho* has been scavenged for parts. Photograph by the author.

vochitos." While it is impossible to verify what people were up to in their little cars, the assertion illustrates how talk of the Beetle readily conjures up recollections and how deeply woven it is into Mexican collective memory as a versatile everyday object. In light of the car's public prominence and private significance, it is no surprise that many drivers regard the Volkswagen as much more than a functional means of transport. Some shower it with declarations of love, and others admire it as a "world wonder." One aficionado who bought twelve VWs throughout his life was moved to say, "I have a *vochito* in my heart."[46]

Even line workers in the factory in Puebla, who regarded their employer's conduct with highly critical eyes, greeted Volkswagen de México's decision to end production of the automotive classic in 2003 with profound regret. While the management continued to manufacture the *vochito* despite falling sales after the mid-nineties, it refrained from further investments in the production line and conducted reviews into the *vochito*'s future every two years. By the turn of the millennium, the old model proved increasingly uncompetitive. Nissan's popular four-door sedan Tsuru in particular had provided drivers with a more highly powered and comfortable alternative to the small rounded VW since the late eighties. On July 30, 2003, the final curtain fell on Beetle production in Puebla as the last model left the assembly line to mournful tunes played by a mariachi band in full regalia. For a worker who had entered the factory in 1979, it was an event that changed the entire character of his workplace. "Without the *vocho,* the factory was like a bakery without rolls," he stated. Rather than include a car among its product range that its own laborers could afford, from that point on Volkswagen de México only made models like the Jetta and the retro vehicle known as the "New Beetle," which were aimed at higher market segments, often for the U.S. market. Some of the sympathy the Beetle still commands undoubtedly derives from memories of the car as an automobile for ordinary Mexicans.[47]

Since 2003, the *vochito* has begun to turn into an object of nostalgia. Workers at the plant in Puebla have expressed their sense of loss by christening the building where the last Beetles had been produced the "hall of tears" *(nave de lágrimas)*. A worker who witnessed the end of Beetle production with sadness acquired a used model soon after. "Adorable, nice" were adjectives he considered suitable for a cherished possession he vowed never to sell. This desire literally to hold on to the *vocho* is by no means restricted to workers. Although the car was still a ubiquitous sight on Mexico's roads half a decade after manufacture ceased, a sense of imminent loss was beginning to surround it. The "última edición" that VW de México designed in 2003 to commemorate the *vochito*'s end not only sold out immediately, but quickly turned into an expensive collector's item. In 2008, well-kept models were advertised for 120,000 pesos or roughly US$12,000, a price only affluent Mexicans could contemplate. The prospect of its absence thus spurred both wealthy and less

The last Beetle rolls off the line in Puebla in July 2003, to the tunes of a mariachi band. As befits a vehicle widely venerated in Mexico as a typically Mexican automobile, the final *vochito* is crowned with a sombrero. Courtesy of Volkswagen Aktiengesellschaft.

wealthy Mexicans alike to buy their own "carro del pueblo" before it eventually disappears from the road.[48]

Despite its foreign origins, the *vochito* has gained a solid reputation as a typically Mexican car. As it met with commercial success across the planet, the Beetle achieved a rare feat in the world of global commodities. In addition to finding its commercial and cultural place in various markets like countless other export articles, it has turned into a vehicle with multiple nationalities. By the time production ended, Mexicans, like West Germans, revered the car as a national icon—and they were not the only ones to do so, because Brazilians cherish the car in a similar vein.

In Mexico, the Beetle's adoption as a national commodity was closely linked to Volkswagen's determination to succeed in international markets. This ambition, which enjoyed active support from the government of the Federal Republic, cast the German car maker as an agent of globalization that tackled the numerous difficulties presented by its Mexican operations with considerable energy and stamina. Deemed a lucrative future market in the mid-sixties, Mexico proved a difficult business environment as the company struggled with limited demand, economic policies favoring state interventionism into the seventies, and subsequent commercial volatility. Although it secured the goodwill of successive Mexican administrations, VW could not prevent an independent trade union from emerging in 1972 that defended workers' rights and wages effectively over the next two decades, at times embroiling VW in deeply contentious labor conflicts. Only in 1992 did the company scale back union influence after seizing the chance to lay off the entire workforce temporarily, a drastic measure buttressing the introduction of post-Fordist production methods as well as Puebla's new strategic position as a manufacturing center for North American markets.

VW's decision to maintain a factory in Puebla proved crucial for the *vochito*'s adoption as an unofficial national treasure. Executives, politicians, and ordinary citizens extolled the vehicle as a quality product "made in Mexico," whose simplicity and sturdiness rendered it ideal for the country's demanding physical terrain. Because of its comparatively low price, the *vocho* gradually extended car ownership beyond the country's elite via the middle class to citizens further along the social scale. Manufactured in the country for fellow countrymen, the *vocho* became the *carro del pueblo* whose technical characteristics appeared to lend the automobile a cultural affinity with everyday life in the country. Viewed from this perspective, it was the *vochito*'s *mexicanidad* that allowed the car to function in Mexico's challenging automotive environment.

Despite its adoption as a Mexican car, the *vocho* nonetheless bears the imprint of capitalism's global inequities. While VW's skilled workers fared exceptionally well, relatively speaking, and commanded good wages and benefits, the *campesinos* on whose erstwhile communal land the factory stands to this day never received financial compensation for losses stemming from the actions of a local government keen to attract a foreign car manufacturer. VW employees in unskilled and semiskilled positions, meanwhile, earned higher incomes than colleagues in other local industries, but their pay levels remained modest by international standards. Twenty years after the factory first opened its gates, the daily wages of VW's Mexican line workers stood nowhere near hourly pay rates in Wolfsburg. Moreover, the laborers at Volkswagen de México did not owe their limited economic gains to their employer's generosity but to trade union activism. In other words, the fact that the incomes at VW in Puebla stood above the Mexican national average cannot be attributed to wealth creation mechanisms inherent in global capitalism but derived from workers' assertiveness.

In light of the repeated ups and downs that have characterized Mexico's recent economic history, the Beetle's ascent to iconic status in Mexico can thus by no means be told as a straightforward success story. Far more fitting is a nonlinear narrative similar to that of the video installation *Rehearsal I* (*Ensayo I*) by Francis Alys, a Belgian artist who has worked in Mexico since 1987. Filmed in Tijuana in 1999, *Rehearsal I* depicts a bright-red Beetle that hurtles up a steep dirt road lined by houses in various states of repair to the sound of a mariachi band rehearsing the tune "El pendulo." Every time the musicians interrupt their play, the car rolls back to the bottom of the slope, only to begin its upward journey afresh when the band resumes the tune. For almost half an hour, the vehicle valiantly reaches for the top, "never succeeding," as the scenario that accompanies the installation states. Alys describes the Sisyphus-like exertion in *Rehearsal I* as "a story of struggle rather than achievement," adding that this work encapsulates "a Latin American development principle" of movement devoid of meaningful progress: "3 steps ahead, 2 steps backwards, 1 step ahead, 2 steps backwards." For Alys, the Beetle's journey up and down the hill ultimately acts as "a metaphor for Mexico's ambiguous affair with modernity."[49]

Four years after Alys first exhibited *Rehearsal I,* the last *vochito* rolled off the line in Puebla. Mexicans greeted this event in 2003 with profound regret, but it did not mark the end of the Beetle's commodity life. On the contrary: five years before the *última edición* left Puebla, VW had presented a new car to the world that openly played on the historical mystique surrounding its first and by now classic model. Sharing no engineering characteristics with its illustrious predecessor, it was nonetheless immediately recognizable as a "New Beetle" that rode a global wave of nostalgia for VW's first model.

8

Of Beetles Old and New

"No chrome. No horsepower. Foreign." Thus runs the dismissive verdict of advertising executive Salvatore Romano upon scrutinizing a Volkswagen ad during a business meeting at fictional agency Sterling Cooper in the contemporary hit television series *Mad Men,* set in 1960. "They did one last year, same kind of smirk. Remember 'Think Small.' It was a half-page ad and full-page buy. You could barely see the product," a colleague shakes his head, only to be contradicted by Pete Campbell, one of the show's personifications of competitive ambition: "Honesty. I think it's a great angle. . . . It's funny. I think it's brilliant." After several further exchanges, Donald Draper, Sterling Cooper's creative supremo, shuts down the meandering discussion. "Say what you want. Love it or hate it. We've been talking about it for the last fifteen minutes."[1]

Focusing on its diminutive, unflashy proportions as well as Doyle Dane Bernbach's noteworthy promotional campaign, the Beetle's cameo appearance in *Mad Men* serves as an indicator of the high cultural profile the car has retained in the United States decades after Volkswagen ended imports. For a 2007 television show whose appeal hinges on evoking the flair of the sixties, the

Beetle provides as much an indispensable period detail as the Cadillac Coupe de Ville Donald Draper acquires in a later episode. In the contemporary United States, the small "Volks" functions as an object of historical memory that allows us to invoke significant aspects of the past. Of course, the Beetle's status as a site of memory is by no means restricted to the United States, extending as it does to other nations where it achieved sales success. A cursory glance at the covers of German-authored books dealing with West German history after 1945 confirms that the Beetle has ranked among the Federal Republic's most prominent symbols well beyond the 1990s. Yet even in countries like Great Britain, where the car never enjoyed major commercial success, the media have used it to conjure up the recent past. The opening scene of the BBC's award-winning series *Life on Mars* (2006) shows its protagonist, who has mysteriously traveled back in time from post-millennium Manchester to the year 1973 after a car accident, stumbling through a thoroughly transformed cityscape in a state of shock before he disbelievingly stares at his own reflection—in the sideview mirror of an olive-green Volkswagen.[2]

The Beetle's popularity as a historical prop in the contemporary media illustrates its continuing cultural salience long after it disappeared from dealerships in Western Europe and the United States. The car once made in Wolfsburg stands tall among the industrial products that have succeeded in extending their commodity lives beyond the period when they were in mass use. Although critics have often berated affluent societies for cultivating throwaway mentalities that derive from the West's fetishization of innovation and economic growth, numerous artifacts have avoided history's dust heaps once they exhausted their apparent usefulness. Frequently safeguarded by collectors, restorers, and others holding old things in high regard, countless objects of mass production ranging from postage stamps, furniture, and corkscrews to beer coasters and birdcages have emerged as items worthy of preservation. This urge to keep the commodities of the past amid a whirlwind of change has accompanied innovation's thrills and upheavals, manifesting itself in what one scholar described as the "transubstantiation of junk." People often ascribe historical significance to objects that link them to their own personal past. The Volkswagen is one of the more prominent historically charged commodities that gesture back in time in present-day Western societies.[3]

The deeply emotive impulses that underlie the first VW's continuing appeal are particularly conspicuous in the international scene of fans who own and drive old Beetles. Investing considerable portions of their disposable income as well as spare time in their small cars, these proprietors of old Volkswagens cherish intimate bonds with their historical automobiles. Their meets and festivals attract large crowds of spectators in Western Europe and the United States during the summer months. The scene of Beetle aficionados testifies to the first Volkswagen's lasting prominence in international popular culture and to its ability to elicit overwhelmingly affectionate reactions and memories among broad sections of the population.

Volkswagen's successful launch of a car under the moniker "New Beetle" in the nineties revealed the extent to which automakers stood to reap substantial commercial benefits from the aura that enshrouds iconic automobiles. While this vehicle shares nothing in terms of engineering features with the original, its shape consciously evokes the classic by copying its distinctive silhouette. The idea to make a new Beetle had to overcome skepticism at the company's German headquarters before turning into the first among a growing number of retro cars that have appeared on auto markets since the turn of the millennium. Beyond highlighting the commercial potential of the past, the new vehicle draws attention to the international profile the first Volkswagen has secured during the second half of the twentieth century. Based on an old classic, this retro vehicle was specifically aimed at a non-German market, developed by an international team of designers and engineers, and manufactured outside Germany. In other words, by the time the New Beetle hit dealerships in 1998, it already possessed a thoroughly international pedigree that reflected the original's global success as well as Volkswagen's evolution into a transnational corporation. Investigating the scene of old Beetle fans in conjunction with the developments behind the New Beetle brings into view multifarious cultural and economic factors that have extended the first Volkswagen's commodity life in affluent societies on both sides of the Atlantic far beyond the period of its everyday mass use.

The profound affection continuing to surround the old Beetle long after sales officially ceased manifests itself most strikingly in the open-air meets that take place across much of Western Europe and North America during summer weekends. Organized by clubs bringing together owners of VWs, these events display lovingly restored specimens of the company's first model as well as versions that have undergone substantial and at times stunning alterations. Just like the cars on display, Beetle meets come in numerous shapes and sizes. While some gatherings remain intimate affairs with fewer than a hundred automobiles, five-figure-strong crowds of spectators flock to the well-publicized, all-weekend festivals that feature speed competitions, awards for show cars, markets for spare parts, and extensive entertainment programs including rock concerts, comedy nights, and discos. In the new millennium's second decade, the historical Beetle scene shows no signs of running out of gas. In 2010, over three thousand drivers of old VWs thronged the "Maikäfer" meet in Hanover, one of the largest German gatherings staged each May. Meanwhile, "Bug Jam," which takes place at the Santa Pod racetrack in Northamptonshire, roughly sixty miles north of London, claims to be the largest Beetle-related event in the world. Having attracted forty thousand spectators in 2007, this three-day festivity fully lived up to its name four years later when the organizers were forced to discourage numerous fans from traveling to the site amid fears of traffic jams in the English countryside.[4]

By 2010, many meets and the societies organizing them had been in existence for well over two decades. Vintage Volkswagen Club of America, an umbrella organization bringing together societies across the nation, goes back to 1976. A year later, a group of West German enthusiasts toyed with the idea of founding a Beetle club, turning it into reality in 1981. Bug Jam first opened its gates in 1987. The emergence of associational networks and festivals dedicated to the Beetle coincided with the end of its production in West Germany and withdrawal from the European and North American markets. To the same degree that the car became dated in the automotive mass markets in Western Europe and North America, a sizable social minority began to treasure it as a precious historical possession.[5]

The people who take to the Beetle as a historical artifact hail from diverse backgrounds. Some fans are wealthy, self-confessed "car nuts" who include their Volkswagens in sizable, expensive car collections, but most owners command far more modest financial resources. A retired auto mechanic, a glazier, a postman, an electrician, a social care worker, a teacher, and a university-educated electrical engineer count among the Beetle aficionados the author encountered at meets in the UK and Germany.[6] While male owners significantly outnumber their female counterparts, the old Beetle scene covers a wide social spectrum spanning the ages of twenty-two to seventy-five. Given its social and generational breadth, the scene is not shaped by a distinctive outlook on life or similar aesthetic preferences that characterize subcultures in a strict sense.[7] Some members form long-lasting friendships at meets, but most enthusiasts are primarily motivated by curiosity about the cars on display. VW aficionados thus form a loosely knit community that revolves around the fascination for an automobile.[8]

The scene's social breadth reflects the comparative affordability that distinguishes the Volkswagen from many highly prized classic cars. In 2006, a British owner offered his "good, solid" and "totally original" vintage vehicle from 1955 for £3,500 (US$5,500), while a well-maintained 1973 model was available for £2,250 (US$3,600). Ten years earlier, a driving school instructor from Nuremberg purchased his 1971 Beetle for a mere DM500 (US$360). Although prices have risen in the first decade of the new millennium as supplies have inevitably begun to dry up, the first Volkswagen continues to count among the more affordable classic automobiles. The ready availability of spare parts at reasonable cost provides another reason why people with comparatively limited means opt for the Beetle. "I have a friend who has an old Mini Cooper," a female German child-care worker explained in 2011. "Of course, the Mini is a nice car, but it is nightmarish and very expensive to get parts for it. The spares for the Beetle are far cheaper." The steady supply of replacement components does not exclusively derive from a highly developed trade in parts scavenged from countless old models that are on offer online and at "swap meets" at Beetle conventions. Rather, REFA Mexicana, a Puebla-based company situated across the road from VW's Mexican factory, continues to turn out thousands of doors, fenders, chassis, bumpers, and more, often using the decades-old stamps once operated in Wolfsburg. While most of REFA's products are aimed at Latin American markets, Volkswagen, which distributes the spares worldwide, ensures that Western European and North American drivers receive what they need. In 2008, an end to inexpensive Beetle parts was nowhere in sight as a REFA manager expected his company to produce them for another twenty years.[9]

Notwithstanding its relatively modest price, an old Volkswagen requires significant investments. Beyond money, fans liberally pour time and emotion into their old Beetles to ensure they remain in working order. Some drivers spend over twenty-five hours a month on their VW. For them, the car's comparatively simple technology is one of its main attractions because it allows owners to perform many repairs and routine maintenance themselves. In addition to helping owners cut down on costs, the first Volkswagen's technological accessibility offers opportunities to display dedication and affection to the car. Although performing repairs can be a frustrating business, lavishing mechanical attention on an aging car also yields considerable satisfaction. "I am happiest when everything has turned out just the way I imagined it," one owner stated. Others consider certain interventions like removing the engine out of curiosity as a rite of passage that marks them as fully paid-up members of the VW drivers' community. "I thought you simply have to do that once as a Beetle owner," stated a German fan.[10]

While proprietors who can't perform their own repairs take their vehicle to a specialized workshop or rely on help from friends, most owners ensure their cars remain in good shape. Practically all men and women with an old Beetle in their garage are vigilantly on the outlook for any new spots of rust. "One permanently has to stay on one's toes or it rusts away under one's hands," said one woman describing her battle against the Beetle's prime enemy all over the world. Concern about symptoms of mechanical decay is widespread among aficionados. "You tend to worry about every little knocking noise you hear," a British devotee noted. "Constantly paranoid that something will go wrong." Statements like these are reminders that many Beetles are over forty years old and, even given the vehicle's fabled stamina and durability, have far outlived their typical life expectancy. Owners shower their automotive treasures with repairs, maintenance, and worry in equal measure.[11]

While the thousands of historical cars congregating at big meets were initially standardized products of Fordism, it is highly unlikely that two vehicles on show at any Beetle convention will be alike. "I like the fact that although there were 20-plus million of them made, you rarely see two of the same. I've never seen a car the same as my own," stated the British driver of a 1965 model featuring a lime and white paint job, lowered suspension, and a "race-spec engine." Over recent decades, a wide variety of styles has sprung up as owners have modified and restored vehicles in accordance with personal tastes. Many proprietors treat the Beetle as a canvas on which they paint in accordance with their aesthetic and technical predilections. In some cases, this process of creation can take years when owners opt to transform a low-priced "wreck" into a shiny display item. To complete this task, which requires skill, knowledge, imagination, discipline, dedication, and cash, many members of the scene at least temporarily make the Beetle the center point of their spare time. Like many hobbies, an old VW is very serious fun. "My other hobby is my daughter," a married German craftsman jokingly replied when asked how much energy he dedicates to his VW. Others self-ironically draw on the language of pathology to characterize their fascination with the car. In the words of an auto mechanic from Connecticut, the old Beetle "is not a hobby—it's an addiction."[12]

For the majority of enthusiasts, the Beetle is anything but an object they wish to display through daily use. To be sure, there are fans who arrive at a Beetle meet with their daily rides, but they are vastly outnumbered by those for whom taking their vehicle out for a spin is a special occasion. One elderly man attending the gathering in Nuremberg in 2011 decided against driving there in his old Beetle because the clouds in the sky suggested rain. Since he was keen to discuss his vehicle with other owners, he had brought a small photo album, of the kind frequently used by young parents for baby snapshots, to give fellow devotees an impression of the treasure he had left safe in his garage. The desire to shield the car from any possible decay motivates such cautious behavior, leading many to drive their automobiles for no more than three hundred to five hundred miles a year.[13]

The types of Beetles that make up the scene are too manifold to be covered in detail here. On one end of the spectrum stand the vehicles that, while retaining the old VW's instantly recognizable body shape, intentionally incorporate alterations that mark them as distinctive creations. The possibilities for modifications are almost limitless. Stripping the car of mudguards and fenders, replacing its original hood, fitting wider front and rear axles, lowering or altogether removing the roof, and decorating the body with new accessories and colorful paint jobs count among the interventions that accelerate many a fan's pulse. Enlarging engine capacity and installing a firmer suspension to enhance the car's speed and performance are alterations dating to the hot-rod scene in Southern California during the 1950s. It is therefore no coincidence that the most popular style for sporty, fast bugs across the world is the "Cal Look" that harks back to the young men who, in the words of an early Californian VW tuner, aimed to make their "car as fast as possible while spending the least amount of money."[14]

Since the mid-seventies, the "Cal Look" has been characterized by informal and at times contested conventions that can include modified front and rear suspensions to lower the car's nose; the removal of chrome detailing, including bumpers; new paint jobs; shiny rims with five- and eight-spoke designs; and the installation of a high-performance engine. With their emphasis on speed, power, and ostentation, Cal Look Beetles contrast starkly in

A "Cal Look" Beetle gleaming at a meeting of British VW enthusiasts in 2012. The vehicle sports a lowered front suspension, five-spoke rims, push bars (instead of bumpers), and virtually no chrome detailing. Photograph by the author.

aesthetic and technical terms with the original's plain appearance, modest acceleration, and slow speed. Cal Look thus displays qualities Susan Sontag long ago identified as "camp." Relying on highly stylized "artifice and exaggeration," a Beetle in full Cal Look drag displays "a strong love of the unnatural" as it seeks to turn itself into a high-performance vehicle that appears more at home on a racing strip than in a suburban driveway.[15]

On the opposite end of the spectrum of the old Beetle collectors operate enthusiasts with the ambition of driving a car whose technical and aesthetic features are identical to those found in the Volkswagen produced in a given model year. VW fanzines are full of admiring articles for those "making sure the car stayed original" by hunting for replacement parts, researching color schemes, and finding fabrics whose patterns exactly resemble those used as far back as the fifties. While an owner's standing among the old Beetle community can be linked to the age of his car, fans of restored VWs are first and foremost motivated by a deeply antiquarian desire for historical authenticity. In some fortuitous cases, faithfully preserving or re-creating a historical car allows aficionados to collapse past and present, thereby stepping back in time. An American devotee who encountered a pristine, "gleaming black" Beetle from 1970 that had hardly left its garage for almost two decades could barely contain himself in 1998: "The car was even better than I had imagined. . . . It was quite literally like looking at a Bug on the dealer showroom floor—in 1970. Restored Bugs will come and restored Bugs will go, but I knew HERE was the REAL DEAL, exactly like Wolfsburg had intended." Seemingly untouched by time, faithfully restored and carefully preserved automobiles thus derive their attractiveness from importing an object from the past into the present. In short, they seem to make history come to life.[16]

No matter whether their cars testify to a predilection for stylized artifice or a desire for historical authenticity, Beetle owners are united by a sense of distinctiveness. Although the scene is a broad church open to numerous aesthetic and technological styles, there exists an article of faith virtually all worshipers adhere to—namely that the engine needs to remain air-cooled and situated in the back. These technical features distinguish the Beetle from other historical as well as contemporary automobiles, as does the Bug's universally recognizable and profoundly idiosyncratic silhouette. "It's different," several German and British owners replied when asked what they liked best about their Beetle, frequently adding that they "love the shape." In a contemporary automotive culture dominated exclusively by far larger, often boxy cars with water-cooled front engines, many agree that the Beetle's form makes it a "cool" and "fun" vehicle.[17]

Numerous owners prize their Beetle because it allows them to step beyond quotidian concerns and worries, not least those related to their professional lives. Like many other hobbies, owning an old VW is a "self-chosen, self-directed activity" that involves "an autonomous kind of 'work' performed for its own sake at a self-determined pace" and offers relief from the alienating pressures permeating the occupational world.[18] An old Volkswagen is not simply a possession; buying, restoring, tuning, and maintaining a Beetle grants a sphere of autonomy foreclosed in other parts of daily life. For some, the car literally functions as an escape pod—and not just when they tinker with it or take it for spin. A German driving school instructor cherishes his vintage VW from 1971 as a shell into which he retreats when the demands of his job threaten to overwhelm him: "When it's been really hectic all day long, I sometimes simply sit in the Beetle when I get home. A quarter of an hour, half an hour. I feel better then. It's such a familiar surrounding. It gives you a sense of being at home, the smell, too." For some owners, the Beetle is an integral component of their domestic comfort zone. The modest tempo at which many first Volkswagens travel enhances the car's regenerative effects. "It's a different kind of driving, far slower. You can really see what's beside the road," a retiree explained his enjoyment before contrasting it with "today's cars, with their engines, you really don't get that." Small wonder, then, that several enthusiasts simply describe a drive in a Beetle as "relaxing."[19]

Beyond providing a temporary escape from everyday cares, old VWs are often regarded by their owners as a token of biographical continuity. In Great Britain, where the Bug never attained a large market share, collectors claim to be attracted primarily by the car's shape or simply state that they had "wanted a Beetle from a young age." In Germany, however, owners frequently identify the long-standing presence of a VW within their family as an important source of their ongoing fascination. The driving school instructor recalled that his father, a mailman, had removed the front passenger seat from the Beetle so that it could simultaneously serve as delivery vehicle, with space for his letter bag, as well as a family car. When the family took a ride in it, "my mum sat on a wooden box next to my dad," he noted. Others attribute their attachment to the fact that a small round VW was their first automobile or that it reminds them of noteworthy personal events. While a man in his late sixties remained unwilling to detail which "memories of youth" the sight of the car brought back, a female child-care worker, who drives her purple specimen on an everyday basis, was considerably more forthcoming: "The Beetle is a big memento of my beautiful, wild youth. I had my first kiss in a Beetle."[20]

The scene devoted to old VWs is permeated by nostalgic, wistful motifs as the car takes owners back to joys that lie irretrievably in the past. Nonetheless, private nostalgia among Beetle enthusiasts tends to stop short of full-blown melancholia. While undoubtedly conjuring up bittersweet memories, Beetles connect proprietors to their individual histories and inject a sense of continuity into their biographies. For those who laboriously restore their cars or tinker with them repeatedly, the vehicle itself becomes an autobiographical fixture about which they can tell numerous anecdotes. For many owners, the Beetle functions as an object that simultaneously articulates nostalgic sentiments and reinforces a sense of autobiographical coherence.[21]

Next to the eminent personal significance Beetles can assume, their roots in the Third Reich pale into significance. British enthusiasts are either not "bothered" by the car's political origins or, while paying no attention to wider ideological issues, state that the vehicle "seems like the only good idea Hitler had." Nor do owners in the Federal Republic of Germany, where the critical memorialization of the Third Reich and the Holocaust has long provided crucial foundations for a democratic public culture, connect the first VW with the Nazi period. Instead, they point out that the Beetle has turned into an "international car" or that it had its "heyday" only in the fifties and sixties. Not even at the 2011 Beetle meet in Nuremberg, which took place, as it had ever since the late 1980s, in front of the main grandstand of the former Nazi Party rallying grounds, did the historically laden location trigger reflections on the car's origins under fascism. That visitors admire historical Beetles less than a hundred yards from where Hitler once delivered speeches ought not to be mistaken for hidden Nazi sympathies. Rather than choose the former party rallying grounds, "Käferteam Nürnberg," the organizer of the occasion, was assigned the site by the city administration, which has used the vast space over the decades for countless outdoor events, including motor racing and rock concerts featuring U2 and Pink Floyd. Käferteam also ensured that a professional historian was available to provide a two-hour, highly informative tour across the party arena.[22]

Although the vast majority of Beetle fans share absolutely no affinities with the extreme Right, one German interviewee nonetheless estimated that around "3 to 6 percent of the scene are made up of political die-hards." Germans are not the only ones who are not altogether immune to the National Socialist appeal. Operating out of Newport, California, Blitzkrieg Racing offers a clothing range, skateboards, parts, and stickers adorned with visual symbols that are unmistakably inspired by National Socialism. With the Iron Cross as the brand's signature ornament, Blitzkrieg Racing also retails adhesive stickers in black and white featuring the eagle of the Wehrmacht as well as SS runes. Meanwhile, one of the oldest and most prestigious tuning clubs in Southern California continues to operate under the name "Der

Beetles parked in front of the grandstand in the former party rally grounds during the 2011 meet in Nuremberg. Aware of the historically charged location, the organizers offered critical tours of the site, which the city of Nuremberg uses for numerous outdoor events. Photograph by the author.

kleiner Panzers" (The Little Tank) and employs as its emblem a Wehrmacht eagle whose claws clutch a VW logo instead of the swastika. Neither Blitzkrieg Racing nor Der kleine Panzers overtly refers to or propagates Nazi ideas. Nonetheless, their use of visual symbols drawing liberally on the militaristic culture of the Third Reich highlights that, for some, the car's ideological roots hold a disturbing aesthetic attraction.[23]

Of course, the vast majority of Beetle lovers are thoroughly repelled by the Third Reich. Most owners view the first VW as a lovable car that generates affection almost everywhere it makes an appearance. Decades after sales ended in Germany, Britain, and the United States, the car continues to command profound sympathy among the Western European and North American public. "My Beetle conjures a smile onto people's faces," a female German driver stated in 2011. "People notice them," a young British owner corroborated. In numerous anecdotes, owners recount how the car's charm extends far beyond their own circles. A man who pursued his restoration project of a twenty-five-year-old Beetle outdoors because he lacked a sizable garage found the friendly attention he received from passers-by too distracting at times: "I really could not bear it anymore." His father, he recalled, contradicted him: "But isn't it nice? Far better than restoring a car that nobody is interested in."[24]

Indeed, personal recollections frequently rear their head when Germans encounter old Volkswagens. More than thirty years after he had sex for the first time in a small cozy VW, a music journalist from Bremen claimed that whenever he saw a Beetle he was still overcome by "the strangely fascinating weakness in the groin" he had felt when his then girlfriend "put [him] through the paces." Reflecting on the end of Beetle production in Puebla in 2003, a former museum director wrote that German collective memory contained "stacks of countless VW tales." If one "loosened the tongues in a mass oral history project and opened Germans' photo albums," he surmised, "a tidal wave of 'remember whens' would flood us." For Germans socialized in the "old" Federal Republic before national reunification in 1989, the first Volkswagen retains enormous memorial significance. Recollections of the "economic miracle" with its high growth rates and full employment, unsurprisingly, underpin the Beetle's mnemonic appeal in a Federal Republic that has been plagued by recalcitrant labor market problems since the mid-seventies and the social costs of the reunification since the nineties. One German fan succinctly stated that for many former West Germans, the vehicle evokes "memories of better days." In Germany, personal memories and collective nostalgia for economic stability intersect in the old Beetle.[25]

While reverence for the first Volkswagen as an object of memory has proved particularly strong in Germany, personal recollections of the car have circulated far beyond the Federal Republic's borders. In the United States, where over five million Beetles had been sold, Volkswagen's decision to remove the car from the showrooms in 1980 did not end the personal attachment drivers had for their cute and unconventional car. For over a decade after VW had stopped sales in the United States, the American market for used Beetles boomed, membership in Volkswagen clubs soared, and meets where owners paraded their bugs mushroomed, leading the *Wall Street Journal* to predict in 1992 that the "trend towards the Beetle" would last for generations.[26]

The Beetle's mystique remained so potent in the United States that Volkswagen was in a position to launch a remarkable comeback by introducing a retro vehicle branded as the "New Beetle" in the late 1990s. This car's crucial role in reenergizing VW's American operations almost two decades after the German automaker had stopped selling the Bug not only demonstrates poignantly how individual and collective memories prolong a commodity's life span; it also highlights how Volkswagen successfully used its global operations to exploit the commercial opportunities afforded by historical memory.

After Beetle sales came to an end in the late seventies, VW failed to find a replacement for its automotive star in the United States. While Volkswagen had a European hit with the Golf, the Rabbit—the American version of this car, produced beginning in 1978 in a new factory in Westmoreland County, Pennsylvania—suffered from problems in quality, as well as a styling that drivers in the United States found unattractive. At the same time, Japanese manufacturers led by Toyota and Honda conquered large parts of the American car market with highly dependable and moderately priced vehicles. As sales slipped throughout the eighties, VW's reputation as a maker of reliable automobiles eroded in the United States. The American recession in the early 1990s affected Volkswagen's position severely, resulting in annual sales dropping below fifty thousand cars in 1993 and fueling speculation about the company's imminent retreat from the world's most lucrative auto market. VW's American weakness brought wider problems into sharp relief: the company posted a global operating deficit of approximately $1.2 billion that year.[27]

Volkswagen developed a multifaceted remedy to these problems. The imposition of stringent new quality controls went hand in hand with a tough cost-cutting program and attempts to coordinate an expanding, increasingly global network of production sites more effectively. Apart from enhancing information flows and lending a stronger voice to regional managers, VW pushed up productivity through internal bidding for production contracts between manufacturing centers across the world. Moreover, the headquarters in Wolfsburg adopted a new strategy of product development that responded more sensitively than before to the specifics of different national markets across the world. While organizing its manufacturing activities around a limited number of chassis "platforms" to control costs, VW expanded its product range, increasingly targeting customers with models designed especially for certain regional markets. In the nineties, Volkswagen morphed from a highly centralized multinational with a rigid global hierarchy between production centers into a transnational company in which the German headquarters retained firm control but whose product range and manufacturing operations took into account far more flexibly global demand and initiatives. Appointed in 1993 after a successful career at sister company Audi, Ferdinand Piëch shaped much of VW's new global strategy during his tenure as CEO.[28]

With regard to the United States, VW initiated a profoundly unorthodox corporate maneuver that proved resoundingly successful. In an industry fixated on innovation, the car maker played the card of historical tradition and decided to resuscitate and repackage its best-known product. Indeed, the impulse to develop a new Beetle came from the United States, not Germany. At the height of the crisis of the early nineties, James Mays and Freeman Thomas, two designers at VW's studio in Southern California, put forward the idea of reviving Volkswagen's first model. They pointed out that American consumers associated the company with that car more than anything else. "We concluded that when the name Volkswagen comes up, all people could talk about was the Beetle," James Mays recalled, adding, "They loved what it stood for." Codenamed "Concept 1," the project met with considerable skepticism at the German headquarters. According to Freeman Thomas, the designers "had to counteract many things in the German attitude to cars and history." Managers in Wolfsburg viewed the Beetle as a symbol of the Federal Republic's "economic miracle," a commodity hailing from a period quickly receding into history in light of Germany's recent unification. VW's German executives saw little need to bring back a car based on the Beetle when its successor, the Golf, was gradually turning into an automotive classic in its own right, enjoying persistently strong sales in Germany and Western Europe.[29]

As a result, Mays and Thomas initially pursued Concept 1 in secrecy before the head of VW's Californian design studio succeeded in securing tentative support from VW's new CEO, who was deeply concerned about VW's American position. A combative power broker, Ferdinand Piëch had absolutely no desire to enter the company's annals as the executive on whose watch Germany's leading car manufacturer had retreated from the United States. He was most certainly drawn to the project for family reasons. Born in 1937, Piëch was Ferdinand Porsche's grandchild and owed his Christian name to anything but a historical accident. The New Beetle thus provided Piëch with a unique opportunity to follow in his ancestor's illustrious footsteps.[30]

To test the waters and provide a boost to Concept 1 within the company, the American management decided to show an early prototype at the Detroit Auto Show in January 1994. With its curved body composed of "three cylindrical shapes, two where the wheels are positioned and one forming the body," Concept 1 emphatically invoked the original's characteristic appearance. As *Road and Track* noted, VW was in the process of bringing back "a familiar shape that people could trust." Concept 1 became a sensation that turned Volkswagen's stand into one of the show's main attractions. While VW enthusiasts gushed that "many Beetle fans" who laid eyes on Concept 1 "thought they had died and gone to heaven," seasoned auto journalists found it equally hard to resist the spell of VW's experimental model. An informal poll conducted by the *Chicago Tribune* asking whether Volkswagen should revive the Beetle triggered an avalanche of affirmative answers and prompted the paper's automotive reporter to pen an open letter to Piëch, who holds a PhD in engineering: "What are you waiting for, Doc? Bring back the Beetle, and hurry." In November 1994, the headquarters eventually gave the green light, clearing funds to develop a car that drew its inspiration from the first VW and targeted American drivers.[31]

The following year, Volkswagen ran an internal competition to determine the site that would manufacture the new vehicle. After vigorous lobbying and collecting more than one million Mexican signatures in favor of the project, Puebla secured the contract and became the first location to produce a new model exclusively outside of Germany. Throughout 1996 and 1997, Mexican staff played a major role in turning the initial design idea from California into a functioning automobile. While German executives retained overall control, the thirty-strong development team overseeing the project in Wolfsburg featured ten technicians and engineers from Puebla. Their brief extended beyond acting as liaison between VW headquarters and the production site. In addition to offering technical advice, the Mexican employees were in a position to issue recommendations on design questions such as the use of interior space. "The engineers in Wolfsburg had no knowledge of these things because they were used to building angular cars. We had the experience with a rounded interior because we were still manufacturing the *vochito* in Puebla, so many things that were strange for our German colleagues were completely normal for us," a technician from Puebla recalled of his two-year posting in Wolfsburg. Encompassing American product designers alongside German and Mexican engineers, the team that transformed Concept 1 into reality lent the car a decidedly international pedigree.[32]

The New Beetle's market launch in March 1998 was a carefully stage-managed affair. Two months before the car reached America's showrooms, VW returned to the Detroit Auto Show to unveil its creation. According to a longtime VW driver who managed to sneak into the event, the fifteen-thousand-square-foot arena prepared by Volkswagen's marketing team was "packed with" hundreds of "jaded auto journalists" at 9 am who were busily sharing "fond memories of old Beetles." After several speeches by Ferdinand Piëch and other executives, the curtain rose and the assembled representatives of the automotive press turned into a "cheering crowd," clamorously greeting seven New Beetles in shades of yellow, red, blue, green, and silver that rolled onto the stage amid flashing lights and catchy music. "It was a lot to take in," the eyewitness wrote, not least since the journalists immediately stormed the podium for a closer look.[33]

When the car eventually reached dealerships, American consumers responded with the rampant enthusiasm that had already gripped the press in Detroit. Having taken the car for a spin, drivers agreed that the "New Beetle is the complete mechanical opposite of the original." With its water-cooled, 122-horsepower front engine, vigorous acceleration, air-conditioned interior, and numerous optional extras ranging from heated seats to high-quality stereo systems, the New Beetle sat on a modified Golf chassis and contrasted strongly with the original's technological simplicity and low power. And with its four airbags, average gas mileage, and price starting at $16,000, the revival car did not call to mind the modest purchase cost and fuel economy that had recommended the first VW to millions of American drivers.[34]

In mechanical and economic terms, the New Beetle differed fundamentally from its predecessor, yet its rounded exterior shape offered the historical quotation that aligned the newcomer aesthetically with the classic. While it looked squatter and featured a less steeply angled windshield in order to reduce air resistance, its rotund hood section, curved roof, and rounded trunk recalled the original's shape. The public greeted the new arrival with an affection reminiscent of the warmth that had characterized the reception of the original in the fifties. "Smiles. They're on the faces of everyone who sees Volkswagen's New Beetle," reported an auto journalist when he returned from a test drive. "It looks like a Beetle," he explained while a colleague diagnosed the car with "terminal cuteness." Demand was so strong that customers agreed to wait several months as VW struggled to raise its annual deliveries to the American market from fifty-six thousand to eighty-three thousand during the first two years after the New Beetle's introduction. The hype surrounding the revival car also alerted customers to other VW products as the

A New Beetle and an older model parked in harmony in Mexico City in 2008. While their profiles are remarkable similar, the New Beetle features a less steeply angled windshield to reduce air resistance, in keeping with its sporty driving credentials. Photograph by the author.

company's overall American sales rose from 133,415 in 1997 to 347,710 in 2000.[35]

While VW claimed that the New Beetle crossed "demographic boundaries" and developed an exceptionally broad appeal, two groups welcomed the car with particular enthusiasm. Dealers reported that many previous owners of the original found their way into VW showrooms, where they circled the new arrival to the American auto scene and reminisced about their youth, exchanging stories from their high school and college days. As one observer explained: "Everyone has a Beetle story. That's because people didn't just own Volkswagens. They had relationships with them." In late March 1998, a manager stated that "about half of our buyers placing orders are older buyers." As the company recouped customers it had previously lost, journalists marveled at how expertly VW rode "the old Bug's huge nostalgia wave."[36]

Sales, however, were not exclusively motivated by personal wistfulness. Volkswagen also noted that the New Beetle attracted a sizable contingent of affluent drivers younger than thirty. As marketing studies before the car's launch had revealed, appealing to this cohort proved crucial to the company's American revival, because potential VW customers overwhelmingly expected "young" and "cool" products from Volkswagen. Parental lore about the original Bug predisposed some youthful drivers toward Volkswagen's newest product. Moreover, well-off urban consumers in their twenties embraced the New Beetle as one of the "retro" objects that have played prominent roles in distinctive generational aesthetics since the 1970s by selectively appropriating and adapting recent yet recognizably historical styles. As a "Janus-faced" commodity "look[ing] both backwards and forwards in time," the new Volkswagen perfectly encapsulated what has been termed "retro-chic," a style that freely imports aesthetic aspects from the past into the present while remaining largely untouched by the bittersweet sense of historical loss that lends nostalgia its serious note. Hiding state-of-the-art automotive equipment underneath its historical silhouette, the New Beetle epitomized retro-chic.[37]

Beyond personal memory and a desire for retro, a powerful sense of nostalgia for the sixties suffused the New Beetlemania of the late nineties. The *New York Times* invited readers to "think back thirty years to a different time when Volkswagen Beetles in every hue took long-haired young men and women in jeans and sandals all over the world. It was a time when lighters weren't just for cigarettes and when a generation of drivers were more worried about war than highway accidents." Similar selective assessments of the sixties, which made light of the decade's racial tensions and political conflicts while idealizing the counterculture's individualistic and hedonistic dimensions, shimmered through more than one explanation of the retro car's appeal. Unlike in the sixties, the culture of the late nineties was "increasingly homogeneous" and valued "emulation . . . more than originality," one Beetle fan opined in a highly conventional critique of his historical present. In this climate, the Volkswagen continued to function as a marker of "one's freedom to be different, unconventional," he asserted.[38]

VW's public relations experts channeled consumer interest toward the original's historical image along these lines. Humor was used just as prominently as in the sixties, but this time there was no focus on quality and economy. Rather it was the Beetle's hippie heritage that played a particularly strong role in these initiatives. At times, ads alluded to Eastern spiritualism to draw on the original's countercultural credentials: "If you were really good in a past life, you come back as something better," ran one tagline that introduced the New Beetle as an automotive reincarnation. Other slogans, however, undercut rather than reinforced countercultural motifs. "Less Flower—More Power," another ad announced, drawing the car-buying public's attention toward the New Beetle's engine and away from the small vase that designers had incorporated into the retro vehicle to allude to Beetle drivers' penchant for floral ornaments. The best-known ad jokingly ironicized the original's association with post-materialism, assuring drivers that "if you sold your soul in the 80s, here's your chance to buy it back." Volkswagen cast the New Beetle as a car for the post-yuppie era and simultaneously caricatured earlier high-minded critiques of consumer culture.[39]

Beneath the ironic and humorous tone accompanying the car's return lurked fundamental differences in the features that marked the old and New Beetle as cute and unconventional automobiles. In the fifties and sixties, a succession of suburbanites, critical

This ad for the New Beetle ironically plays on yuppie guilt and draws on VW's long-standing reputation for humorous marketing. Courtesy of Volkswagen Aktiengesellschaft.

consumers, would-be auto racers, young hedonists, and members of the counterculture played crucial roles in shaping and rejuvenating the original Beetle's eccentric reputation in a market dominated by much larger, ostentatious automobiles. At the time, Volkswagen's prime achievement consisted in adapting the car's technical properties to ensure its long-term attractiveness for a heterogeneous group of consumers in search of a cheap, dependable, and purportedly honest vehicle with an unconventional reputation. Since that reputation was directly related to the car's technical properties, Doyle Dane Bernbach's fabled ad campaign adopted a supporting role in the sixties, sharpening the contours of an iconography that the company never succeeded in controlling fully. In 1998, Volkswagen exerted strict corporate discipline over the vehicle's image through a tightly managed public relations campaign. This shift not only illustrates the professionalization of marketing since the fifties; it also reflects a necessity arising from the revival car's technical features. Since the New Beetle's curved body concealed typical automotive technology, it stood out in terms of its shape and historical reputation, but not its engineering. Put differently, the retro car's distinctiveness was tied far less intimately to its material properties than had been the case with its predecessor.

Combining a historic shape with up-to-date engineering and creature comforts, Volkswagen's latest car product struck *Business Week* as a "postmodern, new Beetle" that harked "back to the future." The new VW responded to shifts in American consumer society that mirrored the biographical trajectory of many of its drivers. "It is really an indication of how the boomers have changed," explained a CEO, who had bought her first Beetle in 1969, to the *New York Times* when she beheld the reincarnation in 1998. While allegedly less interested in consumerism during the late sixties and early seventies, "now the boomers are leading a charge that started in the 80's—we continue to be materialistic and we continue to want things," another interviewee added. As it took up important consumer trends in America, the New Beetle simultaneously invoked the recent past through its distinctive silhouette as well as parts of its interior. One journalist was struck by the "bud vase for a daisy plucked straight from the 1960s" that sat "right next to . . . a high-tech multi-speaker stereo." With its back wheels in the sixties and its front wheels pointing to the future, the New Beetle was an updated piece of history tailored to late-nineties American society. Given its intentionally ambiguous historical properties, it was ideally situated to articulate and ironically ridicule "yuppie guilt," as one observer remarked. As such, it exhibited the very "incredulity toward metanarratives" that Jean-François Lyotard singled out as one of postmodernism's defining hallmarks.[40]

A New Beetle in its postmodern habitat in Johnson City, Tennessee, in 2006.
Photograph by the author.

As the New Beetle's launch frequently revolved around selective, at times ironic references to the sixties, the American media anchored the retro vehicle in the nation's recent past. Numerous reports cast the New Beetle as an American car rather than a foreign automobile that encountered American success. As a matter of fact, several observers ventured to state that Volkswagen had brought back nothing less than an "American icon." The original Beetle's long-standing presence in the country, the new car's wistful reception by American consumers, nostalgia for the sixties, and VW's promotional strategy in the United States partly account for the New Beetle's status as a quintessential American cultural symbol. Thoroughly naturalized and domesticated by the turn of the twenty-first century through protracted processes of cultural appropriation, the New Beetle found its place within the American commodity landscape with the ease of many a second-generation descendant of immigrants.[41]

The New Beetle's inclusion within the pantheon of American commodities was furthered by the silence with which most media treated the vehicle's transnational historical background and origins in National Socialism. More than fifty years after the end of World War II, a considerable number of Americans had moved beyond viewing twentieth-century Germany as well as its companies and their products primarily through a Nazi prism. To be sure, the late nineties witnessed a series of vociferous and effective legal initiatives under the aegis of the Jewish Conference on Material Claims to press for compensation for the forced laborers whom German businesses had exploited during World War II. In the summer of 1998, Volkswagen counted among the German companies named in a series of class-action suits that sought payments for erstwhile forced laborers through courts in the United States.[42]

Yet these interventions did not adversely affect the retro vehicle. Volkswagen, which had initially adopted a hard-line stance opposing compensation claims, ultimately diffused the issue by signaling its willingness to contribute funds toward a foundation for forced laborers that American and German negotiators began to discuss in 1998.[43] Moreover, the Beetle's long-standing presence in the United States insulated its successor from negative associations with its National Socialist origins. In fact, even those who were sensitive to the crimes of the Third Reich found ways of detaching the New Beetle from the Nazi past. Half a year after the New Beetle's launch, a half-Jewish columnist for the *New York Times,* who had repeatedly criticized friends and acquaintances driving German automobiles, recounted how he overcame his deep-seated resentment about VW and acquired one of the retro cars because he found it "thoroughly captivating." When he told his mother-in-law—a "conservative Jew"—of his purchase, she replied: "Congratulations darling. Maybe the war is finally over." One reader of his column considered the whole issue far-fetched: "No one would hope that the atrocities of the 1940s be forgotten. At the same time, it becomes increasingly tiresome when people . . . continue to obsess on historic injustices from five decades ago."[44]

The majority of American observers not only neglected the original's German origins but also marginalized the New Beetle's transnational background. Based on a design idea generated in Southern California, developed by an international engineering team in Wolfsburg, and manufactured by Mexican workers, the New Beetle possessed a quintessentially global pedigree. Unlike the original Beetle, which began as a commodity for the domestic market made in Germany, the postmodern Beetle was initially conceived as an export item for the United States. That the management in Wolfsburg decided to make this car exclusively in Puebla was not just a reflection of Mexico's geographical proximity to the New Beetle's main market but also highlighted VW's ongoing efforts to globalize its manufacturing operations. Crucially, Puebla recommended itself as a comparatively low-wage production site, especially after the management's effective crackdown on local trade unions in the early nineties. Although full-time VW workers in Puebla received solid wages by Mexican standards, they nonetheless enjoyed far less effective union representation as well as considerably lower remuneration than their German colleagues in the late nineties. Bearing the imprint of global capitalism's social and economic inequities, the New Beetle's instant anointment as an American icon depended to a considerable extent on a media silence that kept the car's global dimensions conveniently out of sight.

VW's decision to launch the retro car in the United States rather than in Germany bolstered the New Beetle's status as an American automotive star because it allowed observers to focus exclusively on its American dimensions. Indeed, VW introduced the car in Germany only a full eight months after it had hit showrooms in the United States. The New Beetle partially reversed a global commodity flow that had started in Germany decades earlier, effectively reimporting an updated version of an erstwhile export classic into the Federal Republic. While this move further mirrored Volkswagen's internal reorganization into a transnational company throughout the nineties, it demonstrates the car maker's role in stoking the increasingly intensive international commodity exchanges that characterized late twentieth-century globalization.

When the retro vehicle eventually reached Germany, it did not create the same stir that propelled it to instant fame across the Atlantic. Although the German public had eagerly awaited the New Beetle's launch, sales figures in the Federal Republic did not even remotely approach the heights that stemmed from New Beetlemania in the United States. In its home country, VW did not need to stage a comeback, since the corporation enjoyed lavish sales with models such as the Golf. Characteristically, the automotive correspondent for a Berlin daily newspaper categorized the New Beetle as a largely inconsequential fun car: "Nobody needs it, but the mass producer Volkswagen lacked it." This assessment cast the new arrival as a somewhat frivolous complement, taking its place among a product range that had underpinned VW's domestic reputation as a purveyor of eminently reasonable automobiles. Moreover, the German setting proved far less conducive to playful, ironic, and humorous marketing that bolstered VW's success in the United States. In a Germany plagued by high unemployment as a result of structural economic change as well as the socioeconomic fallout of recent political reunification, the old Beetle evoked an "economic miracle" that the citizens of the Federal Republic undoubtedly regarded with nostalgia but could not treat as a laughing matter. When linked with its automotive ancestor, the New Beetle certainly did not point "back to the future" in Germany in 1998. As a result, observers did not attempt to argue that VW had managed to bring back a German icon. That the New Beetle triggers enthusiasm about the original as an American icon but fails to prompt similar feelings in Germany provides a final reminder of its status as a global commodity, which, by the turn of the millennium, had developed resolutely plural and nationally specific histories.[45]

Decades after having been withdrawn from the market, the VW Beetle has retained a multifarious presence in Western Europe and North America. As a comparatively inexpensive classic car, the old Beetle stands at the center of a socially diverse, loosely knit, international scene of fans who cherish it either as a platform for fanciful automotive creations or as an antiquarian object worthy of faithful historical preservation. Frequently valuing their Beetle as a hobby that takes them beyond everyday cares and obligations, owners regard their small, rounded, air-cooled Volkswagens as badges of individuality that lend them distinctiveness within automotive cultures dominated by larger, angular cars with water-cooled engines. In countries like Germany where the old Beetle once enjoyed sustained commercial success, the vehicle also functions as a highly valued source of nostalgia.

In the United States, the car's endearing shape and mnemonic salience allowed Volkswagen to stake a corporate comeback on the launch of the New Beetle in the late 1990s. The retro vehicle offered a potent demonstration of the commercial opportunities that lie dormant in unconventionality and nostalgia. By bringing out a vehicle whose silhouette recalled the original's iconic shape, VW tapped markets among young affluent drivers drawn toward retro objects, as well as among the baby boomers harboring nostalgia for the sixties. Whatever the New Beetle's conventional technical characteristics, American auto fans recognized it as the postmodern reincarnation of the classic. Some commentators went so far as to congratulate VW for having revived an American icon, a reading that disregarded the original's historical roots and the newcomer's international design and production history. While the old Beetle provides enthusiasts with a history-laden car that allows them to step outside prosaic concerns, the New Beetle is a contemporary vehicle steeped in history that owners integrate into their everyday lives. Both original and New Beetle show how firmly the unique Bug first designed by Ferdinand Porsche has lodged itself within personal and collective memory at home and abroad. With the original model still in production at the same Mexican factory that began to produce the retro version for the American market in 1998, the Volkswagen Beetle had evolved into a thoroughly global commodity despite its deep roots in German twentieth-century history.

Epilogue

The Volkswagen Beetle as a Global Icon

"Germany has become unfaithful to itself. For a long time, we have neglected the recipe of success that brought the Federal Republic optimism and affluence, stability and prestige after the war. Those were the times when no one yet spoke of globalization, but the Beetle ran all over the world—and ran and ran and ran. At that time, the Federal Republic was characterized by an order that encouraged achievement and social progress." With these words President Horst Köhler opened a high-profile speech that intervened in a contentious debate about welfare reform in Germany in 2005. As a former head of the International Monetary Fund, Köhler unsurprisingly sided with those advocating social cuts to remedy stubbornly high unemployment. In doing so, he called upon the small car as an inspiration to build a better future for (re)unified Germany on the virtuous foundations of the past. Invoking the Beetle in the midst of heated exchanges on how to recapture growth and restore the country to what it had supposedly once been amounted to a profoundly conventional move. As he insinuated that German workers and employees should be prepared to adopt the tough, reliable, and undemanding Beetle as a social role model, Köhler did what thousands had done before him: he held up the first Volkswagen as an icon of the Federal Republic.[1]

Köhler's speech provides only one of many illustrations that the Volkswagen Beetle exhibited no sign of diminishing resonance after the turn of the millennium. Although production of the model based on Ferdinand Porsche's design ceased in 2003, both an international collectors' scene and the New Beetle testified to the continuing appeal of a vehicle that has come to rank highly among the world's classic cars. In a global commodity culture replete with goods circulating beyond national boundaries, the first VW has achieved a rare feat over the second half of the twentieth century. In addition to securing a degree of international visibility and recognition arguably on a par with Coca-Cola and McDonald's, the Volkswagen Beetle has been adopted as a national icon in countries as diverse as Germany, the United States, and Mexico.

The first Volkswagen owes its status as an icon with multiple nationalities to far more than superior technology and manufacturing routines. To be sure, its material properties provided an important element of the Beetle's global progress, but this car also had the capacity to embody and articulate a bewildering range of ideas during its long commodity life. Tracing the Beetle's history uncovers how, irrespective of false starts, crises, and unexpected twists, a broad range of actors within Germany and beyond have lent this automobile prominence in an increasingly globalized culture of goods after 1945. In this process, the car's unique rounded silhouette, which remained largely unchanged over the decades, became one of the world's best-known shapes. Drivers and owners played particularly important roles in propelling the car to stardom, for they not only regarded it as a convenient mode of transport but often treated it as a personal treasure. Its ability to become an intimate individual possession provided the indispensable foundation for the Beetle's enduring presence. Given its character as a multinational commodity supercharged with private and public significance, the first Volkswagen highlights the complex dynamics that have fueled a globalizing commodity culture from the 1950s until today.

In Germany, the first Volkswagen owes its outstanding rank to the dream of universal car ownership that was first cautiously aired during the Weimar Republic. The Nazi regime's propaganda campaigns turned desire for an automobile into a prominent aspiration, ideologically incorporating mass motorization into its racist vision of a modern "people's community." Although the Third Reich never put the vehicle into production, it bequeathed a crucial legacy in the form of a technically sophisticated prototype, Europe's largest auto works, and echoes of repeated proclamations that individual car ownership amounted to a realistic expectation. When the Volkswagen became a mass commodity after 1945, its proliferation eventually fulfilled a desire that had long been frustrated, thereby retroactively lending legitimacy to the Third Reich's motorization initiatives. As it inextricably intermeshed the recent past with the postwar historical present, the Beetle provides a potent example of how National Socialism shaped twentieth-century Germany far beyond the crimes and military disasters that remain the regime's starkest hallmarks. In this light, the VW draws attention to prominent cultural and economic continuities that overarched deep political ruptures from the Weimar Republic to the Federal Republic.

At the same time, the Beetle underlines the yawning chasms in twentieth-century German history. While the factory's commercial success highlighted West Germany's transformation into a car-producing country during the fifties and sixties, the vehicle made in Wolfsburg embodied several aspects of the country's new order. By delivering on what had remained an empty promise before 1945, the Volkswagen's proliferation offered contemporaries proof of the Federal Republic's superiority over the Third Reich.

By enhancing individual mobility on an unprecedented scale, the car helped substantiate the ubiquitous Cold War rhetoric of freedom that the West German public initially viewed with considerable political skepticism. And through its intimate link with prosperity, the Beetle crystallized West Germany's "economic miracle" and growing stability, heralding the advent of a new affluent era supposedly based on an ethos of achievement, hard work, cooperative industrial relations, full employment, and, crucially, high wages. While the VW's rapid proliferation and the joys of car ownership underlined West Germany's consolidation into an affluent society, its modest appearance as well as technical dependability reassured contemporaries that the attractive postwar order rested on solid foundations. Simultaneously portraying the postwar settlement as both thoroughly exceptional yet altogether normal, the Volkswagen Beetle drove forward a West German success story that has repeatedly given rise to nostalgic longings ever since the "economic miracle" ended in the seventies. While these yearnings underline the car's continuing personal appeal, they also signal an awareness that the exceptional conditions of capitalism's "golden age" are unlikely to return in the foreseeable future—occasional erstwhile presidential exhortations like Köhler's notwithstanding.

Although German commentators have overwhelmingly cast the Volkswagen as a tale of self-made success, it became the Federal Republic's foremost national icon only because it bore an international imprint from the outset. Beyond the fact that Ferdinand Porsche—like Hitler—was a native Austrian who pursued his ambitions in Germany, his prototype borrowed heavily from Czech designs and ideas published in the French motoring press. The huge production site at Wolfsburg could never have been built without Italian construction workers in the late thirties. During the war, forced laborers from all over occupied Europe toiled in its vast halls as the management preserved the Volkswagen corporation as an independent enterprise within the German war economy. After the war, it was the British occupational authorities who put the prototype into production. Italian laborers returned to Wolfsburg in the early sixties, this time as so-called "guest workers." In addition to European impulses, stimuli from the United States provided crucial momentum to the Volkswagen after Henry Ford pioneered not only a "universal car" but also the cost-effective mode of production that bore his name. Having caught Hitler's eye for his anti-Semitic leanings early on, the American tycoon offered Porsche organizational advice on large-scale automobile production in the thirties. America also influenced Heinrich Nordhoff, who drew on a novel managerial model first developed at General Motors in postwar Detroit and shaped VW into a corporation renowned for its collaborative industrial relations and high-wage policies. As such, the Volkswagen Beetle ranks as a national icon whose decidedly international pedigree extends from its technical features to production procedures to its workforce at the plant in Wolfsburg.

Beyond giving rise to production sites in Brazil, Mexico, Australia, and South Africa, Nordhoff's early decision to position Volkswagen as an international player laid the foundations for the Beetle's global success. While the vehicle owed its international appeal to the same material properties that attracted West German drivers, foreign owners often beheld a different car when they laid eyes on a Beetle. As is frequently the case, transferring a commodity into new cultural environments endowed it with meanings it did not possess at home. In the Beetle's case, the effects of transfers were particularly dramatic because they lent the car altogether new national identities. In finding an enthusiastic customer base in many countries, VW made an atypical contribution to the Federal Republic's "export miracle." Unlike most West German companies, the corporation from Wolfsburg secured only limited European market shares in the fifties and sixties because of high import duties on cars, as well as popular memory that associated the Beetle with its Third Reich origins. This was particularly true in European countries with strong domestic auto industries such as Great Britain, where a significant portion of the public ostracized the Volkswagen as an unwelcome competitor.

Against Nordhoff's initial expectations, the United States became the Volkswagen's most important sales territory, where the little German car secured a sizable market niche owing to its high quality, low price, and moderate maintenance costs. The car's Nazi heritage presented few obstacles in 1950s America, as West Germany came to be seen much more as an important Cold War ally than a former enemy. Nor did the American auto industry regard the VW as a serious competitor, since it targeted a market segment in which Detroit displayed little interest. Despite being regarded as a niche product, the Beetle quickly secured a high profile as a cute and unconventional product on the margins of America's consumer landscape. In an automotive culture dominated by far larger, more ornate, and more expensive cars, the Beetle attracted numerous female drivers from suburban backgrounds in search of a second family car. At the same time, the German export allowed disgruntled white middle-class consumers to register their discontent with Detroit. Its unchanging appearance, humorous advertising, suitability to technical modifications, and resourceful owners and drivers consolidated the car's reputation as an individualistic product that elicited profound affection, from countercultural circles to the world of commercial entertainment. In the sixties, the VW moved from a position as a highly visible yet culturally marginal foreign commodity into the American cultural mainstream. The love of the Beetle survived the car's withdrawal from the American market in the late seventies, renewing itself in the rapturous reception accorded to the New Beetle two decades later. A completely different vehicle in technical terms, the New Beetle sported the original's familiar shape. By the late nineties, the deeply nostalgic enthusiasm triggered by memories of the "Bug" led several observers in the United States to declare the Beetle an American icon. Similar to countless descendants of immigrants, the Beetle became fully American in the second generation.

In the meantime, Mexicans had also incorporated the vehicle into their national culture. The decision to build a comprehensive production facility in Puebla paved the way for subsequent portrayals of the *vochito* as a product of Mexican quality labor and Volkswagen as a company with firm economic roots in the country. Its presence over decades and its wide social proliferation turned the Volkswagen into a permanent feature of the Mexican landscape, where it has come to function as an important site of private memory. With its comparatively uncomplicated, sturdy technology, the car struck numerous Mexicans as eminently suited to the demanding conditions on the country's roads—a dependable vehicle with characteristics reminiscent of a tough Mexican ready to confront the challenges of everyday life in a nation characterized by recurring economic instability. In the absence of a popular, affordable automobile not only made but also initially designed in the country, the *vochito* filled a cultural vacuum and was gradually adopted as a national icon because of its success in navigating the vagaries of Mexican everyday life.

In the Federal Republic, the Beetle's international profile enhanced a deeply flattering self-image of West Germany's place in the wider world. West German commentators read the vehicle's commercial success in the United States as evidence that their young country was moving beyond a pariah status in international affairs. While steering clear of triumphalism and emphasizing the VW's marginal market position in the United States, West German coverage tended to underplay the fact that commercial expansion turned Volkswagen into a powerful corporate player in other parts of the world. Only trade-union friendly, left-leaning publications covered the confrontational tactics VW adopted vis-à-vis its Mexican workforce when the company accelerated its organizational transformation into a transnational car producer in the late eighties and early nineties. Beyond helping VW preserve its reputation as an exemplary employer at home, silence about heavy-handed practices in Puebla underpinned depictions of the Federal Republic as a country whose corporations did not exacerbate the inequities that are deeply etched into globalization. Small, peaceful, cute, likable—these antonyms of the country's previous international reputation provide the dominant motifs in self-portraits of Germany's global role that the Beetle has conveyed. The Volkswagen ideally complemented a pronounced trend among West German politicians to conduct international affairs with demonstrative restraint in the hope of overcoming the country's flawed global reputation.

Volkswagen's reluctance to strike overtly national poses during the Beetle's worldwide run of commercial success played an important part in elevating the car into an icon with multiple nationalities. Many international customers were undoubtedly aware of the vehicle's main production site, and their appreciation contributed to the postwar rehabilitation of the label "made in Germany." Nonetheless, the company was hesitant to cast its best seller as a typically German product, because promotional strategies along those lines risked dragging the car's Nazi origins into public view. As its high quality marked the car as an unconventional commodity in American contexts, VW refrained from branding its technical properties as explicitly "German." In Mexico, relative silence about the *vochito* as a typically German car went hand in hand with playing up its manufacturing history in Puebla. Allowing the corporation to sidestep the Beetle's earlier history, an international public relations strategy that deliberately underplayed the prominent role of "the people's car" in German history before 1945 loosened the vehicle's link with its national background. As a result, the Volkswagen posed fewer obstacles to a thorough cultural assimilation abroad than export hits that flaunted their national origins. Compared with commodities like McDonald's and Coca-Cola, whose global allure (as well as the occasional animosity they elicit) has hinged directly on their ability to evoke distinctively American lifestyles, the Beetle projected its place of origin far less confidently on the international stage. In a deeply ironic twist, National Socialism's ultranationalist ideology indirectly contributed to the car's subsequent ascent to multinational icon in the long run.[2]

The Beetle's international prominence serves as a potent reminder of the numerous sources that have fed an ever-expanding international commodity culture since World War II. American consumer goods, practices, and styles have undoubtedly played particularly important roles in shaping this global consumer culture since 1945, as the proliferation of rock 'n' roll, jazz, Hollywood movies, supermarkets, informal clothing items such as jeans, and countless other examples indicates. Nonetheless, a steady stream of artifacts and practices from Western Europe, ranging from French cuisine and cinematography to British pop music to Italian design, also internationalized the global commodity landscape—with a cumulative impact. In this context, the United States' relative cultural openness during the Cold War sheds light not only on the transnational dimensions of American postwar culture but the cultural dynamics that underpinned America's international alliances. For West German commentators in the fifties and sixties, the U.S. embrace of the Beetle proved psychologically important because they could read American enthusiasm for the vehicle as evidence of their new republic's acceptance by the West's leading power. At the same time, the Beetle gained prominence in a global economy that continued to exhibit deep international inequality, as developments in Mexico demonstrated. To be sure, the employees of VW de México fared better than many colleagues in other Mexican corporations, but they owed their advantage to determined trade unions rather than wealth-distribution mechanisms often alleged to be inherent in capitalism. The Beetle demonstrates how globalization promotes the proliferation of a colorful international commodity culture under the conditions of persistent worldwide inequality.[3]

The Beetle's return as a postmodern retro vehicle in 1998 provides the best illustration of the global contours that Volkswagen's best-known product had accrued over previous decades. Designed in California with the intention of reviving Volkswagen of America, developed in Wolfsburg, and produced in Puebla by workers of comparatively low wages, the New Beetle was first launched in the United States eight months before being offered to German drivers. In a move that mirrored a new international division of labor within the corporation and partially reversed the initial commodity flow from Europe across the Atlantic, Volkswagen effectively reimported an updated and foreign-made version of its classic to Germany. Sustained by cultural and economic links that the first Volkswagen had initially helped establish between disparate regions, the retro vehicle embodied its manufacturer's transformation into a transnational corporation and owed its existence directly to the original's worldwide iconic allure. In terms of its commodity history, production record, and consumer appeal, the New Beetle epitomizes the international reach of the car initially designed by Ferdinand Porsche.

Nonetheless, the first Volkswagen's global properties display clear limits. Commercial success undoubtedly lent the original Beetle global prominence. Yet despite its ability to cross borders, the original Beetle remained firmly rooted in national frames. Rather than adopting a hybrid or fully fledged transnational identity, which would have rendered the ascription of a distinct national identity impossible, the first VW developed into an icon with multiple nationalities. Globalization did not divest the car of national resonance. While the Beetle highlights how processes of reception have incorporated objects from elsewhere into new cultural landscapes, it simultaneously draws attention to the resilience of national categories in the era of globalization since World War II.

It is unclear how long the first Volkswagen is destined to retain its iconic shine. As the decision to end its production in 2003 indicates, the car eventually became obsolete after a production run of almost sixty years. By the turn of the millennium, the Beetle was set to turn into a purely historical artifact, a museum piece that lends itself to ironic, retro cultural, and nostalgic citation. All over the world, including the countries that still await mass motorization, customer demand has moved on. Nonetheless, the Beetle's appeal persists. "Will we ever let the Beetle die?" a VW ad asked in the 1960s. Five decades and several relaunches of the New Beetle later, we now know that Wolfsburg's answer has been no.

Notes

Translations are by the author unless otherwise credited.

Prologue

1. *Unter dem Sonnenrad: Ein Buch von Kraft durch Freude* (Berlin: Verlag der Deutschen Arbeitsfront, 1938), 182.

2. Tom McCarthy, *Auto Mania: Cars, Consumers, and the Environment* (New Haven, CT: Yale University Press, 2007), 30–76; Douglas Brinkley, *Wheels for the World: Henry Ford, His Company, and a Century of Progress, 1903–2003* (New York: Penguin, 2003), 90–179, 199–206.

3. Mary Nolan, *Visions of Modernity: American Business and the Modernization of Germany* (New York: Oxford University Press, 1994); Stefan Link, "Rethinking the Ford-Nazi Connection," *Bulletin of the German Historical Institute, Washington DC* 49 (Fall 2011): 135–150.

4. David Edgerton, *The Shock of the Old: Technology and Global History since 1900* (London: Profile, 2006).

5. Roland Barthes, *Mythologies* (New York: Vintage, 1994), 88.

6. *The Marx-Engels Reader: Second Edition,* ed. Robert C. Tucker (New York: Norton, 1978), 319–329, esp. 319–321. On Marx, see Hartmut Böhme, *Fetischismus und Kultur: Eine andere Theorie der Moderne* (Reinbeck: Rowohlt, 2006), 283–372, esp. 326; Arjun Appadurai, "Introduction: Commodities and the Politics of Value," in *The Social Life of Things: Commodities in Cultural Perspective,* ed. Arjun Appadurai (Cambridge: Cambridge University Press, 1986), 1–63, here 7.

7. Leora Auslander, "Beyond Words," *American Historical Review* 110 (2005): 1015–1045, here 1016; Sherry Turkle, "What Makes an

Object Evocative?" in *Evocative Objects: Things We Think With* (Cambridge, MA: MIT Press, 2007), 307–326; Igor Kopitoff, "The Cultural Biography of Things: Commoditization as Process," in Appadurai, *Social Life of Things,* 64–93; Donald A. Norman, *The Design of Everyday Things* (Cambridge, MA: MIT Press, 1998); Harvey Molotch, *Where Stuff Comes From: How Toasters, Toilets, and Many Other Things Come to Be as They Are* (New York: Routledge, 2005); Roger-Pol Droit, *How Are Things? A Philosophical Experience* (London: Faber, 2005); Daniel Miller, *The Comfort of Things* (London: Polity, 2008); Randy O. Frost and Gail Stekete, *Stuff: Compulsive Hoarding and the Meaning of Things* (New York: Mariner Books, 2011).

8. C. A. Baily, *The Birth of the Modern World, 1780–1914* (Oxford: Blackwell, 2004); Jürgen Osterhammel and Niels P. Petersson, *Globalization: A Short History* (Princeton, NJ: Princeton University Press, 2005); Victoria de Grazia, *Irresistible Empire: America's Advance through Twentieth-Century Europe* (Cambridge, MA: Harvard University Press, 2005); Thomas Bender, *Nation among Nations: America's Place in World History* (New York: Hill & Wang, 2006); Andrei S. Markovits, *Uncouth Nation: Why Europe Dislikes America* (Princeton, NJ: Princeton University Press, 2007); Denis Lacorne and Tony Judt, eds., *With Us or against Us: Studies in Global Anti-Americanism* (New York: Palgrave Macmillan, 2005).

9. Gunilla Budde, Sebastian Conrad, and Oliver Janz, eds., *Transnationale Geschichte: Themen, Tendenzen und Theorien* (Göttingen: Vandenhoeck & Ruprecht, 2006); Heinz-Gerhard Haupt and Jürgen Kocka, eds., *Comparative and Transnational History: Central European Approaches and New Perspectives* (New York: Berghahn Books, 2009); C. A. Bayly et al., "AHR Conversation: On Transnational History," *American Historical Review* 111 (2006): 1441–1464; Elizabeth Buettner, "'Going for an Indian': South Asian Restaurants and the Limits of Multiculturalism in Britain," *Journal of Modern History* 80 (2008): 865–901; Priscilla Parkhurst Ferguson, *Accounting for Taste: The Triumph of French Cuisine* (Chicago: Chicago University Press, 2004); James L. MacDonald, ed., *Golden Arches East: McDonald's in East Asia* (Stanford, CA: Stanford University Press, 1997).

1 Before the People's Car

1. "Der Verein," *Mein Kleinauto,* October 1927, 1–2.

2. Jean-Pierre Bardou et al., *The Automobile Revolution: The Impact on Industry* (Chapel Hill: University of North Carolina Press, 1982),

112; Christoph Maria Merki, *Der holprige Siegeszug des Automobils 1895–1930: Zur Motorisierung des Straßenverkehrs in Frankreich, Deutschland und der Schweiz* (Vienna: Böhlau, 2002), 115.

3. Wolfgang König and Wolfhard Weber, *Netzwerke, Stahl und Strom* (Berlin: Propyläen, 1997), 449–453.

4. "Nachruf Carl Benz," *Das Auto,* April 15, 1929, 292.

5. Bardou et al., *Automobile Revolution,* 112.

6. Susan Carter et al., *Historical Statistics of the United States: Earliest Times to the Present,* vol. 4, *Economic Sectors* (Cambridge: Cambridge University Press, 2006), 288, 635; B. R. Mitchell, *European Historical Statistics, 1750–1975* (London: Macmillan, 1975), 384–389, 420–422.

7. Robert E. Gallman, "Economic Growth and Structural Change," in *The Cambridge Economic History of the United States,* vol. 2, *The Long Nineteenth Century,* ed. Stanley L. Engerman and Robert E. Gallman (Cambridge: Cambridge University Press, 2000), 1–55; Naomi R. Lamoreaux, "Entrepreneurship, Business Organization, and Economic Concentration," in Engerman and Gallman, *Cambridge Economic History of the United States,* 2:403–434; Gary Cross, *An All-Consuming Century: Why Commercialism Won in Modern America* (New York: Columbia University Press, 1999), 24–38.

8. Anton Erkelenz, *Amerika von heute: Briefe von einer Reise* (Berlin: Weltgeist-Bücher, 1925), 29.

9. Henry Ford with Samuel Crowther, *My Life and Work* (Garden City, NY: Doubleday, 1922), 67. On the book's authors, see Stefan Link, "Rethinking the Ford-Nazi Connection," *Bulletin of the German Historical Institute, Washington, DC* 49 (Fall 2011): 135–150, esp. 139.

10. Tom McCarthy, *Auto Mania: Cars, Consumers, and the Environment* (New Haven, CT: Yale University Press, 2007), 32; James J. Flink, *The Automobile Age* (Cambridge, MA: MIT Press, 2001), 37; Douglas Brinkley, *Wheels for the World: Henry Ford, His Company, and a Century of Progress, 1903–2003* (New York: Penguin, 2003), 101–104, 120.

11. Ford, *My Life and Work,* 13–14; Brinkley, *Wheels for the World,* 120.

12. Ford, *My Life and Work,* 145; Brinkley, *Wheels for the World,* 111, 116, 129, 236; McCarthy, *Auto Mania,* 36.

13. Ford, *My Life and Work,* 145; Brinkley, *Wheels for the World,* 77–80.

14. Reynold M. Wik, *Henry Ford and Grass-Roots America* (Ann Arbor: University of Michigan Press, 1972), 33; Ronald R. Kline,

Consumers in the Country: Technology and Social Change in Rural America (Baltimore: Johns Hopkins University Press, 2000), 72–79.

15. Quoted in Brinkley, *Wheels for the World,* 118. On women, see Cotton Seiler, *Republic of Drivers: A Cultural History of Automobility in America* (Chicago: Chicago University Press, 2008), 50–60; Virginia Scharff, *Taking the Wheel: Women and the Coming of the Motor Age* (New York: Free Press, 1991), 15–34, 67–88. The percentages are from McCarthy, *Auto Mania,* 37; Kline, *Consumers in the Country,* 63–65.

16. Kathleen Franz, *Tinkering: Consumers Reinvent the Early Automobile* (Philadelphia: University of Pennsylvania Press, 2005), 26–31; Orvar Löfgren, *On Holiday: A History of Vacationing* (Berkeley: University of California Press, 1999), 58–71; McCarthy, *Auto Mania,* 35–36.

17. Quoted in Franz, *Tinkering,* 20. Ford's quip is from Ford, *My Life and Work,* 72. For nicknames, see Brinkley, *Wheels for the World,* 122. On animosity, see Brian Ladd, *Autophobia: Love and Hate in the Automotive Age* (Chicago: Chicago University Press, 2008), 13–41; Kline, *Consumers in the Country,* 63–65.

18. Stephen Meyer III, *The Five Dollar Day: Labor Management and Social Control at Fort Motor Company, 1908–1921* (Albany: SUNY Press, 1981), 2; Adam Smith, *An Inquiry into the Nature and Causes of the Wealth of Nations,* books 1–3 (Harmondsworth, UK: Penguin, 1986), 109–121. On Smith, see Emma Rothschild, *Economic Sentiments: Adam Smith, Condorcet, and the Enlightenment* (Cambridge, MA: Harvard University Press, 2001).

19. David A. Hounshell, *From the American System to Mass Production, 1800–1932: The Development of Manufacturing Technology in the United States* (Baltimore: Johns Hopkins University Press, 1982), esp. 67–124, 189–216; Carroll Pursell, *The Machine in America: A Social History of Technology* (Baltimore: Johns Hopkins University Press, 1995), 90–93; König and Weber, *Netzwerke, Stahl und Strom,* 427–441; Jonathan Zeitlin and Jonathan Sabel, eds., *Worlds of Possibilities: Flexibility and Mass Production in Western Industrialization* (Cambridge: Cambridge University Press, 1997).

20. Horace Lucien Arnold and Fay Leone Faurote, *Ford Methods and the Ford Shops* (New York: Engineering Magazine, 1919), 5.

21. Wilson J. Warren, *Tied to the Great Packing Machine: The Midwest and Meatpacking* (Iowa City: University of Iowa Press, 2007); Rick Halpern, *Down on the Killing Floor: Black and White Workers in Chicago's Packinghouses, 1904–1954* (Champaign: University of Illinois Press, 1997), 7–42.

22. Brinkley, *Wheels for the World,* 155; Frederico Buccci, *Albert Kahn: Architect of Ford* (New York: Princeton Architectural Press, 2002), 37–47.

23. Ford, *My Life and Work,* 79; Meyer, *Five Dollar Day,* 77.

24. Brinkley, *Wheels for the World,* 281–282.

25. Ibid., 170–171.

26. Meyer, *Five Dollar Day,* 123–148.

27. Flink, *Automobile Age,* 114; David L. Lewis, *The Public Image of Henry Ford: An American Folk Hero and His Company* (Detroit: Wayne State University Press, 1976), esp. 69–113. The folksy truism is in Ford, *My Life and Work,* 77. On the ill-fated rubber plantation, see Greg Grandin, *Fordlandia: The Rise and Fall of Henry Ford's Forgotten Jungle City* (London: Icon, 2010).

28. Brinkley, *Wheels for the World,* 259–264, 288–290; Flink, *Automobile Age,* 231–235; McCarthy, *Auto Mania,* 81–84, 87–89. On GM, see David Farber, *Sloan Rules: Alfred P. Sloan and the Triumph of General Motors* (Chicago: Chicago University Press, 2002); Sally H. Clarke, *Trust and Power: Consumers, the Modern Corporation, and the Making of the United States Automobile Market* (Cambridge: Cambridge University Press, 2007), 109–138, 175–204.

29. Egbert Klautke, *Unbegrenzte Möglichkeiten: "Amerikanisierung" in Deutschland und Frankreich, 1900–1930* (Stuttgart: Steiner, 2003), 191.

30. Eric D. Weitz, *Weimar Germany: Promise and Tragedy* (Princeton, NJ: Princeton University Press, 2007); Bernd Widdig, *Culture and Inflation in Weimar Germany* (Berkeley: University of California Press, 2001).

31. Detlev J. K. Peukert, *Die Weimarer Republik* (Frankfurt: Suhrkamp, 1987), 179. On film, see Thomas J. Saunders, *Hollywood in Berlin: American Cinema in Weimar Germany* (Berkeley: University of California Press, 1994); Katharina von Ankum, ed., *Women and the Metropolis: Gender and Modernity in Weimar Germany* (Berkeley: University of California Press, 1997).

32. Mary Nolan, *Visions of Modernity: American Business and the Modernization of Germany* (New York: Oxford University Press, 1994), 30–57; Joachim Radkau, *Technik in Deutschland: Vom 18. Jahrhundert bis heute* (Frankfurt: Suhrkamp, 2008), 188–196, 286–300; Klautke, *Unbegrenzte Möglichkeiten,* 196–199; Erkelenz, *Amerika von heute,* 61; Irene Witte, *Taylor, Gilbreth, Ford: Gegenwartsfragen der amerikanischen und europäischen Arbeitswissenschaften* (Munich: Oldenbourg, 1924), 74; Gustav Winter, *Der falsche Messias Henry Ford: Ein*

Alarmsignal für das gesamte deutsche Volk (Leipzig: Freie Meinung, 1924), 19; Carl Köttgen, *Das wirtschaftliche Amerika* (Berlin: VDI-Verlag, 1925).

33. Franz Westermann, *Amerika, wie ich es sah: Reiseskizzen eines Ingenieurs* (Halberstadt: Meyer, 1925), 18–19.

34. Merki, *Der holprige Siegeszug,* 115, 342; Benjamin Ziemann, "Weimar Was Weimar: Politics, Culture, and the Emplotment of the German Republic," *German History* 28 (2010): 542–571.

35. Heidrun Edelmann, *Vom Luxusgut zum Gebrauchsgegenstand: Die Geschichte der Verbreitung von Personenkraftwagen in Deutschland* (Frankfurt: VDA, 1989), 83, 87; Anita Kugler, "Von der Werkstatt zum Fließband: Etappen der frühen Automobilproduktion in Deutschland," *Geschichte und Gesellschaft* 13 (1987): 304–339, esp. 329–332.

36. Karl August Kroth, *Das Werk Opel* (Berlin: Schröder, 1928), 117, 119. On Opel, see Edelmann, *Vom Luxusgut,* 88; Rainer Flik, *Von Ford lernen? Automobilbau und Motorisierung in Deutschland bis 1933* (Cologne: Böhlau, 2001), 222.

37. Paul Thomes, "Searching for Identity: Ford Motor Company in the German Market (1900–2003)," in *Ford, 1903–2003: The European History,* vol. 2, ed. Hubert Bonin, Yannick Lung, and Steven Tolliday (Paris: PLAGE, 2003), 151–193, esp. 157–158; Sabine Saphörster, "Die Ansiedelung der Ford-Motor-Company 1929/30 in Köln," *Rheinische Vierteljahresblätter* 53 (1989): 178–210; *Frankfurter Zeitung,* March 14, 1929, evening edition, 3; *Neue Preußische Kreuz-Zeitung,* March 19, 1929, 2; *Der Abend,* March 19, 1929, 1.

38. Merki, *Der holprige Siegeszug,* 18–19.

39. Hans-Ulrich Wehler, *Deutsche Gesellschaftsgeschichte*, vol. 4 (Munich: Beck, 2003), 276–279, 313, 333–334.

40. Merki, *Der holprige Siegeszug,* 116, 120–125.

41. See Wehler, *Deutsche Gesellschaftsgeschichte,* vol. 4, 284–285, 294–304; Dietmar Petzina, Werner Abelshauser, and Anselm Faust, eds., *Sozialgeschichtliches Arbeitsbuch,* vol. 3 (Munich: Beck, 1978), 101–102; David Landes, *The Unbound Prometheus: Technological Change and Industrial Development in Western Europe from 1750 to the Present* (Cambridge: Cambridge University Press, 1969), 429, 450–451. The marketing study is Josef Bader, *Einkommen und Kraftfahrzeughaltung in Deutschland* (Berlin: Verlag der Wirtschaftsgesellschaft des Automobilhändler-Verbandes, 1929), 1, 9.

42. Edelmann, *Vom Luxusgut,* 95; "Was die Opelwerke über ihre neuen Wagen sagen," *Kleinauto-Sport,* May 1930, 4; Richard Hofmann, *Das Klein-Auto für den Selbstfahrer* (Berlin: Volckmann, 1925), 185–188.

43. R. J. Wyatt, *The Austin Seven: The Motor for the Million, 1922–1939* (Newton Abbot: David & Charles, 1982), 78–79, 117; "Der neue BMW-Kleinwagen," *Das Auto,* July 15, 1929, 560.

44. *Das Auto,* June 10, 1929, title page; "Von unseren Kleinen," *Kleinauto-Sport,* April 1930, 12; "Der 3/20 PS BMW 1932," *Das Auto,* April 30, 1932, 64–65.

45. Richard Hofmann and Fritz Wittekind, *Motorrad und Kleinauto* (Braunschweig: Westermann, 1925), 188. The rhyme is in Flik, *Von Ford lernen?* 155–156; "Hanomag jetzt Viersitzer!" *Kleinauto-Sport,* August 1930, 2–4, here 4.

46. "Aus dem Wirtschaftsbuch eines BMW-Kleinautos," *Kleinauto-Sport,* March 1930, 6–8; Hofmann, *Klein-Auto für den Selbstfahrer,* 24–25; Merki, *Der holprige Siegeszug,* 110, 375–403; Edelmann, *Vom Luxusgut,* 104–105; Flik, *Von Ford lernen?* 62–70, 300. The enraged outcry is from "Das gefesselte Auto," *Das Auto,* December 30, 1932, 183.

47. Hofmann and Wittekind, *Motorrad und Kleinauto,* 173–174, 178, 180; "Wintersorgen," *Kleinauto-Sport,* December 1929, 1–4; "Aus dem Wirtschaftsbuch eines BMW-Kleinautos," 8; "Das Hanomag-Kabriolett, ein Wagen für die Dame," *Das Auto,* May 30, 1929, 431.

48. Hofmann and Wittekind, *Motorrad und Kleinauto,* 109. See also "Hanomag-Kameraden," *Der Hanomagfahrer,* August 1929, 1–4, esp. 3; Merki, *Der holprige Siegeszug,* 111; Edelmann, *Vom Luxusgut,* 93–95.

49. Hofmann and Wittekind, *Motorrad und Kleinauto,* 109; "Hanomag-Kameraden," 1.

50. "Jedem sein Kleinauto," *Mein Kleinauto,* October 1927, 4; Hofmann and Wittekind, *Motorrad und Kleinauto,* 109; "Mit 16 PS an den Busen der Natur," *Der Hanomagfahrer,* August 1929, 4–8, esp. 6.

51. "Unsere Sonntagsfahrten," *Kleinauto-Sport,* September 1929, 13–15; *Kleinauto-Sport,* October 1929, 13–15; Rudy Koshar, "Germans at the Wheel: Cars and Leisure Travel in Interwar Germany," in *Histories of Leisure,* ed. Rudy Koshar (Oxford: Berg, 2002), 215–230.

52. Merki, *Der holprige Siegeszug,* 178–180, 194–196.

53. "ADAC-Avus-Rennen 1932," *Das Auto,* May 31, 1932, 83–84; *Der Abend,* May 23, 1932, 6; *Berliner Tageblatt,* May 23, 1932, evening edition, 2. See also "Der große Preis der Nationen," *Das Auto,* July 15, 1929, 588–589.

54. On technology and modernity, see Bernhard Rieger, *Technology and the Culture of Modernity in Britain and Germany, 1890–1945* (Cambridge: Cambridge University Press, 2005), esp. 20–30. On veneration of the automobile in Germany, see Wolfgang Ruppert, "Das Auto: Herrschaft

über Raum und Zeit," in *Fahrrad, Auto, Fernsehschrank: Zur Kulturgeschichte der Alltagsdinge,* ed. Wolfgang Ruppert (Frankfurt: Fischer, 1993), 119–161; Wolfgang Sachs, *For Love of the Automobile: Looking Back into the History of Our Desires* (Berkeley: University of California Press, 1992), 32–46.

55. *Berliner Tageblatt,* December 14, 1924, morning edition, 17; *Vorwärts,* December 14, 1924, 6; *Berliner Tageblatt,* February 17, 1929, morning edition, 9. See also *Tempo,* March 18, 1929, 9. For a particularly polemical intervention, see L. Betz, *Das Volksauto: Rettung oder Untergang der deutschen Automobilindustrie?* (Stuttgart: Petri, 1931), esp. 29, 62–63, 73–74. On driving cultures, see Rudy Koshar, "Cars and Nations: Anglo-German Perspectives on Automobility in the Interwar Period," *Theory, Culture and Society* 21:4/5 (2004): 121–144, esp. 137–139.

2 A Symbol of the National Socialist People's Community?

1. Neil Baldwin, *Henry Ford and the Jews: The Mass Production of Hate* (New York: Public Affairs, 2001), 284–285; *New York Times,* August 1, 1938, 5.

2. *New York Times,* August 4, 1938, 13; August 7, 1938, 13. On the wider context, see Philipp Gassert, *Amerika im Dritten Reich: Ideologie, Propaganda und Volksmeinung, 1933–1945* (Stuttgart: Steiner, 1997); Timothy W. Ryback, *Hitler's Private Library: The Books That Shaped His Life* (London: Vintage, 2010), 69.

3. *The Jewish Question: A Selection of Articles (1920–1922) Published by Mr. Henry Ford's Paper* (London: MCP Publications, 1927), 20, 40. On Ford's anti-Semitic publishing activities, see Douglas Brinkley, *Wheels for the World: Henry Ford, His Company, and a Century of Progress* (New York: Penguin, 2003), 257–268. Henry Ford's anti-Semitism differed from the National Socialism's racial variant. See Henry Ford with Samuel Crowther, *My Life and Work* (Garden City, NY: Doubleday, 1922), 251–253; Helmut Walser Smith, *The Continuities of German History: Nation, Religion, and Race in the Long Nineteenth Century* (Cambridge: Cambridge University Press, 2008); Hermann Graml, *Anti-Semitism in the Third Reich* (Oxford: Oxford University Press, 1992), esp. 33–86; Saul Friedländer, *Nazi Germany and the Jews: The Years of Persecution, 1933–1939* (New York: HarperPerennial, 1997), 73–112. The book list is reproduced in Ryback, *Hitler's Private Library,* 57.

4. Hans Mommsen, "Cumulative Radicalisation and Progressive Self-Destruction as Structural Determinants of the Nazi Dictatorship," in

Stalinism and Nazism: Dictatorships in Comparison, ed. Ian Kershaw and Moshe Lewin (Cambridge: Cambridge University Press, 1997), 75–87. The quote is from *Kraft des Motors, Kraft des Volkes: Sechs Reden zur Internationalen Automobil- und Motorrad-Ausstellung Berlin 1937* (Berlin: RDA, 1937), 17–18.

5. Gerhard L. Weinberg, "Foreign Policy in Peace and War," in *The Short History of Germany: Nazi Germany,* ed. Jane Caplan (Oxford: Oxford University Press, 2008), 196–218; Wolfgang Benz, *A Concise History of the Third Reich* (Berkeley: University of California Press, 2006), 155–170. For contrary assessments of popular enthusiasm, see Richard J. Evans, *The Third Reich in Power* (London: Penguin, 2006), 708–709; Peter Fritzsche, *Life and Death in the Third Reich* (Cambridge, MA: Harvard University Press, 2009); Michael Wildt, *Volksgemeinschaft als Selbstermächtigung: Gewalt gegen Juden in der deutschen Provinz, 1919–1939* (Hamburg: Hamburger Edition, 2007); Frank Bajohr, *Aryanisation in Hamburg: The Economic Exclusion of Jews and the Confiscation of Their Property in Nazi Germany* (New York: Berghahn, 2002).

6. Ian Kershaw, *Hitler: 1889–1936 Hubris* (London: Penguin, 2001), 435.

7. Otto Dietrich, *Mit Hitler an die Macht: Persönliche Erlebnisse mit meinem Führer* (Munich: Eher, 1934), 13. Good summaries of Nazi ideology are Richard J. Evans, "The Emergence of Nazi Ideology," in Caplan, *Short History of Germany,* 26–47; Lutz Raphael, "Die nationalsozialistische Weltschanschauung: Profil, Verbreitungsformen und Nachleben," *Forum Politik* 24 (2006): 27–42.

8. "Zum Geleit," in *Volk ans Gewehr! Das Buch vom neuen Deutschland,* ed. Walter Gruber (Wiesbaden: Heinig, 1934), 7. On the "people's community," see Frank Bajohr and Michael Wildt, eds., *Volksgemeinschaft: Neuere Forschungen zur Gesellschaft des Nationalsozialismus* (Frankfurt: Fischer, 2009); Norbert Frei, "Volksgemeinschaft: Erfahrungsgeschichte und Lebenswirklichkeit der Hitler-Zeit," in *1945 und wir: Das Dritte Reich im Bewusstsein der Deutschen* (Munich: Beck, 2009), 121–142. On expansionism, see Mark Mazower, *Hitler's Empire: Nazi Rule in Occupied Europe* (London: Penguin, 2008), esp. 31–52.

9. Jochen Helbeck, *Revolution on My Mind: Writing a Diary under Stalin* (Cambridge, MA: Harvard University Press, 2006); David L. Hoffmann, *Stalinist Values: The Cultural Norms of Modernity, 1917–1941* (Ithaca, NY: Cornell University Press, 2003), 57–87; Peter Fritzsche and Jochen Helbeck, "The New Man in Stalinist Russia and Nazi

Germany," in *Beyond Totalitarianism: Stalinism and Nazism Compared,* ed. Michael Geyer and Sheila Fitzpatrick (Cambridge: Cambridge University Press, 2009), 302–341.

10. Dietrich, *Mit Hitler an die Macht,* 72. For a call for productivism, see the speech by the minister of transport von Eltz-Rübenach in 1934 in *Vollgas voraus! Drei Reden, gehalten aus Anlass der Internationalen Automobil- und Motorradausstellung* (Berlin: RDA, 1934), 14–20, esp. 15. On science, see Robert D. Proctor, *The Nazi War on Cancer* (Princeton, NJ: Princeton University Press, 1999); Paul Weindling, *Health, Race and German Nation between National Unification and National Socialism* (Cambridge: Cambridge University Press, 1989). On technology, see Bernhard Rieger, *Technology and the Culture of Modernity in Britain and Germany, 1890–1945* (Cambridge: Cambridge University Press, 2005), esp. 243–263.

11. *Kraftfahrt tut not! Zwei Reden zur Eröffnung der Internationalen Automobil- und Motorradausstellung in Berlin am 11. Februar 1933* (Berlin: RDA, 1933), 9–10; Anette Gudjons, *Die Entwicklung des "Volksautomobils" von 1904 bis 1945 unter besonderer Berücksichtigung des "Volkswagens"* (PhD diss., Technical University Hanover, 1988), 151–154; Adam Tooze, *The Wages of Destruction: The Making and Breaking of the Nazi Economy* (London: Penguin, 2007), 100.

12. *Kraftfahrt tut not!* 10; "Rosemeyers großer Sieg," *Motor und Sport* (hereafter *MuS*), June 21, 1936, 16; *Völkischer Beobachter,* July 7, 1937, 1; *NSKK-Mann,* July 2, 1938, 4; "Großer Preis von Monaco," *MuS,* April 26, 1936, 38–41; "Der moderne Rennwagen," *MuS,* July 26, 1936, 12–15. Scholarship includes Dorothee Hochstetter, *Motorisierung und "Volksgemeinschaft": Das Nationalsozialistische Kraftfahrkorps (NSKK) 1931–1945* (Munich: Oldenbourg, 2005), 277–329; Eberhard Reuß, *Hitlers Rennschlachten: Silberpfeile unterm Hakenkreuz* (Berlin: Aufbau Verlag, 2006).

13. *Kraftfahrt tut not!* 10; *Völkischer Beobachter,* May 20, 1935, 1; Erhard Schütz and Eckhard Gruber, *Mythos Reichsautobahn: Bau und Inszenierung der "Straßen des Führers," 1933–1941* (Berlin: Christoph Links, 2000), 51; Thomas Zeller, *Driving Germany: The Landscape of the German Autobahn* (New York: Berghahn, 2007), 47–78, 127–180.

14. Karl Gustav Kaftan, "Die Reichsautobahnen: Marksteine des Dritten Reiches," in Gruber, *Volk ans Gewehr!* 308–314, here 310; *Völkischer Beobachter,* May 20, 1935, 2. See also *Kraftfahrt tut not!* 10; *Parole: Motorisierung—Ein Jahr nationalsozialistischer Kraftverkehrsförderung* (Berlin: RDA, 1934), 3; Peter Reichel, *Der schöne Schein des*

Dritten Reiches: Gewalt und Faszination des deutschen Faschismus (Hamburg: Ellert & Richter, 2006), 361.

15. Waldemar Wucher, ed., *Fünf Jahre Arbeit an den Straßen Adolf Hitlers* (Berlin: Volk und Reich, 1938), 19; *Völkischer Beobachter,* September 15, 1933, 5. For similar celebrations, see Archive, Institut für Zeitgeschichte, Munich, file "Autobahnen," *Völkischer Beobachter,* February 3, 1935; August 18, 1936; *Der Oberbayrische Gebirgsbote,* November 26, 1934.

16. On this perception, see Fritzsche, *Life and Death in the Third Reich,* 58; Günter Morsch, *Arbeit und Brot: Studien zu Lage, Stimmung, Einstellung und Verhalten der deutschen Arbeiterschaft, 1933–1936* (Frankfurt: Lang, 1993).

17. Tooze, *Wages of Destruction,* 43–46, 60–63; esp. 62; Schütz and Gruber, *Mythos Reichsautobahn,* 10–12; Reichel, *Der schöne Schein,* 358.

18. The full text of the code is in "Aber noch fehlt der Volkswagen," *MuS,* June 17, 1934, 10–11, 40–42, here 41. The comment is "Einige Bemerkungen," *MuS,* October 7, 1934, 5. For a rare exploration of this measure, see Hochstettter, *Motorisierung und "Volksgemeinschaft,"* 376–379. On earlier speed restrictions, see Christoph Maria Merki, *Der holprige Siegeszug des Automobils, 1895–1930: Zur Motorisierung des Straßenverkehrs in Frankreich, Deutschland und der Schweiz* (Vienna: Böhlau, 2002), 355.

19. Joe Moran, *On Roads: A Hidden History* (London: Profile, 2010), 97. See also Sean O'Connell, *The Car in British Society: Class, Gender and Motoring, 1896–1939* (Manchester: Manchester University Press, 1998), 123–136.

20. Gustav Langenscheidt, "Nationalsozialismus und Kraftfahrwesen," in Gruber, *Volk ans Gewehr!* 315–329, here 325–326; *Völkischer Beobachter,* June 21, 1937, 2; "Der Führer eröffnet die Ausstellung," *MuS,* February 26, 1939, 21–25, here 25.

21. On "discipline" and "chivalry," see Langenscheidt, "Nationalsozialismus und Kraftfahrwesen," 326; *Völkischer Beobachter,* June 21, 1937, 2; Hochstetter, *Motorisierung und "Volksgemeinschaft,"* 383.

22. Langenscheidt, "Nationalsozialismus und Kraftfahrwesen," 327; "Wichtiges vom Volkswagen," *MuS,* July 3, 1938, 6. Joseph Goebbels drew on the term in 1938. See "Eröffnung der Automobil- und Motorradausstellung," *MuS,* February 27, 1938, 21–24. See also Karl Krug and Hans Kindermann, *Das neue Straßenverkehrsrecht* (Stuttgart and Berlin: Kohlhammer, 1938), x; Johannes Floegel, *Straßenverkehrsrecht* (Munich: Beck, 1939), 10; *NSKK-Mann,* May 6, 1939, 1.

23. Langenscheidt, "Nationalsozialismus und Kraftfahrwesen," 327; Krug and Kindermann, *Das neue Straßenverkehrsrecht,* xi; Hochstetter, *Motorisierung und "Volksgemeinschaft,"* 377, 387–393.

24. *Kraftfahrt tut not!* 7–8. The passage from Hitler's speech was cited repeatedly. See Wilfried Bade, *Das Auto erobert die Welt: Biographie des Kraftwagens* (Berlin: Zeitgeschichte Verlag, 1938), 311; Langenscheidt, "Nationalsozialismus und Kraftfahrwesen," 322. The legal opinion is Krug and Kindermann, *Das neue Straßenverkehrsrecht,* x. On individualism, see Moritz Foellmer, "Was Nazism Collectivistic? Redefining the Individual in Berlin, 1930–1945," *Journal of Modern History* 82 (2010): 61–99.

25. Victor Klemperer, *Tagebücher 1937–1939,* ed. Walter Nowojski (Berlin: Aufbau, 1999), 118. Himmler's decree is in *Völkischer Beobachter,* December 5, 1938, 5. On violence in November 1938, see Friedländer, *Nazi Germany and the Jews,* 269–279; Richard J. Evans, *The Third Reich in Power* (New York: Penguin, 2005), 580–610. On its automotive dimension, see Hochstetter, *Motorisierung und "Volksgemeinschaft,"* 202–206, 405–412.

26. "Der Führer eröffnet," 25. The revised highway code in "Die neue Straßenverkehrsordnung," *MuS,* November 28, 1937, 22–23; December 5, 1937, 21–22; December 12, 1937, 28–29, 34. See also Hochstetter, *Motorisierung und "Volksgemeinschaft,"* 374–379.

27. Heidrun Edelmann, *Vom Luxusgut zum Gebrauchsgegenstand: Die Geschichte der Verbreitung von Personenkraftwagen in Deutschland* (Frankfurt: VDA, 1989), 132, 160–165, 171; Bade, *Das Auto erobert,* 326; *Vollgas voraus!* 3; Hochstetter, *Motorisierung und "Volksgemeinschaft,"* 363. On Opel, see Henry Ashby Turner Jr., *General Motors and the Nazis: The Struggle for Control of Opel, Europe's Biggest Car Maker* (New Haven, CT: Yale University Press, 2005).

28. *Vollgas voraus!* 7–8, 10–12.

29. Gert Selle, *Design im Alltag: Thonetstuhl zum Mikrochip* (Frankfurt: Campus, 2007), 99–109; Tooze, *Wages of Destruction,* 147–149; Wolfgang König, *Volkswagen, Volksempfänger, Volksgemeinschaft: "Volksprodukte" im Dritten Reich* (Paderborn: Schöningh, 2004), 25–99.

30. Hartmut Berghoff, "Träume und Alpträume: Konsumpolitik im nationalsozialistischen Deutschland," in *Die Konsumgesellschaft in Deutschland, 1890–1990,* ed. Heinz-Gerhard Haupt and Claudius Torp (Frankfurt: Campus, 2009), 268–288; Hartmut Berghoff, "Gefälligkeitsdiktatur oder Tyrannei des Mangels? Neue Kontroversen zur Konsumgeschichte des Nationalsozialismus," *Geschichte in Wissenschaft und Unterricht* 58 (2007): 502–518.

31. "Wer den Volkswagen bauen soll," *MuS,* March 25, 1934, 7.

32. Hans Mommsen and Manfred Grieger, *Das Volkswagenwerk und seine Arbeiter im Dritten Reich* (Düsseldorf: Econ, 1997), 63–66.

33. Edelmann, *Vom Luxusgut,* 179.

34. Fabian Müller, *Ferdinand Porsche* (Berlin: Ullstein, 1999), 11–34; Heidrun Edelmann, *Heinz Nordhoff und Volkswagen: Ein deutscher Unternehmer im amerikanischen Jahrhundert* (Göttingen: Vandenhoeck & Ruprecht, 2003), 37.

35. Ulrich Kubisch and Hermann-J. Pölking, eds., *Allerweltswagen: Die Geschichte eines automobilen Wirtschaftswunders, von Porsches Volkswagen-Vorläufer zum Käfer-Ausläufer-Modell* (Berlin: Elefanten-Press, 1986), 16–20; Mommsen and Grieger, *Das Volkswagenwerk,* 87; Müller, *Ferdinand Porsche,* 39–40.

36. Mommsen and Grieger, *Das Volkswagenwerk,* 66, 104–105. Hitler's reminiscences are from Henry Picker, *Hitlers Tischgespräche im Führerhauptquartier* (Stuttgart: Seewald, 1977), 374. On the engineer as a "man of action," see Kees Gispen, *Poems in Steel: National Socialism and the Politics of Inventing* (New York: Berghahn, 2002).

37. *Schrittmacher der Wirtschaft: Vier Reden gehalten zur Internationalen Automobil- und Motorrad-Ausstellug* (Berlin: RDA, 1936), 16; Mommsen and Grieger, *Das Volkswagenwerk,* 94–99.

38. Mommsen and Grieger, *Das Volkswagenwerk,* 96–97, 167.

39. Ibid., 71. Porsche joined the NSDAP only in 1937.

40. Ibid., 74–75. On design issues, see Wilhelm Hornbostel and Nils Jockel, eds., *Käfer—der Ervolkswagen: Nutzen, Alltag, Mythos* (Munich: Prestel, 1999), 31; Selle, "Ein Auto für alle," 113–114; "Den Volkswagen erfunden," *Der Spiegel,* April 23, 1952, 10; Steven Tolliday, "Enterprise and State in the West German Wirtschaftswunder: Volkswagen and the Automobile Industry, 1939–1962," *Business History Review* 69 (1995): 272–350, 281.

41. Bade, *Das Auto erobert,* 356–357. See also Deutsches Museum, Munich, archive, Firmenschriften, file Volkswagen, *Dein KdF-Wagen* (Berlin: KdF, 1938), esp. 6–9; *Volkswagenwerk GmbH* (Berlin: KdF, 1939).

42. James J. Flink, *The Automobile Age* (Cambridge, MA: MIT Press, 2001), 213; Paul Atterbury, "Travel, Transport and Art Deco," in *Art Deco, 1910–1939,* ed. Charlotte Benton, Tim Benton, and Ghislaine Wood (London: V&A Publications, 2003), 315–323; Selle, "Ein Auto für alle," 121.

43. Mommsen and Grieger, *Das Volkswagenwerk,* 148–154. On Citroën and Fiat, see Omar Calabrese, "L'utilitaria," in *I luoghi della*

memoria: Simboli e miti dell'Italia unita, ed. Mario Isnenghi (Rome: Laterza, 1998), 537–557, esp. 543–545; Valerio Castronovo, *Fiat: Una storia del capitalismo italiano* (Milan: Rizzoli, 2005), 248–250; Dominique Pagneux, *La 2CV de 1939 à 1990* (Paris: Hermé, 2005), 8–12.

44. DAF had twenty-three million members in 1939. See Gerhard Starcke, *Die Deutsche Arbeitsfront: Eine Darstellung über Zweck, Leistungen und Ziele* (Berlin: Verlag für Sozialpolitik, 1940), 144. On DAF, see Tilla Siegel, *Industrielle Rationalisierung unter dem Nationalsozialismus* (Frankfurt: Campus, 1991); Matthias Frese, *Betriebspolitik im "Dritten Reich": Deutsche Arbeitsfront, Unternehmer und Staatsbürokratie in der westdeutschen Großindustrie* (Paderborn: Schöningh, 1991). On disorderly planning, see Mommsen and Grieger, *Das Volkswagenwerk,* 117–128, 156–176, 268–276; Marie-Luise Recker, *Die Großstadt als Wohn- und Lebensbereich im Nationalsozialismus: Zur Gründung der Stadt des KdF-Wagens* (Frankfurt: Campus, 1981).

45. Shelley Baranowski, *Strength through Joy: Consumerism and Mass Tourism in the Third Reich* (Cambridge: Cambridge University Press, 2004), 40–41, 118–161; Heinz Schön, *Hitlers Traumschiffe: Die "Kraft-durch Freude" Flotte, 1934–1939* (Kiel: Arndt, 2000); Jürgen Rostock, *Paradiesruinen: Das KdF-Seebad der Zwanzigtausend auf Rügen* (Berlin: Christoph Links, 1995). The propaganda statements are from *Unter dem Sonnenrad: Ein Buch von Kraft durch Freude* (Berlin: Deutsche Arbeitsfront, 1938), 93; Starcke, *Die Deutsche Arbeitsfront,* 8.

46. Institut für Zeitgeschichte und Stadtpräsentation, Wolfsburg, EB 1, Sigrid Barth, "Wie ich den Führer traf," photocopy of school exercise book, 3; *Völkischer Beobachter,* May 27, 1938, 1. On the ceremony, see Mommsen and Grieger, *Das Volkswagenwerk,* 182–186.

47. For PR initiatives, see "Mein Auto und ich," *MuS,* February 5, 1939, 6; *MuS,* April 30, 1939, 9; Karen Peters, ed., *NS-Presseanweisungen der Vorkriegszeit, vol. 6: 1938* (Munich: Saur, 1999), 872; "Probefahrt im KdF-Wagen," *MuS,* September 18, 1938, 32; *Völkischer Beobachter,* December 4, 1938, 5; "Wir fuhren den KdF-Wagen," *Arbeitertum,* September 1, 1938, 5–9; *New York Times,* July 3, 1938, 112. Technical information was provided in *Dein KdF-Wagen,* 21. Similar claims can be found in "Das Wichtigste über die Ausstellung," *MuS,* February 19, 1939, 51; "Dein KdF-Wagen," *Arbeitertum,* November 1, 1938, 9–10.

48. "Der KdF-Wagen kommt," *MuS,* June 5, 1938, 13. Ley's categorical statement is in *Völkischer Beobachter,* August 2, 1938, 2.

49. Starcke, *Das Auto erobert,* 348; "Wir fuhren den KdF-Wagen," 9. *Dein KdF-Wagen,* front cover. See also *Volkswagenwerk GmbH,* 3, 10.

50. Postcard, author's collection. On advertising, see Kristin Semmens, *Tourism in the Third Reich* (Basingstoke: Palgrave Macmillan, 2005), esp. 72–97.

51. Quoted in Monika Uliczka, *Berufsbiographie und Flüchtlingsschicksal: VW-Arbeiter in der Nachkriegszeit* (Hanover: Hahn, 1993), 179; *Deutschland-Berichte der Sozialdemokratischen Partei Deutschlands (Sopade), 1934–1940: Sechster Band 1939,* ed. Klaus Behnken (Salzhausen: Nettelbeck, 1980), 488; "Der Volkswagen," *MuS,* June 19, 1938, 39–40. For the figures, see Mommsen and Grieger, *Das Volkswagenwerk,* 197; Hochstetter, *Motorisierung und "Volksgemeinschaft,"* 185.

52. Deutsches Tagebucharchiv, Emmendingen, 1614/I, letter, Helmut Hartmann to his parents, April 14, 1935.

53. Mommsen and Grieger, *Das Volkswagenwerk,* 198.

54. Philipp Kratz, "Sparen für das kleine Glück," in *Volkes Stimme: Skepsis und Führervertrauen im Nationalsozialismus,* ed. Götz Aly (Frankfurt: Fischer, 2007), 59–79.

55. Tooze, *Wages of Destruction,* 141–143, 195–197; König, *Volkswagen,* 186–190; Mommsen and Grieger, *Das Volkswagenwerk,* 201; Gudjons, *Die Entwicklung des "Volksautomobils,"* 60–61; Hochstetter, *Motorisierung und "Volksgemeinschaft,"* 185.

56. *Meldungen aus dem Reich: Die geheimen Lageberichte des Sicherheitsdienstes der SS, 1938–1945,* vol. 2., ed. Heinz Boberach (Herrsching: Pawlack, 1984), 177.

57. Mommsen and Grieger, *Das Volkswagenwerk,* 250–311, 1032.

58. Ibid., 338–382, 477–496, 601–624, 677–710.

59. Ulrich Herbert, *Hitler's Foreign Workers: Enforced Foreign Labor in the Third Reich* (Cambridge: Cambridge University Press, 1997); Neil Gregor, *Daimler-Benz in the Third Reich* (New Haven, CT: Yale University Press, 1998), 150–217; Constanze Werner, *Kriegswirtschaft und Zwangsarbeit im Nationalsozialismus* (Munich: Oldenbourg, 2006); Turner, *General Motors and the Nazis,* 145–146. On the sadistic cook, see Institut für Zeitgeschichte und Stadtpräsentation, Wolfsburg, EB2, Gespräch Dr. Gericke mit Wilhelm Mohr, January 21, 1970, 20–21; EB1, Gespräch Dr. Siegfried und Hugo Bork, August 30, 1979, 8–11; Julian Banaś, *Abfahrt ins Ungewisse: Drei Polen berichten über ihre Zeit als Zwangsarbeiter im Volkswagenwerk von Herbst 1942 bis Sommer 1945* (Wolfsburg: Volkswagen, 2007), 25–60, here 28. See also Henk 't Hoen, *Zwei Jahre Volkswagenwerk: Als niederländischer Student im "Arbeitseinsatz" im Volkswagenwerk von Mai 1943 bis zum Mai 1945* (Wolfsburg: Volkswagen, 2005).

60. National Archives, London, WO 235/236, Proceedings of a Military Court for the Trial of War Criminals Held at Helmstedt, Germany, May 20 and 21, 1946, 10. For a love story against all odds, see *Olga und Piet: Eine Liebe in zwei Diktaturen* (Wolfsburg: Volkswagen, 2006). On prisoner abuse, see Mommsen and Grieger, *Das Volkswagenwerk,* 516–599, 713–799; Karl Ludvigsen, *Battle for the Beetle* (Cambridge, MA: Bentley, 2000), 61–75.

61. Mommsen and Grieger, *Das Volkswagenwerk,* 320–335, 383–405, 488–495, 1032.

62. Ibid., 624–649, 876–902.

63. *Motor-Schau* 5 (1941), 729, quoted, alongside other propaganda works, in König, *Volkswagen,* 172; Joseph Goebbels, "Wofür?" in Joseph Goebbels, *Das eherne Herz: Reden und Aufsätze aus den Jahren 1941/1942* (Munich: Eher, 1943), 329–335, here 334–335. (This article first appeared in *Der Angriff* on May 31, 1942.)

64. Postcard in the author's possession.

3 "We Should Make No Demands"

1. Hans Mommsen and Wolfgang Grieger, *Das Volkswagenwerk und seine Arbeiter im Dritten Reich* (Düsseldorf: Econ, 1997), 880–885, 926–927.

2. Institut für Zeitgeschichte und Stadtpräsentation, Wolfsburg (hereafter IZS), EB 1, Erlebnisbericht Hermann Chall, October 17, 1982, 33; Simone Neteler, "Besetzt und doch frei: Wolfsburg unter alliierter Herrschaft," in *Die Wolfsburg-Saga,* ed. Christoph Stölzl (Stuttgart: Theiss, 2009), 92. For an overview, see Richard Bessel, *Germany 1945: From War to Peace* (New York: Simon & Schuster, 2009).

3. Arthur Maier, *Wahlen, Wahlverhalten und Sozialstruktur in Wolfsburg von 1945 bis 1960* (Göttingen: n.p., 1979), 38; I. D. Turner, *British Occupation Policy and Its Effects on the Town of Wolfsburg and the Volkswagenwerk, 1945–1949* (PhD diss., University of Manchester, 1984), 71–78; Neteler, "Besetzt und doch frei," 93.

4. Tony Judt, *Postwar: A History of Europe since 1945* (London: Pimlico, 2007), 104–106; Mommsen and Grieger, *Das Volkswagenwerk,* 1031.

5. Christina von Hodenberg, *Konsens und Krise: Eine Geschichte der westdeutschen Medienöffentlichkeit 1945–1973* (Göttingen: Wallstein, 2006), 103–229. The newsreel *Wochenschau* reported on Wolfsburg both in 1946 and in 1948. See www.wochenschau.de (accessed July 21, 2011).

6. Ian Turner, "The British Occupation and Its Impact on Germany," in *Reconstruction in Post-War Germany: British Occupation Policy and*

the Western Zones, 1945–1955, ed. Ian Turner (Oxford: Berg, 1989), 3–14, esp. 4–5; John W. Cell, "Colonial Rule," in *The Oxford History of the British Empire,* vol. 4, *The Twentieth Century,* ed. Judith M. Brown and Wm. Roger Louis (Oxford: Oxford University Press, 1999), 232–254.

7. Ian Turner, "British Policy Towards German Industry, 1945–1949," in Turner, *Reconstruction in Post-War Germany,* 67–91, esp. 70–71; Turner, "British Occupation and Its Impact," 5–6; Werner Plumpe, "Wirtschaftsverwaltung und Kapitalinteresse im britischen Besatzungsgebiet 1945/6," in *Wirtschaftspolitik im britischen Besatzungsgebiet,* ed. Dietmar Petzina and Walter Euchner (Düsseldorf: Schwann, 1984), 121–152, esp. 128–130.

8. Markus Lupa, *Das Werk der Briten: Volkswagenwerk und Besatzungsmacht* (Wolfsburg: Volkswagen, 2005), 6–8; Ralf Richter, *Ivan Hirst: Britischer Offizier und Manager des Volkswagenaufbaus* (Wolfsburg: Volkswagen, 2003), 35–39; IZS, EB 16, *Fragen an Ivan Hirst, Januar/Februar 1996,* 6.

9. Christoph Kleßmann, *Die doppelte Staatsgründung: Deutsche Geschichte 1945–1955* (Bonn: Bundeszentrale für politische Bildung, 1986), 67; National Archives, London (hereafter NA), FO 1039/797, Zonal Executive Offices Economic Sub-Commission, Minden, Minutes of Meeting, July 15, 1946, 2.

10. Richter, *Ivan Hirst,* 38–41; Mommsen and Grieger, *Das Volkswagenwerk,* 952–953; IZS, *Fragen an Ivan Hirst,* 6.

11. IZS, *Fragen an Ivan Hirst,* 1; Richter, *Ivan Hirst,* 38; Ronald Hyam, "Bureaucracy and 'Trusteeship' in the Colonial Empire," in Brown and Louis, *Oxford History of the British Empire*, vol. 4, 255–279; Kenneth Robinson, *The Dilemmas of Trusteeship: Aspects of British Colonial Policy between the Wars* (London: Oxford University Press, 1965).

12. Richter, *Ivan Hirst,* 51; Lupa, *Das Werk der Briten,* 10, 25; Mommsen and Grieger, *Das Volkswagenwerk,* 1031.

13. *Times* (London), June 27, 1946, 3; Lutz Niethammer, *Die Mitläuferfabrik: Die Entnazifizierung am Beispiel Bayerns* (Berlin: Dietz, 1982); Ian Turner, "Denazification in the British Zone," in Turner, *Reconstruction in Post-War Germany*, 239–270.

14. Turner, *British Occupation Policy and Its Effects on the Town of Wolfsburg,* 257, 267–278; Lupa, *Das Werk der Briten,* 16–18; NA, FO 1039/797, Minutes of the Fifth Meeting of the Board of Control of the Volkswagenwerk, June 13, 1946.

15. Turner, *British Occupation Policy and Its Effects on the Town of Wolfsburg,* 278–298; Peter Reichel, *Vergangenheitsbewältigung in*

Deutschland: Die Auseinandersetzung mit der NS-Diktatur von 1945 bis heute (Munich: Beck, 2001), 37; Konrad Jarausch, *Die Umkehr: Deutsche Wandlungen 1945–1995* (Munich: Deutsche Verlags-Anstalt, 2004), 68–75; Edgar Wolfrum, *Die geglückte Demokratie: Geschichte der Bundesrepublik von ihren Anfängen bis zur Gegenwart* (Stuttgart: Klett-Cotta, 2006), 26–27; Axel Schildt and Detlef Siegfried, *Deutsche Kulturgeschichte: Die Bundesrepublik von 1945 bis zur Gegenwart* (Munich: Hanser, 2009), 46–48.

16. Lupa, *Das Werk der Briten,* 40; IZS, EB 1, Gespräch Dr. Gericke mit Horst Bischof, October 10, 1966, 5, 10, 13; NA, FO 1032/1379, Minutes of the Seventh Board of Control Meeting of the Volkswagenwerk, August 12, 1946; Minutes of the Eighth Board of Control Meeting of the Volkswagenwerk, September 12, 1946.

17. Quoted in Monika Uliczka, *Berufsbiographie und Flüchtlingsschicksal: VW-Arbeiter in der Nachkriegszeit* (Hanover: Hahn, 1993), 220.

18. Maier, *Wahlen, Wahlverhalten und Sozialstruktur,* 44–45; Jessica Reinisch and Elizabeth White, eds., *The Disentanglement of Populations: Migration, Expulsion and Displacement in Postwar Europe, 1945–1949* (Basingstoke: Palgrave Macmillan, 2011); Andreas Kossert, *Kalte Heimat: Die Geschichte der Vertriebene nach 1945* (Berlin: Pantheon, 2009), esp. 43–87; Pertti Ahonen, *After the Expulsion: West Germany and Eastern Europe, 1945–1990* (Oxford: Oxford University Press, 2003); Ralf Richter, "Die Währungs- und Wirtschaftsreform 1948 im Spiegel unternehmerischer Personalpolitik—Volkswagen, 1945–1950," *Zeitschrift für Unternehmensgeschichte* 48 (2003): 215–238, esp. 222.

19. Richter, "Die Währungs- und Wirtschaftsrefrom," 227–228; Turner, *British Occupation Policy and Its Effects on the Town of Wolfsburg,* 118, 127; Kleßmann, *Die doppelte Staatsgründung,* 48; A. J. Nicholls, *Freedom with Responsibility: The Social Market Economy in Germany, 1918–1963* (Oxford: Oxford University Press, 2000), 127.

20. Quoted in Uliczka, *Berufsbiographie und Flüchtlingsschicksal,* 235; Ian Connor, "The Refugees and the Currency Reform," in Turner, *Reconstruction in Post-war Germany,* 301–324, esp. 302; Turner, *British Occupation Policy and Its Effects on the Town of Wolfsburg,* 144; Richter, *Ivan Hirst,* 68; NA, SUPP 14/397, Report, The Volkswagenwerk Complex in Control Under Law 52, June 1947, 4–5, 8–9; FO 1046/193, Report, Head of DAF Section, August 26, 1947.

21. NA, SUPP 14/397, The Volkswagenwerk Complex in Control Under Law 52, 10. Unternehmensarchiv Volkswagen AG, Wolfsburg

(hereafter UVW), 69/150/2, memorandum for Major Hirst, July 31, 1947; UVW, 69/150/81, memorandum, Dr. Münch to Major Hirst, November 16, 1946; Lupa, *Das Werk der Briten,* 21; IZS, *Fragen an Ivan Hirst,* 10–11; UVW, 69/150/159, 160, note, Dr. Kemmler (Kaufmännische Leitung) an die britische Werksleitung, August 9, 1946.

22. Lupa, *Das Werk der Briten,* 30; Günter J. Trittel, *Hunger und Politik: Die Ernährungskrise in der Bizone, 1945–1949* (Frankfurt: Campus, 1990), esp. 81–126.

23. UVW, 69/150/32, memo (translation), Januar Produktion, December 27, 1946; UVW, 69/196/1/2, Bericht über die Tätigkeit der Technischen Leitung im Volkswagenwerk bis einschließlich Oktober 1947, [no date], 7, 21, 23–24; UVW, 69/149/21, Lagebericht für den Monat Juni 1947; UVW, 69/149/33, Lagebericht für den Monat Mai 1947; Richter, "Die Währungs- und Wirtschaftsreform," 220; Lupa, *Das Werk der Briten,* 64–65.

24. British Intelligence Objectives Sub-Committee, *Investigation of the Developments in the German Automobile Industry during the War Period: BIOS Final Report No. 300* (London: HMSO, 1945), 72; British Intelligence Objectives Sub-Committee, *The German Automobile Industry: BIOS Final Report 768* (London: HMSO, 1946), 12; Karl Ludvigsen, ed., *People's Car: A Facsimile Reprint of B.I.O.S. Final Report No. 998 Investigation into the Design and Performance of the Volkswagen or German People's Car: First Published in 1947* (London: Stationery Office, 1996), 79–80.

25. Ludvigsen, *People's Car,* 68–69, 115, 117, 118; *BIOS Final Report 300,* 97; *BIOS Final Report 768,* 31.

26. Ludvigsen, *People's Car,* 118, 69, 85; *BIOS Final Report 768,* 31. See Ministry of Supply, *National Advisory Council for the Motor Manufacturing Industry: Report on Proceedings* (London: HMSO, 1948), 15. Some historians have criticized British industry for missing a supposedly golden opportunity. See Martin Adeney, *The Motor Makers: The Turbulent History of Britain's Car Makers* (London: Fontana, 1989), 209; James Laux, *The European Automobile Industry* (New York: Twayne, 1992), 170.

27. Judt, *Postwar,* 90–99; Geir Lundestad, *The United States and Western Europe since 1945* (Oxford: Oxford University Press, 2003), 55–58; Alan Milward, *The Reconstruction of Western Europe, 1945–1951* (Berkeley: University of California Press, 1984).

28. Steven Tolliday, "Enterprise and State in the West German Wirtschaftswunder: Volkswagen and the Automobile Industry, 1939–1962," *Business History Review* 69 (1995): 272–350, here 296; Henry

Walter Nelson, *Small Wonder: The Amazing Story of the Volkswagen Beetle* (Boston: Little, Brown, 1970), 104–112.

29. Heidrun Edelmann, *Heinz Nordhoff und Volkswagen: Ein deutscher Unternehmer im amerikanischen Jahrhundert* (Göttingen: Vandenhoeck & Ruprecht, 2003), 9–63.

30. Paul Erker, "Industrie-Eliten im 20. Jahrhundert," in *Deutsche Unternehmer zwischen Kriegswirtschaft und Wiederaufbau: Studien zur Erfahrungsbildung von Industrie-Eliten,* ed. Paul Erker and Toni Pierenkämper (Munich: Oldenbourg, 1999), 3–18, esp. 5–6, 8–9; Edelmann, *Heinz Nordhoff und Volkswagen,* 68.

31. Heidrun Edelmann, "Heinrich Nordhoff: Ein deutscher Manager in der Automobilindustrie," in Erker and Pierenkämper, *Deutsche Unternehmer zwischen Kriegswirtschaft und Wiederaufbau,* 19–52, esp. 35–37, 39–44; Christoph Buchheim, "Unternehmen in Deutschland und NS-Regime 1933–1945: Versuch einer Synthese," *Historische Zeitschrift* 282 (2006): 351–390. For two far more egregious offenders, see Peter Hayes, *Industry and Ideology: IG Farben in the Nazi Era* (Cambridge: Cambridge University Press, 1987), 319–376; Norbert Frei et al., *Flick: Der Konzern. Die Familie. Die Macht* (Munich: Blessing, 2009), 327–368.

32. Nina Grunenberg, *Die Wundertäter: Netzwerke in der deutschen Wirtschaft, 1942–1966* (Munich: Pantheon, 2007); Tim Schanetzky, "Unternehmer: Profiteure des Unrechts," in *Karrieren im Zwielicht: Hitlers Eliten nach 1945,* ed. Norbert Frei (Frankfurt: Campus, 2001), 73–130; Erker, "Industrie-Eliten im 20. Jahrhundert," 12–13.

33. "Unproduktive Botschaft," *Der Spiegel,* June 19, 1948, 16–17.

34. Edelmann, *Heinz Nordhoff und Volkswagen,* 76, 89, 95–97; UVW, CH 4920/10, *Volkswagen Informationsdienst,* issue 3, December 16, 1948, 2; *Volkswagen Informationsdienst,* issue 1, August 1, 1948, 1; issue 2, October 5, 1948, 3–5; Karsten Line, ". . . bisher nur Sonnentage . . . Der Aufbau der Volkswagen-Händlerorganisation 1948 bis 1967," *Zeitschrift für Unternehmensgeschichte* 53 (2008): 5–32, esp. 9–12; Turner, *British Occupation Policy and Its Effects on the Town of Wolfsburg,* 604; Lupa, *Das Werk der Briten,* 64–71.

35. Wendy Carlin, "Economic Reconstruction in Western Germany, 1945–1955: The Displacement of 'Vegetative Control,'" in Turner, *Reconstruction in Post-War Germany,* 67–92.

36. Nicholls, *Freedom with Responsibility,* 178–233.

37. Harold James, "Die D-Mark," in *Deutsche Erinnerungsorte: Eine Auswahl,* ed. Etienne François and Hagen Schulze (Munich: Beck, 2005), 367–384; Christoph Buchheim, "Die Währungsreform 1948 in

Westdeutschland," *Vierteljahrshefte für Zeitgeschichte* 36 (1988): 189–231, esp. 217–220.

38. Nicholls, *Freedom with Responsibility,* 217; Buchheim, "Die Währungsreform," 220–223.

39. See Richter, "Die Währungs- und Wirtschaftsreform 1948," 233–235; Mommsen and Grieger, *Das Volkswagenwerk,* 1031; UVW, CH 4920/10, *Volkswagen Informationsdienst,* no. 3, December 16, 1948, 13–15; Richter, "Die Währungs- und Wirtschaftsreform 1948," 234; "Unproduktive Botschaft," 17; *Die Welt,* May 1, 1949, 14; Institut für Zeitgeschichte, Munich, archive, file Automobilindustrie, *Neue Zeitung,* January 27, 1949.

40. UVW, Presse 1948/1949, untitled speech manuscript, October 5, 1948, 2; Bernd Wiersch, *Volkswagen Typenkunde* (Bielefeld: Delius Klasing, 2010), 20–22.

41. Wiersch, *Volkswagen Typenkunde,* 28–29; Heinrich Nordhoff, "Presse-Empfang am 28. Juni 1949," in *Reden und Aufsätze: Zeugnisse einer Ära* (Düsseldorf: Econ, 1992), 73–87, here 76.

42. Edelmann, *Heinz Nordhoff und Volkswagen,* 102. UVW, Presse 1948/1949, *Tagesspiegel,* April 3, 1949.

43. *Die Welt,* November 30, 1948, 1; "Immer schwächer als die Männer," *Der Spiegel,* December 4, 1948, 8; Maier, *Wahlen, Wahlverhalten und Sozialstruktur,* 101–105; Turner, *British Occupation Policy and Its Effects on the Town of Wolfsburg,* 715; Günther Koch, *Arbeitnehmer steuern mit: Belegschaftsvertretung bei VW ab 1945* (Cologne: Bund, 1987), 61–62; Uliczka, *Berufsbiographie und Flüchtlingsschicksal,* 286.

44. NA, FO 1005/1869, Special Report No. 185, June 25, 1948; Public Opinion Research Office, Special Report No. 222, July 21, 1948; Reichel, *Vergangenheitsbewältigung in Deutschland,* 35–36.

45. The verdict is in NA, WO 235/779, War Crimes Group to Legal Division, memorandum, August 13, 1948; NA, WO 235/518, Minutes of Trial, August 13, 1948, Ferdinand Porsche, Erklärung an Eidesstatt, December 1, 1947; Georg Tyrolt, Eidesstattliche Erklärung, January 5, 1948.

46. NA, WO 311/523, Deposition by Ernst Lütge, May 20, 1947; NA, WO 309/202, Judge Advocate General's Office, memorandum, July 14, 1947.

47. Neteler, "Besetzt und doch frei," 95; UVW, file Presse 1948/1949, *Tagesspiegel,* April 3, 1949; *Niedersächsische Landeszeitung,* January 28, 1948; *Tagesspiegel,* April 3, 1949.

48. Klaus J. Bade and Jochen Oltmer, "Einführung: Einwanderungsland Niedersachsen—Zuwanderung und Integration seit dem Zweiten

Weltkrieg," in *Zuwanderung und Integration in Niedersachsen seit dem Zweiten Weltkrieg*, ed. Klaus J. Bade and Jochen Oltmer (Osnabrück: Rasch, 2002), 11–36, here 14–15; Buchheim, "Die Währungsreform," 229; Paul Erker, *Ernährungskrise und Nachkriegsgesellschaft: Bauern und Arbeiterschaft in Bayern* (Stuttgart: Klett-Cotta, 1990), 284; Connor, "The Refugees and the Currency Reform," 305–306, 318–323.

49. Nordhoff, "Ansprache an die Belegschaft am 6. Dezember 1948 über den Werkfunk," in *Reden und Aufsätze,* 63–66, here 64; Edelmann, *Heinz Nordhoff und Volkswagen,* 79.

50. Robert G. Moeller, *War Stories: The Search for a Usable Past in the Federal Republic of Germany* (Berkeley: University of California Press, 2001); Mary Nolan, "Air Wars, Memory Wars," *Central European History* 38 (2005): 7–40, esp. 17–19; Neil Gregor, *Haunted City: Nuremberg and the Nazi Past* (New Haven, CT: Yale University Press, 2008), 135–186.

51. "Ansprache an die Belegschaft am 6. Dezember 1948," 64; Heinrich Nordhoff, "Werkfunk-Ansprache an die Belegschaft am 25. Juni 1948," in *Reden und Ausätze,* 54–55, here 55; Heinrich Nordhoff, "Werkfunk-Ansprache an die Belegschaft anlässlich der Fertigstellung des 30 000. Volkswagens am 9. September 1948," in ibid., 57–58, here 58; Institut für Zeitgeschichte, Munich, archive, file Ferdinand Porsche, *Der Angriff,* May 5, 1942. For scholarship on the rhetoric of achievement, see S. Jonathan Wiesen, *Creating the Nazi Marketplace: Commerce and Consumption in the Third Reich* (Cambridge: Cambridge University Press, 2011), 28–33; Moritz Foellmer, "Was Nazism Collectivistic? Redefining the Individual in Berlin, 1930–1945," *Journal of Modern History* 82 (2010): 61–100, esp. 88–90.

52. *Neue Zeitung,* January 27, 1949; Günter Neliba, *Die Opelwerke im Konzern von General Motors (1929–1948) in Rüsselsheim und Brandenburg* (Frankfurt: Brandes & Apsel, 2000), 152–164; Paul Thomes, "Searching for Identity: Ford Motor Company in the German Market, 1903–2003," in *Ford, 1903–2003: The European History,* vol. 2, ed. Hubert Bonin, Yannick Lung, and Steven Tolliday (Paris: PLAGE, 2003), 151–193, esp. 160–161; *Die Daimler-Benz AG 1916–1948: Schlüsseldokumente zur Konzerngeschichte,* ed. Karl-Heinz Roth and Michael Schmid (Nordingen: Greno, 1987), 403–405.

4 Icon of the Early Federal Republic

1. "In König Nordhoffs Reich," *Der Spiegel,* August 20, 1955, 16–26, here 16–17. See also www.youtube.com/watch?v=tNDeowQA_Jk (accessed May 27, 2011).

2. "Erlkönige in Detroit," *Der Spiegel,* February 18, 1959, 47–49, here 47; Heidrun Edelmann, *Heinz Nordhoff und Volkswagen: Ein deutscher Unternehmer im amerikanischen Jahrhundert* (Göttingen: Vandenhoeck & Ruprecht, 2003), 179–180; *Süddeutsche Zeitung,* July 11/12, 1953; Günter Riederer, "Das Werk im Kornfeld: Der Industriefilm *Aus eigener Kraft* (1954), Volkswagen und die Stadt Wolfsburg," in *Die Wolfsburg-Saga,* ed. Christoph Stölzl (Stuttgart: Theiss, 2009), 148–151.

3. Gert Selle, *Design im Alltag: Vom Thonetstuhl zum Mikrochip* (Frankfurt: Campus, 2007), 112; Erhard Schütz, "Der Volkswagen," in *Deutsche Erinnerungsorte: eine Auswahl,* ed. Etienne François and Hagen Schulze (Munich: Beck, 2005), 351–368, esp. 353.

4. Konrad Jarausch, *Die Umkehr: Deutsche Wandlungen* (Munich: Deutsche Verlags-Anstalt, 2004), 64–66, 76–96; Konrad H. Jarausch and Michael Geyer, *Shattered Past: Reconstructing German Histories* (Princeton, NJ: Princeton University Press, 2003), 235–237; Friedrich Kießling and Bernhard Rieger, "Einleitung: Neuorientierung, Tradition und Transformation in der Geschichte der alten Bundesrepublik," in *Mit dem Wandel leben: Neuorientierung und Tradition in der Bundesrepublik der 1950er und 60er Jahre,* ed. Friedrich Kießling and Bernhard Rieger (Cologne: Böhlau, 2011), 7–28, esp. 20–23; Axel Schildt and Detlef Siegfried, *Deutsche Kulturgeschichte: Die Bundesrepublik von 1945 bis zur Gegenwart* (Munich: Hanser, 2009), 124, 131–132.

5. Axel Schildt, *Ankunft im Westen: Ein Essay zur Erfolgsgeschichte der Bundesrepublik* (Frankfurt: Fischer, 1999), 93; Rudolf Oswald, *"Fußball-Volksgemeinschaft": Ideologie, Politik und Fanatismus im deutschen Fußball, 1919–1964* (Frankfurt: Campus, 2008), 300–303.

6. Hans-Ulrich Wehler, *Deutsche Gesellschaftsgeschichte, Fünfter Band: Bundesrepublik Deutschland und DDR, 1949–1990* (Munich: Beck, 2008), 54–58; Werner Abelshauser, *Deutsche Wirtschaftsgeschichte seit 1945* (Munich: Beck, 2004), 300–301.

7. Abelshauser, *Deutsche Wirtschaftsgeschichte seit 1945;* A. J. Nicholls, *Freedom with Responsibility: The Social Market Economy in Germany, 1918–1963* (Oxford: Oxford University Press, 2000); James C. Van Hook, *Rebuilding Germany: The Creation of the Social Market Economy, 1945–1957* (Cambridge: Cambridge University Press, 2004); J. Adam Tooze, "Reassessing the Moral Economy of Post-war Reconstruction: The Terms of the West German Settlement in 1952," in *Post-war Reconstruction in Europe: International Perspectives, 1945–1949,* ed. Mark Mazower et al. (Oxford: Oxford University Press, 2011), 47–70; Christoph Buchheim, *Die Wiedereingliederung Westdeutschlands in die*

Weltwirtschaft, 1945–1958 (Munich: Oldenbourg, 1990); Jeffry R. Frieden, *Global Capitalism: Its Fall and Rise in the Twentieth Century* (New York: Norton, 2006), 278–300; Charles S. Maier, *Among Empires: American Ascendancy and Its Predecessors* (Cambridge, MA: Harvard University Press, 2006), 198–228.

8. Manfred Grieger et al., *Volkswagen Chronik* (Wolfsburg: Volkswagen, 2004), 25, 61; Volker Wellhöner, *"Wirtschaftswunder"—Weltmarkt—westdeutscher Fordismus: Der Fall Volkswagen* (Münster: Westfälisches Dampfboot, 1996), 74, 85; Arthur Maier, *Wahlen, Wahlverhalten und Sozialstruktur in Wolfsburg von 1945 bis 1960* (Göttingen: n.p., 1979), 38; Ortwien Reichold, ed., *Erleben, wie eine Stadt entsteht; Städtebau, Architektur und Wohnen in Wolfsburg, 1938–1968* (Braunschweig: Meyer, 1998), 63.

9. Horst Mönnich, "Eine Stadt von morgen," in *Merian,* July 1958, iii–x, here iii; *Hannoversche Allgemeine Zeitung,* July 2, 1953; *Süddeutsche Zeitung,* July 11/12, 1953.

10. *Hamburger Freie Presse,* October 1951; *Frankfurter Neue Presse,* July 12, 1954; *Süddeutsche Zeitung,* August 8, 1958; "In König Nordhoffs Reich," 25; Edelmann, *Heinz Nordhoff und Volkswagen,* 80–84, 162. For critical letters, see *Der Spiegel,* October 7, 1959, 3, 6, 8; *Stuttgarter Zeitung,* December 8, 1959; Roland Marchand, *Creating the Corporate Soul: The Rise of Public Relations and Corporate Imagery in American Big Business* (Berkeley: University of California Press, 2001).

11. Heinrich Nordhoff, "Rede zur Betriebsversammlung am 23. März 1956," in *Reden und Aufsätze: Zeugnisse einer Ära* (Düsseldorf: Econ, 1992), 184–193, here 185; Wellhöner, *"Wirtschaftswunder,"* 110.

12. David Noble, *Forces of Production: A Social History of Industrial Automation* (New York: Knopf, 1984); Amy Sue Bix, *Inventing Ourselves Out of Jobs? America's Debate over Technological Unemployment, 1929–1981* (Baltimore: Johns Hopkins University Press, 2000); Christian Kleinschmidt, *Der produktive Blick: Wahrnehmung amerikanischer und japanischer Management- und Produktionsmethoden durch deutsche Unternehmer 1950–1985* (Berlin: Akademie Verlag, 2002), 159–161; Edelmann, *Heinz Nordhoff und Volkswagen,* 183–189; Wellhöner, *"Wirtschaftswunder,"* 109–135.

13. *Industriekurier,* June 23, 1959; Heinrich Nordhoff, "Ansprache bei der Pressekonferenz aus Anlaß der Produktion des 500000. Volkswagens am 4. Juli 1953," in *Reden und Aufsätze,* 146–164, here 156; Institut für Zeitgeschichte und Stadtpräsentation, Wolfsburg (hereafter IZS), EB 16, "Zeitzeugen-Interview zur Geschichte Wolfsburgs mit Eberhard Anlauf," September 12, 1995, transcript, 1; IZS, EB 18,

"Gespräch am 10.6.1996 mit Herren Amtenbrinck, Ziegler, Hondke und Kagelmann," transcript, 10–12; *Süddeutsche Zeitung,* January 31, 1956; Monika Uliczka, *Berufsbiographie und Flüchtlingsschicksal: VW-Arbeiter in der Nachkriegszeit* (Hanover: Hahn, 1993), 246.

14. Wellhöner, "*Wirtschaftswunder,*" 138–140; Stephen Meyer III, *The Five Dollar Day: Labor Management and Social Control in the Ford Motor Company, 1908–1921* (Albany: SUNY Press, 1981), 48–50.

15. Heinrich Nordhoff, "Vortrag vor der schwedischen Handelskammer in Stockholm am 13. März 1953," in *Reden und Aufsätze,* 129–145, here 131, 137; Günther Koch, *Arbeitnehmer steuern mit: Belegschaftsvertretung bei VW ab 1945* (Cologne: Bund, 1987), 59, 89–91.

16. Nordhoff, "Vortrag vor der schwedischen Handelskammer," 139–141.

17. Kevin Boyle, *The UAW and the Heyday of American Liberalism, 1945–1960* (Ithaca, NY: Cornell University Press, 1995), 61–106; Tom Sugrue, *The Origins of the Urban Crisis: Race and Inequality in Postwar Detroit: With a New Preface by the Author* (Princeton, NJ: Princeton University Press, 2005), 91–152; Volker Berghahn, *Otto A. Friedrich, ein politischer Unternehmer: Sein Leben und seine Zeit, 1902–1975* (Frankfurt: Campus, 1993), 230, 328.

18. Nordhoff, "Vortrag vor der schwedischen Handelskammer," 139, 142.

19. Koch, *Arbeitnehmer steuern mit,* 81–91; Steven Tolliday, "Enterprise and State in the West German Wirtschaftswunder: Volkswagen and the Automobile Industry, 1939–1962," *Business History Review* 69 (1995): 272–350, esp. 318–319; Werner Conze, *Die Suche nach der Sicherheit: Eine Geschichte der Bundesrepublik Deutschland von 1949 bis zur Gegenwart* (Munich: Siedler, 2009), 165–168; Anselm Doering-Manteuffel, *Wie westlich sind die Deutschen? Amerikanisierung und Westernisierung im 20. Jahrhundert* (Göttingen: Vandenhoeck & Ruprecht, 1999), 90–102; Julia Angster, *Konsenskapitalismus und Sozialdemokratie: Die Westernisierung von SPD und DGB* (Munich: Oldenbourg, 2003).

20. Tolliday, "Enterprise and State," 319; Wellhöner, "*Wirtschaftswunder,*" 146; Koch, *Arbeitnehmer steuern mit,* 78–80; Ralf Rytlewski and Manfred Opp de Hipt, *Die Bundesrepublik Deutschland in Zahlen 1945/49–1980: Ein sozialgeschichtliches Arbeitsbuch* (Munich: Beck, 1989), 119.

21. *Statistisches Jahrbuch für die Bundesrepublik Deutschland 1959* (Stuttgart: Kohlhammer, 1959), 445; *Neue Rhein Zeitung,* June 29,

1957; *Neue Rhein Zeitung,* January 5, 1956; Tolliday, "Enterprise and State," 321–323.

22. Mönnich, "Eine Stadt von morgen," III; *Die Welt,* April 11, 1957; Erich Kuby, "Der bürgerliche Arbeiter," in *Das ist des Deutschen Vaterland* (Reinbeck: Rowohlt, 1959 [1957]), 408–434, here 428.

23. Josef Mooser, "Abschied von der 'Proletarität': Sozialstruktur und Lage der Arbeiterschaft in der Bundesrepublik in historischer Perspektive," in *Sozialgeschichte der Bundesrepublik Deutschland: Studien zum Kontinuitätsproblem,* ed. Werner Conze and M. Rainer Lepsius (Stuttgart: Klett-Cotta, 1983), 143–186; Andreas Kossert, *Kalte Heimat: Geschichte der deutschen Vertriebenen nach 1945* (Munich: Siedler, 2008), 92–138; Helmut Schelsky, "Die Bedeutung des Schichtungsbegriffs für die Analyse der gegenwärtigen Gesellschaft," in *Auf der Suche nach der Wirklichkeit* (Düsseldorf: Diederichs, 1965), 331–336, here 332; Paul Nolte, *Die Ordnung der deutschen Gesellschaft: Selbstentwurf und Selbstbeschreibung im 20. Jahrhundert* (Munich: Beck, 2001), 330–335.

24. *Die Welt,* April 11, 1957; Reichold, *Erleben, wie eine Stadt entsteht,* 41–55; Simone Neteler, "Die Stadtmaschine springt an," in Stölzl, *Die Wolfsburg-Saga,* 106–113; Dietrich Kautt, "Wolfsburg im Wandel städtebaulicher Leitbilder," in *Aufbau West, Aufbau Ost: Die Planstädte Wolfsburg und Eisenhüttenstadt in der Nachkriegszeit,* ed. Rosemarie Baier (Berlin: DHM, 1997), 99–109.

25. *Stuttgarter Zeitung,* June 30, 1956; *Die Welt,* April 6, 1957; *Rheinische Post,* February 6, 1958; Postcard, author's private collection; Klaus-Jörg Siegfried, "Die 'Autostadt': Zur Selbstdarstellung Wolfsburgs in der Nordhoff-Ära," in Baier, *Aufbau West,* 239–247; Neil Gregor, *Haunted City: Nuremberg and the Nazi Past* (New Haven, CT: Yale University Press, 2008); Bernhard Rieger, "Was Roland a Nazi? Victims, Perpetrators, and Silences during the Restoration of Civic Identity in Postwar Bremen," *History and Memory* 17:2 (2007): 75–112; Rudy Koshar, *From Monuments to Traces: Artifacts of German Memory* (Berkeley: University of California Press, 2000), 143–173.

26. Joachim Käppner, *Berthold Beitz: Die Biographie* (Berlin: Berlin Verlag, 2010), 207–220; Frank Bajohr, *Hanseat und Grenzgänger: Erik Blumenfeld—eine politische Biographie* (Göttingen: Wallstein, 2010), 72–84; Constantin Goschler, *Schuld und Schulden: Die Politik der Wiedergutmachung für NS-Verfolgte seit 1945* (Göttingen: Wallstein, 2008), 125–254; S. Jonathan Wiesen, *West German Industry and the Challenge of the Nazi Past* (Chapel Hill: University of North Carolina Press, 2001).

27. Horst Mönnich, *Die Autostadt* (Munich: Andermann, 1951), 87, 89. For Nordhoff's praise, see Unternehmensarchiv Volkswagen AG, Wolfsburg (hereafter UVW), 319/10226, *VW-Informationen: Mitteilungsblatt für die VW-Organisation,* November 1951, 56; Peter Reichel, *Vergangenheitsbewältigung in Deutschland: Die Auseinandersetzung mit der NS-Diktatur von 1945 bis heute* (Munich: Beck, 2001), 66–72; Norbert Frei, *Vergangenheitspolitik: Die Anfänge der Bundesrepublik und die NS-Vergangenheit* (Munich: dtv, 1999); Hartmut Berghoff, "Zwischen Verdrängung und Aufarbeitung: Die bundesdeutsche Gesellschaft und ihre nationalsozialistische Vergangenheit in den fünfziger Jahren," *Geschichte in Wissenschaft und Unterricht* 49:2 (1998): 96–114; Robert G. Moeller, "Remembering the War in a Nation of Victims: West German Pasts in the 1950s," in *The Miracle Years: A Cultural History of West Germany, 1949–1968,* ed. Hanna Schissler (Princeton, NJ: Princeton University Press, 2001), 83–109.

28. Heinz Todtmann and Alfred Trischler, *Kleiner Wagen auf großer Fahrt* (Offenbach: Verlag Dr. Franz Burda, 1949), 52; UVW, Presse 1948/49, letter, Heinz Todtmann to Heinrich Nordhoff, January 22, 1949; letter, Heinrich Nordhoff to Heinz Todtmann, January 27, 1949.

29. *Süddeutsche Zeitung,* July 11/12, 1953; *Augsburger Zeitung,* April 6, 1957; *Industriekurier,* June 23, 1959; *Lübecker Nachrichten,* September 17, 1950; Heinrich Nordhoff, "Ansprache am 1. Juli 1961 anläßlich der ersten Hauptversammlung nach Umwandlung in eine Aktiengesellschaft," in *Reden und Aufsätze,* 276–286, here 277; Wehler, *Deutsche Gesellschaftsgeschichte, Fünfter Band,* 48; Schildt and Siegfried, *Deutsche Kulturgeschichte,* 98.

30. *Frankfurter Allgemeine Zeitung,* December 23, 1955; *Süddeutsche Zeitung,* July 11/12, 1953; Heidrun Edelmann, "Privatisierung als Sozialpolitik: 'Volksaktien' und Volkswagenwerk," *Jahrbuch für Wirtschaftsgeschichte* 1(1999): 55–72; Edelmann, *Heinz Nordhoff und Volkswagen,* 206–225. The Erhard quote is from "Vom Volkswagen zum Volkskapitalismus," *Der Spiegel,* February 20, 1957, 26–31, here 27.

31. *Christ und Welt,* July 11, 1957; *Frankfurter Allgemeine Zeitung,* December 23, 1955.

32. *Tatsachen und Zahlen aus der Kraftverkehrswirtschaft 1963/64* (Frankfurt: VDA, 1964), 148–149, 154, 195, 350; *Tatsachen und Zahlen aus der Kraftverkehrswirtschaft 1957/58* (Frankfurt: VDA, 1958), 159; Arnold Sywottek, "From Starvation to Excess? Trends in the Consumer Society from the 1940s to the 1970s," in Schissler, *Miracle*

Years, 341–358; Michael Wildt, *Vom kleinen Wohlstand: Eine Konsumgeschichte der fünfziger Jahre* (Frankfurt: Fischer, 1996).

33. Rytlewski and Opp de Hipt, *Die Bundesrepublik Deutschland in Zahlen,* 123; *Frankfurter Neue Presse,* July 7, 1953; *Tatsachen und Zahlen 1957/58,* 119; Dietmar Klenke, *Bundesdeutsche Verkehrspolitik und Motorisierung: Konfliktträchtige Weichenstellungen in den Jahren des Wiederaufstiegs* (Stuttgart: Steiner, 1993), 119, 124–132.

34. Wehler, *Deutsche Gesellschaftsgeschichte, Fünfter Band,* 155; *Tatsachen und Zahlen 1957/58,* 119; *Tatschen und Zahlen 1963/64,* 160; Thomas Südbeck, *Motorisierung, Verkehrsentwicklung und Verkehrspolitik in der Bundesrepublik der 1950er Jahre: Umrisse der allgemeinen Entwicklung und zwei Beispiele—Hamburg und Emsland* (Stuttgart: Steiner, 1994), 37–43.

35. Manfred Caroselle, "Die Düsenjäger des kleinen Mannes," in *Mein erstes Auto: Erinnerungen und Geschichten,* ed. Franz-Josef Oller (Frankfurt: Fischer, 1999), 67–83; Südbeck, *Motorisierung,* 34–36, 53–62.

36. Ulrich Kubisch and Volker Janssen, *Borgward: Ein Blick zurück auf Wirtschaftswunder, Werkalltag und einen Automythos* (Berlin: Elefanten Press, 1984), 92; Südbeck, *Motorisierung,* 34–35, 46; *Tatsachen und Zahlen 1963/64,* 247; Siegfried Rauch, *DKW—Die Geschichte einer Weltmarke* (Stuttgart: Motorbuch-Verlag, 1988).

37. Volkswagenwerk, *Bericht der Geschäftsführung für die Jahre 1951 bis 1953* (Wolfsburg: Volkswagenwerk, 1955), 6; Volkswagenwerk, *Bericht der Geschäftsführung für das Jahr 1956* (Wolfsburg: Volkswagenwerk, 1956), 12; Volkswagenwerk, *Bericht der Geschäftsführung für das Jahr 1961* (Wolfsburg: Volkswagenwerk, 1962), 20; *Tatsachen und Zahlen 1963/64,* 253; Tolliday, "Enterprise and State," 329; "In König Nordhoffs Reich," 18; "Der Kunde als Kreditgeber," *Der Spiegel,* October 15, 1958, 22–23; *Weißenburger Tageblatt,* August 24, 1957.

38. Walter Henry Nelson, *Small Wonder: The Amazing Story of the Volkswagen* (Boston: Little, Brown, 1970), 345–350; Bernd Wiersch, *Volkswagen Typenkunde, 1945–1974* (Bielefeld: Delius Klasing, 2009), 23–28, 32–55.

39. "Volkswagen," *Auto Motor und Sport* (hereafter *AMS*), no. 19, 1951, 649–651, here 649. Further test reports praising the car's quality include "Volkswagen 1958," *AMS,* no. 21, 1957, 13–16; "Ist der Volkswagen veraltet?" *Stern,* no. 43, 1957, 54–61; Arthur Westrup, *Besser fahren mit dem Volkswagen: Ein Handbuch* (Bielefeld: Delius Klasing, 1950), 12, 26–27, 61; *Süddeutsche Zeitung,* July 11/12, 1953; "Volkswagen 1956," *AMS,* no. 20, 1955, 22–25.

40. See letters to the editor, *Gute Fahrt* (hereafter *GF*), no. 5, 1955, 32; *GF*, no. 10, 1951, 12; "Was kostet Dich Dein Auto," *AMS*, no. 15, 1955, 18–19; Maiken Umbach, "Made in Germany," in François and Schulze, *Deutsche Erinnerungsorte*, 244–257; Helmuth Trischler, "'Made in Germany': Die Bundesrepublik als Wissensgesellschaft und Innovationssystem," in *Modell Deutschland: Erfolgsgeschichte oder Illusion?* ed. Thomas Hertfelder and Andreas Rödder (Göttingen: Vandenhoeck & Ruprecht, 2007), 44–60.

41. "Wie sieht der VW der Zukunft aus?" *GF*, no. 11, 1957, 16–17; UVW, 319/10226, *VW-Informationsdienst: Mitteilungsblatt für die VW-Organisation* 13, November 1951, 8; *Industriekurier*, June 23, 1959; UVW, 174/406/5, minutes, Hauptabteilungsleiter-Besprechung, September 27, 1954, 2–3.

42. "Wie sieht der VW der Zukunft aus," 16–17; Westrup, *Besser fahren*, 91; "Ist der VW veraltet?" *Der Spiegel*, September 30, 1959, 40–48.

43. Karsten Linne, "'Bisher nur Sonntage': Der Aufbau der Volkswagen-Händlerorganisation 1948 bis 1967," *Zeitschrift für Unternehmensgeschichte* 53 (2008): 5–32, esp. 12–22; "Der brave Wolfsburger," *Constanze*, no. 11, 1960, 72–73, here 72.

44. "Volkswagen," *AMS*, no. 19, 1951, 651; *Augsburger Allgemeine*, April 2, 1961.

45. For a reprint of this 1949 review, see Westrup, *Besser fahren*, 61–62, here 61; Bernhard Rieger, "Schulden der Vergangenheit? Der Mammutprozess der Volkswagensparer 1949–1961," in Kießling and Rieger, *Mit dem Wandel leben*, 185–209, esp. 204–205.

46. *Die Welt*, April 11, 1957; *Industriekurier*, June 23, 1959; *Das Auto*, no. 18, 1950, 587; *Bremer Nachrichten*, January 31, 1951; *Hannoversche Allgemeine Zeitung*, July 2, 1953; *Süddeutschen Zeitung*, October 29, 1961; Fritz Kölling, *Ein Auto zieht Kreise: Herkunft und Zukunft des Volkswagens* (Reutlingen: Bardtenschlager, 1962), 6; "In König Nordhoffs Reich," 17.

47. "Porsche von Fallersleben," *Der Spiegel*, May 18, 1950, 21–27, here 24; "Warum kaufen Sie eine Limousine," *GF*, no. 2, 1951, 17; *Industriekurier*, June 23, 1959; note, *GF*, no. 1, 1952, 22–23; letter to the editor, *GF*, no. 8, 1955, 37; note, *GF*, no. 3, 1957, 3.

48. Mönnich, *Die Autostadt*, 238–240; Thomas Kühne, "Zwischen Vernichtungskrieg und Freizeitgesellschaft: Veteranenkultur in der Bundesrepublik (1945–1995)," in *Nachkrieg in Deutschland*, ed. Klaus Naumann (Hamburg: Hamburger Editon, 2001), 90–113; Karsten Wilke, "Organisierte Veteranen der Waffen-SS zwischen Systemopposition und

Integration," *Zeitschrift für Geschichtswissenschaft* 53:2 (2005): 149–166; Detlef Bald, Johannes Klotz, and Wolfram Wette, *Mythos Wehrmacht: Nachkriegsdebatten und Traditionspflege* (Berlin: Aufbau, 2001).

49. Paul Betts, *The Authority of Everyday Objects: A Cultural History of West German Industrial Design* (Berkeley: University of California Press, 2004).

50. "Von großen und von kleinen Wagen," *GF,* no. 1, 1951, 16–17; *Industriekurier,* June 23, 1959; "Ein Auto ist kein Damenhut," *GF,* no. 9, 1952, 3; UVW, file 319/10226, *VW-Informationen: Mitteilungsblatt für die VW-Organisation,* no. 14, 1953, 45.

51. "Die unsichtbaren Verbesserungen am Volkswagen," *AMS,* no. 12, 1960, 16–17; "Ist der Volkswagen veraltet," *Stern,* no. 43, 1957, 54–61; "Ist der Volkswagen veraltet," *Der Spiegel,* September 30, 1959, 40–58. For praise, see the letters to the editor, *Der Spiegel,* October 7, 1959, 12; October 14, 1959, 16. On the culture of security, see Eckart Conze, "Sicherheit als Kultur: Überlegungen zu einer 'modernen Politikgeschichte' der Bundesrepublik Deutschland," *Vierteljahrshefte für Zeitgeschichte* 53 (2005): 357–380, here 366.

52. Gerhard Kießling, interview with author, Erlangen, June 17, 2010; Alon Confino, "Traveling as a Culture of Remembrance: Traces of National Socialism in West Germany, 1845–1960," in *Germany as a Culture of Remembrance: Promises and Limits of Writing History* (Chapel Hill: University of North Carolina Press, 2006), 235–254, esp. 249, 251.

53. Hasso Spode, "Der Aufstieg des Massentourismus im 20. Jahrhundert," in *Die Konsumgesellschaft in Deutschland 1890–1990: Ein Handbuch,* ed. Heinz-Gerhard Haupt and Claudius Torp (Frankfurt: Campus, 2009), 114–128, esp. 127; Axel Schildt, *Sozialgeschichte der Bundesrepublik Deutschland bis 1989/90* (Munich: Oldenbourg, 2007), 46.

54. *Gute Fahrt in Italien: Ein Reiseführer für motorisierte Menschen* (Bielefeld: Delius Klasing, 1954), 15; "Reisetips," *ADAC Motorwelt,* no. 6, 1953, 25; "Gute Tipps," *Constanze Reisetips 1955,* 4–5; "Falls Sie nach Italien fahren," *Constanze,* no. 10, 1956, 80.

55. Hanna Schissler, "'Normalization' as Project: Some Thoughts on Gender Relations in West Germany during the 1950s," in Schissler, *Miracle Years,* 359–375, esp. 362; Elizabeth Heineman, "The Hour of the Woman: Memories of Germany's 'Crisis Years' and West German National Identity," in Schissler, *Miracle Years,* 21–56; Dagmar Herzog, "Desperately Seeking Normality: Sex and Marriage in the Wake of the War," in *Life after Death: Approaches to the Cultural and Social His-*

tory of Europe during the 1940s and 1950s, ed. Richard Bessel and Dirk Schumann (Cambridge: Cambridge University Press, 2003), 161–192; Till van Rahden, "Wie Vati die Demokratie lernte: Religion, Familie und die Frage der Autorität in der frühen Bundesrepublik," in *Demokratie im Schatten der Gewalt: Private Geschichten im deutschen Nachkrieg,* ed. Daniel Fulda et al. (Göttingen: Wallstein, 2010), 122–151.

56. Michael Wildt, "'Wohlstand für alle': Das Spannungsfeld von Konsum und Politik in der Bundesrepublik," in Haupt and Torp, *Die Konsumgesellschaft in Deutschland,* 305–316, esp. 310–311; Conze, *Die Suche nach der Sicherheit,* 186; Schildt and Siegfried, *Deutsche Kulturgeschichte,* 105–108.

57. Erika Spiegel, *Soziologische Bedingtheiten von Verkehrsunfällen: Ein Beitrag zur Soziologie des PKW-Fahrers und des PKW-Verkehrs* (Frankfurt: Institut für Sozialforschung, 1963), 10, 102; Paul Reibestahl, *Das Buch vom Volkswagen: Eine aktuelle Plauderei in Wort und Bild über eine geniale Konstruktion* (Braunschweig: Schmidt, 1951), 162–164; Marianne Ludorf, "Ferien mit 80 Mark in der Tasche," in *Deutschland—Wunderland: Neubeginn, 1950–1960: 44 Erinnerungen aus Ost und West,* ed. Jürgen Kleindienst (Berlin: Zeitgut, 2003), 317–320.

58. Josef Heinrich Darchinger, *Wirtschaftswunder: Deutschland nach dem Krieg, 1952–1967* (Cologne: Taschen, 2008), 97.

59. Ludger Claußen, "Volltanken mit Obenöl, oder: Die Operation Katwijk," in Oller, *Mein erstes Auto,* 190–194, here 190; Alexander Spoerl, *Mit dem Auto auf du* (Munich: Piper, 1957), 11.

60. Westrup, *Besser fahren,* 187–205; Carl Otto Windecker, *Besinnliches Autobuch: Eine gedruckte Liebeserklärung* (Bielefeld: Delius Klasing, 1953), 50.

61. Westrup, *Besser fahren,* 198–199, 209–215; Spoerl, *Mit dem Auto auf du,* 242.

62. Letter to the editor, *GF,* no. 2, 1955, 27–28; Westrup, *Besser fahren,* 63; Kurt Möser, *Geschichte des Autos* (Frankfurt: Campus, 2002), 332; Hanns-Peter von Thyssen, "Mausi mit Familienanschluss," in Oller, *Mein Erstes Auto,* 161–163, 162; Spoerl, *Mit dem Auto auf du,* 239.

63. Spoerl, *Mit dem Auto auf du,* 76, 259–271. See also Helmut Dillenburger, *Das praktische Autobuch* (Gütersloh: Bertelsmann, 1957), 288–304, 370–373; Westrup, *Besser fahren,* 153.

64. Windecker, *Besinnliches Autobuch,* 6–7.

65. Interview with Gerhard Kießling; "Ein Deutscher am Steuer verwandelt sich," *Der Spiegel,* October 28, 1964, 65–72, here 70; Thyssen,

"Mausi mit Familienanschluss," 161; Max Reisch, *Mit "Fridolin" nach Indien* (Munich: Ehrenwirth, 1960); Peter Fischer, "Und zahlt und zahlt und zahlt," in Oller, *Mein erstes Auto,* 149–151, here 150; Jutta Aurahs, "Erst mein Fünfter war ein Kerl," in ibid., 157–160, 158.

66. Harold James, "Die D-Mark," in François and Schulze, *Deutsche Erinnerungsorte,* 367–384.

67. Dillenburger, *Das praktische Autobuch,* 17, 142–287, 377; Westrup, *Besser fahren,* 87, 139–140; Möser, *Geschichte des Autos,* 307; "Autoreise," *Constanze,* no. 11, 1957, 30–31.

68. Letter to the editor, *GF,* no. 2, 1951, 31; Ursula Eyermann and Heidemarie Bade, interview by author, Erlangen, June 19, 2010; "Der brave Wolfsburger," 73; letter to the editor, *GF,* no. 6, 1963, 3.

69. Spiegel, *Soziologische Bedingtheiten von Verkehrsunfällen,* 42; *Tatsachen und Zahlen aus der Kraftverkehrswirtschaft 1963/64,* 192; "Sie erobern sich einen Männerberuf," *Constanze,* no. 12, 1961, 40–42; "Sonntagsschule der Frauen," *GF,* no. 5, 1963, 14.

70. Marlies Schröder, "Heimlich zur Fahrschule," in Kleindienst, *Deutschland—Wunderland,* 330–334, here 333. For the figures, see Schildt, *Sozialgeschichte der Bundesrepublik Deutschland,* 18; Christina von Oertzen, *Teilzeitarbeit und die Lust am Zuverdienen: Geschlechterpolitik und gesellschaftlicher Wandel in Westdeutschland, 1948–1969* (Göttingen: Vandenhoeck & Ruprecht, 1999); Detlef Siegfried, *Time Is on My Side: Konsum und Politik in der westdeutschen Jugendkultur der 60er Jahre* (Göttingen: Wallstein, 2006), 45–50.

71. Response to letter, *GF,* no. 2, 1951, 31; letter to the editor, *GF,* no. 5, 1951, 8; note, *Constanze,* no. 8, 1950, 6; "Frauen fahren besser," *Constanze,* no. 4, 1951, 10–11. See also "Tipps für die Frau am Steuer," *Constanze,* no. 12, 1959, 60; letters to the editor, *GF,* no. 5, 1955, 37; *GF,* no. 9, 1957, 3.

72. Letter to the editor, *GF,* no. 6, 1963, 3; "Für Zuwiderhandelnde wird gebetet," *GF,* no. 4, 1963, 14.

73. "Die Steuerlast in der Ehe," *GF,* no. 2, 1963, 26–29, here 27.

74. Ludwig Erhard, "Verführt Wohlstand zum Materialismus?" in *Wohlstand für alle* (Düsseldorf: Econ, 1957), 232–245, here 236–237. Heinrich Nordhoff repeatedly attacked collectivism. See Heinrich Nordhoff, "Ansprache bei der Pressekonferenz aus Anlaß der Produktion des 500000. Volkswagens am 4. Juli 1953," in *Reden und Aufsätze,* 157–158; Nordhoff, "Rede anlässlich der Verleihung des Elmer A. Sperry-Preises am 13. November 1958," ibid., 226–239, esp. 238; Wilhelm Röpke, "Die Abstimmung von Straße und Schiene," *Der Volks-*

wirt, April 30, 1954, 9–19, here 9. On Röpke, see Alexander Nützenadel, *Die Stunde der Ökonomen: Wissenschaft, Politik und Expertenkultur in der Bundesrepublik, 1949–1974* (Göttingen: Vandenhoeck & Ruprecht, 2005), 37–43, 57–59.

75. Windecker, *Besinnliches Autobuch,* 85; Spoerl, *Mit dem Auto auf du,* 239; Schildt and Siegfried, *Deutsche Kulturgeschichte,* 130–131; Schildt, *Ankunft im Westen,* 93–95.

76. "Gelegenheit macht Liebe," *GF,* no. 7, 1955, 28; Spoerl, *Mit dem Auto auf du,* 83; Sybille Steinbacher, *Wie der Sex nach Deutschland kam: Der Kampf um Sittlichkeit und Anstand in der frühen Bundesrepublik* (Munich: Siedler, 2011), esp. 124–133; Elizabeth D. Heineman, "The Economic Miracle in the Bedroom: Big Business and Sexual Consumption in Reconstruction West Germany," *Journal of Modern History* 78 (2006): 846–877; Dagmar Herzog, *Sex after Fascism: Memory and Morality in Twentieth-Century Germany* (Princeton, NJ: Princeton University Press, 2005), esp. 101–104.

77. Letter to the editor, *GF,* no. 2, 1951, 31; interview with Gerhard Kießling.

78. Heinrich Popitz et al., *Technik und Industriearbeit: Soziologische Untersuchungen in der Hüttenindustrie* (Tübingen: Mohr, 1957), 112–119, here 118; interview with Gerhard Kießling; Spoerl, *Mit dem Auto auf du,* 81.

79. Georg Heinrich Spornberger, "Im Dschungel des Verkehrs," *Magnum,* no. 12 (1957), 39–40, here 40; Dillenburger, *Das praktische Autobuch,* 19–20. For scholarship, see David W. Plath, "My Car-isma: Motorizing the Showa Self," *Daedalus* 119:3 (1990): 229–244, esp. 231; Tim Dant, "The Driver-Car," *Theory, Culture & Society* 21:4/5 (2004): 61–79; Nigel Thrift, "Driving in the City," *Theory, Culture & Society* 21:4/5 (2004): 41–59.

80. Dietmar Klenke, *Freier Stau für freie Bürger: Die Geschichte der bundesdeutschen Verkehrspolitik, 1949–1994* (Darmstadt: Wissenschaftliche Buchgesellschaft, 1995), 50–59; Axel Schildt, "Vom Wohlstandsbarometer zum Belastungsfaktor—Autovision und Autoängste in der westdeutschen Presse," in *Geschichte der Zukunft des Verkehrs: Verkehrskonzepte von der Frühen Neuzeit bis zum 21. Jahrhundert,* ed. Hans-Liudger Dienel and Helmuth Trischler (Frankfurt: Campus, 1997), 289–309, esp. 297–300.

81. *Verhandlungen des Deutschen Bundestages,* 1. Wahlperiode 1949, vol. 9 (Bonn: n.p., 1951), 7049; *Verhandlungen des Deutschen Bundestages,* 1. Wahlperiode 1949, vol. 14 (Bonn: n.p., 1953), 11572; *Der Spiegel,* October 17, 1956, 25.

82. *Statistisches Jahrbuch für die Bundesrepublik Deutschland 1952* (Stuttgart: Kohlhammer, 1952), 306; *Statistisches Jahrbuch für die Bundesrepublik Deutschland 1955* (Stuttgart: Kohlhammer, 1955), 340; *Statistisches Jahrbuch für die Bundesrepublik Deutschland 1963* (Stuttgart: Kohlhammer, 1963), 373.

83. Windecker, *Besinnliches Autobuch,* 74, 77; Martin Beheim-Schwarzbach, *Der geölte Blitz: Aus den Aufzeichnungen eines Volkswagens* (Hamburg: Dulk, 1953), 37; "Psychologie des Überholens," *AMS,* no. 21, 1957, 11; "Nächstes Jahr langsamer," *Der Spiegel,* October 17, 1956, 22–31, esp. 28; "Die Menschen versagen," *Constanze,* no. 1, 1955, 15; "So ist das heute," *GF,* no. 6, 1955, 14–15; Bundesarchiv, Koblenz (hereafter BAK), 108/2638, Cardinal Frings, public statement, Cologne, January 18, 1958; BAK, B108/2202, memorandum, Auswärtiges Amt an Bundesministerium für Verkehr, July 24, 1961. On debates about accidents, see Dietmar Klenke, *Bundesdeutsche Verkehrspolitik und Motorisierung,* 145–161; Helmut Vogt, "'Das schaurige Schlachtfeld der Straße': Mobilitätskonflikte in der Frühzeit der Bundesrepublik," *Geschichte im Westen* 16 (2001): 38–46.

84. Windecker, *Besinnliches Autobuch,* 75; Spornberger, "Im Dschungel des Verkehrs," 39; "Psychologie des Überholens," 11; "Ein Deutscher am Steuer," 65.

85. *Verhandlungen des Deutschen Bundestages,* 1. Wahlperiode 1949, vol. 9, 7049; BAK, B108/2638, minutes, 11. Sitzung des Straßenverkehrssicherheitsausschusses, March 14, 1957; Das Schwerpunktprogramm für die Zeit vom März 1958 bis März 1959, typescript, Bonn, no date. On the introduction of the award "Kavalier der Straße" (Gentleman of the Road) in 1959, see BAK, B108/ 2677, Kuratorium "Wir und die Straße," copy, November 11, 1965. On manners, see Paul Betts, "Manners, Morality, and Civilization: Reflections on Postwar German Etiquette Books," in *Histories of the Aftermath: The Legacies of the Second World War in Europe,* ed. Frank Bies and Robert G. Moeller (New York: Berghahn, 2010), 196–214, here 198–199; Jarausch, *Die Umkehr,* esp. 26–30. On road rage, see Tom Vanderbilt, *Traffic: Why We Drive the Way We Do (and What It Says about Us)* (London: Penguin, 2008), 19–39; Mike Michael, "The Invisible Car: The Cultural Purification of Road Rage," in *Car Cultures,* ed. Daniel Miller (Oxford: Berg, 2001), 59–80; Jack Katz, *How Emotions Work* (Chicago: Chicago University Press, 2000), 18–86.

86. *Statistisches Jahrbuch für die Bundesrepublik Deutschland 1953,* 341; *Statistisches Jahrbuch für die Bundesrepublik Deutschland, 1959* (Stuttgart: Kohlhammer, 1959), 315; *Statistisches Jahrbuch für die*

Bundesrepublik Deutschland, 1963, 375. On lobbying, see *Süddeutsche Zeitung,* January 31, 1956; "Nächstes Jahr langsamer."

87. Letters to the editor, *GF,* no. 4, 1952, 29; *GF,* no. 3, 1955, 3. See also "Psychologie des Überholens"; letters to the editor *GF,* no. 8, 1955, 33; *GF,* no. 1, 1957, 6; *GF,* no. 5, 1957, 3.

5 An Export Hit

1. "The Beetle Does Float," *Sports Illustrated,* August 19, 1963, 58–67.

2. "A Volkswagen Runaway," *Business Week,* April 9, 1955, 140–144; "Volkswagen May Not Be a Big Car," *Popular Mechanics,* October 1956, 155–159, 304–313, esp. 155.

3. C. A. Bayly et al., "AHR Conversation: On Transnational History," *American Historical Review* 111 (2006): 1441–1464, here 1444.

4. Heinrich Nordhoff, "Vortag vor dem neuernannten Beirat am 22. Mai 1951," in *Reden und Aufsätze: Zeugnisse einer Ära* (Düsseldorf: Econ, 1992), 110–118, here 117; Heidrun Edelmann, *Heinz Nordhoff und Volkswagen: Ein deutscher Unternehmer im amerikanischen Jahrhundert* (Göttingen: Vandenhoeck & Ruprecht, 2003), 135–138.

5. Volker Wellhöner, *"Wirtschaftswunder"—Weltmarkt—westdeutscher Fordismus: Der Fall Volkswagen* (Münster: Westfälisches Dampfboot, 1996), 181. Lothar Gall, "Von der Entlassung Alfried Krupp von Bohlen und Halbachs bis zur Errichtung seiner Stiftung, 1951–1967," in *Krupp im 20. Jahrhundert: Die Geschichte des Unternehmens vom Ersten Weltkrieg bis zur Gründung der Stiftung,* ed. Lothar Gall (Berlin: Siedler, 2002), 473–590, here 526.

6. Hans-Ulrich Wehler, *Deutsche Gesellschaftsgeschichte, Fünfter Band: Bundesrepublik Deutschland und DDR 1949–1990* (Munich: Beck, 2008), 52; Nina Grunenberg, *Die Wundertäter: Netzwerke in der deutschen Wirtschaft, 1942–1966* (Berlin: Pantheon, 2007), 148–150; Werner Abelshauser, *Deutsche Wirtschaftgeschichte seit 1945* (Munich: Beck, 2004), 258–262.

7. Wellhöner, *"Wirtschaftswunder,"* 181–182; Markus Lupa, *Das Werk der Briten: Volkswagenwerk und Besatzungsmacht, 1945–1949* (Wolfsburg: Volkswagen, 2005), 72–77; Society of Motor Manufacturers and Traders, *Monthly Statistical Review March 1963* (London: The Society, 1963), 7, 22.

8. James Foreman-Peck, Sue Bowden, and Alan McKinley, *The British Motor Industry* (Manchester: Manchester University Press, 1995), 94; Roy Church, *The Rise and Fall of the British Motor Industry* (Cambridge: Cambridge University Press, 1999), 47.

9. John Ramsden, *Don't Mention the War: The British and the Germans since 1890* (London: Little, Brown, 2006), esp. 212–294. Aaron L. Friedberg, *The Weary Titan: Britain and the Experience of Relative Decline* (Princeton, NJ: Princeton University Press, 1988); Maiken Umbach, "Made in Germany," in *Deutsche Erinnerungsorte: Eine Auswahl,* ed. Etienne François and Hagen Schulze (Munich: Beck, 2005), 244–257.

10. "How Now, Mr. Ostrich," *Autocar,* March 13, 1953, 1; "Foreign Sales," *Autocar,* May 11, 1956, 515. On the sales network, see "Volkswagen Abroad," *Autocar,* August 10, 1951, 937; "Volkswagen Production," *Autocar,* April 9, 1954, 494; "Volkswagen de Luxe Saloon," *Autocar,* March 19, 1954, 401–403; letter to the editor, *Autocar,* May 10, 1957, 660.

11. *Daily Telegraph,* February 1, 1956; *Guardian,* June 27, 1960, 10.

12. *Daily Telegraph,* March 26, 1956; *Daily Telegraph,* February 1, 1956; *Observer,* September 16, 1956, 7.

13. On the lack of adequate vehicles, see letter to the editor, *Autocar,* February 12, 1954, 225; David Kynaston, *Austerity Britain, 1945–1951* (London: Bloomsbury, 2008), 497; "An Assessment of German Competition," *Motor Business,* September 1955, 1–12, esp. 6; *Daily Telegraph,* February 1, 1956.

14. *Daily Mail,* June 25, 1953; *Daily Mail,* November 8, 1954; *Observer,* September 16, 1956; letter to the editor, *Autocar,* May 24, 1957, 724.

15. Edelmann, *Heinz Nordhoff und Volkswagen,* 137. For the figures, see Henry Walter Nelson, *Small Wonder: The Amazing Story of the Volkswagen Beetle* (New York: Little, Brown, 1970), 333; Steven Tolliday, "From 'Beetle Monoculture' to the 'German Model': The Transformation of the Volkswagen, 1967–1991," *Business and Economic History* 24 (1995): 111–132, here 112–113.

16. U.S. Bureau of the Census, *Statistical Abstracts of the United States: 1956* (Washington, DC, 1956), 550. On stamina, see "A Volkswagen Runaway," 141, 144; "Volkswagen May Not," 157, 304, 313; *New York Times,* January 30, 1955, X25; "Herr Tin Lizzie," *Nation,* December 3, 1955, 475–476; "Volkswagen Races 858 Miles," *Popular Science,* September 1956, 145–149, 296–298, 310; "Will Success Spoil Volkswagen?" *Popular Mechanics,* February 1958, 160–184, 254, 282–283, here 254; "Big Forever," *Time,* August 13, 1965, 71; *"Road and Track" on Volkswagen* (Cobham, UK: Brooklands Books 1986), 4–7.

17. "A Volkswagen Runaway," 144; "Volkswagen May Not," 154; "Why People Buy Bugs," *Sales Management,* July 19, 1963, 33–39. On

two-car households, see Sally Clarke, *Trust and Power: Consumers, the Modern Corporation, and the Making of the United States Automobile Market* (Cambridge: Cambridge University Press, 2007), 239; Lizabeth Cohen, *A Consumers' Republic: The Politics of Mass Consumption in Postwar America* (New York: Vintage, 2004), 195; Maggie Walsh, *At Home at the Wheel? The Woman and Her Automobile in the 1950s* (London: British Library, 2007), esp. 3; Maggie Walsh, "Gendering Mobility: Women, Work and Automobility in the United States," *History* 93 (2008): 376–395, esp. 383–387; Tom McCarthy, *Auto Mania: Cars, Consumers, and the Environment* (New Haven, CT: Yale University Press, 2007), 101.

18. "Volkswagen May Not," 154, 155, 159, 306, 313; *New York Times,* January 30, 1955, X25; "Herr Tin Lizzie," 475. On cuteness, see Gary S. Cross, *The Cute and the Cool: Wondrous Innocence and Modern American Children's Culture* (Oxford: Oxford University Press, 2004); Anne Higonnet, *Pictures of Innocence: The History and Crisis of Ideal Childhood* (London: Thames and Hudson, 1998).

19. *New York Times,* January 30, 1955, X25.

20. McCarthy, *Auto Mania,* 101; Clarke, *Trust and Power,* 249; Robert Baldwin, "The Changing Nature of U.S. Trade Policy since World War II," in *The Structure and Evolution of Recent Trade Policy,* ed. Robert E. Baldwin and Anne O. Krueger (Chicago: Chicago University Press, 1984), 5–32, esp. 7–13; Raymond Bauer, *American Business and Public Policy: The Politics of Foreign Trade* (New York: Atherton Press, 1964), 251–264.

21. McCarthy, *Auto Mania,* 99–109; David Gartman, *Auto Opium: A Social History of American Automobile Design* (London: Routledge, 1994), 136–181; Lawrence J. White, *The Automobile Industry since 1945* (Cambridge, MA: Harvard University Press, 1971), 92–176.

22. "The Badness of Bigness," *Consumer Reports,* April 1959, 206–209; "Volkswagen May Not," 159; "Herr Tin Lizzie," 476; letter to the editor, *New York Times Magazine,* October 16, 1955, 6; letter to the editor, *Consumer Reports,* May 1957, 210.

23. Karal Ann Marling, *As Seen on TV: The Visual Culture of Everyday Life in the 1950s* (Cambridge, MA: Harvard University Press, 1994); Andrew Hurley, *Diners, Bowling Alleys, and Trailer Parks: Chasing the American Dream in Postwar Consumer Culture* (New York: Basic Books, 2001); Shelley Nickels, "More Is Better: Mass Consumption, Gender and Class Identity in Postwar America," *American Quarterly* 54 (2002): 581–622; Alison J. Clarke, *Tupperware: The*

Promise of Plastic in 1950s America (Washington, DC: Smithsonian Institution Press, 1999); Daniel Horowitz, *The Anxieties of Affluence: Critiques of American Consumer Culture, 1939–1979* (Amherst: University of Massachusetts Press, 2004), 101–128; Cohen, *Consumers' Republic,* 347–357.

24. *New York Times,* October 20, 1956, 28; letter to the editor, *New York Times Magazine,* October 2, 1955, 4. On the wider phenomenon, see Giles Slade, *Made to Break: Technology and Obsolescence in America* (Cambridge, MA: Harvard University Press, 2006), 151–186.

25. Martin Mayer, *Madison Avenue, U.S.A.* (New York: J. Lane, 1958), 26; Lola Clare Bratten, "Nothin' Could Be Finah: The Dinah Shore Chevy Show," in *Small Screens, Big Ideas: Television in the 1950s,* ed. Janet Thumin (London: I. B. Tauris, 2002), 88–104; Christopher Innes, *Designing Modern America: Broadway to Main Street* (New Haven, CT: Yale University Press, 2005), 120–143, 156–169; Thomas E. Bonsall, *Disaster in Dearborn: The Story of the Edsel* (Stanford, CA: Stanford University Press, 2002).

26. *New York Times,* October 20, 1956, 28; "Volkswagen Runaway," 141; "Volkswagen May Not," 155.

27. Grace Elizabeth Hale, *A Nation of Outsiders: How the White Middle Class Fell in Love with Rebellion in Postwar America* (New York: Oxford University Press, 2011), esp. 13–48; Robert Bruegman, *Sprawl: A Compact History* (Chicago: Chicago University Press, 2005), 121–136; Elaine Tyler May, *Homeward Bound: American Families in the Cold War Era* (New York: Basic Books, 1999 [1988]); "Comeback in the West," *Time,* February 15, 1954, 84–91, esp. 85; "Herr Tin Lizzie," 474; "Not since the Model T," *Forbes,* July 15, 1964, 20–21.

28. Christina von Hodenberg, "Of German Fräuleins, Nazi Werewolves, and Iraqi Insurgents: The American Fascination with Hitler's Last Foray," *Central European History* 41 (2008): 71–92; Petra Goedde, *GIs and Germans: Culture, Gender and Foreign Relations, 1945–1949* (New Haven, CT: Yale University Press, 2003); Geir Lundestad, *The United States and Western Europe since 1945* (Oxford: Oxford University Press, 2005). The following sections follow my line of reasoning in Bernhard Rieger, "From People's Car to New Beetle: The Transatlantic Journeys of the Volkswagen Beetle," *Journal of American History* 97 (2010): 91–115, esp. 96–99.

29. "Germany—Report on a Perplexing People," *New York Times Magazine,* April 3, 1955, 9, 68–71; Charles Thayer, *The Unquiet Ger-*

mans (London: Michael Joseph, 1957), 43, 57–59; "Germany and the West," *Nation,* June 18, 1960, 537–538. On the importance of work, see "Hans Schmidt Lives to Work," *New York Times Magazine,* May 25, 1959, 15, 81–83.

30. "Comeback in the West," 84; "The Volkswagen: A Success Story," *New York Times Magazine,* October 2, 1955, 14, 63–64, esp. 14.

31. "Comeback in the West," 86, 88; *New York Times,* November 16, 1958, F3; "The Volkswagen: A Success Story," 63; "Will Success Spoil Volkswagen?" 180–184.

32. Jeffrey Louis Decker, *Made in America: Self-Styled Success from Horatio Alger to Oprah Winfrey* (Minneapolis: University of Minnesota Press, 1997); David E. Shi, *The Simple Life: Plain Living and High Thinking in American Culture* (Athens: University of Georgia Press, 1985).

33. Peter Novick, *The Holocaust in American Life* (Boston: Houghton Mifflin, 1999), 98; Brian C. Etheridge, "*The Desert Fox,* Memory Diplomacy, and the German Question in Early Cold War America," *Diplomatic History* 32:2 (2008): 207–231, esp. 223–232, 235–236; Shlomo Shafir, *Ambiguous Relations: The Jewish American Community and Germany since 1945* (Detroit: Wayne State University Press, 1999); Hans Koningsberger, "Should a Jew Buy a Volkswagen?" *fact* 2:1 (1965): 40–43.

34. Nelson, *Small Wonder,* 333; Wellhöner, "*Wirtschaftswunder,*" 217; *Advertising Age,* April 18, 1960, 178; *Advertising Age,* October 5, 1959, 28.

35. "Import Revival," *Time,* November 24, 1961, 77–78; Nelson, *Small Wonder,* 349–354; Edelmann, *Heinz Nordhoff und Volkswagen,* 202–203; Dana Frank, *Buy American: The Untold Story of Economic Nationalism* (Boston: Beacon Press, 1999).

36. Robert Jackall and Janice M. Hirota, *Image Makers: Advertising, Public Relations, and the Ethos of Advocacy* (Chicago: Chicago University Press, 2000), 67–89; Thomas Frank, *The Conquest of Cool: Business Culture, Counterculture, and the Rise of Hip Consumerism* (Chicago: Chicago University Press, 1997), 52–73; Daniel Pope and William Toll, "We Tried Harder: Jews in American Advertising," *Jewish American History* 72 (1982): 26–51, esp. 41–50.

37. Nelson, *Small Wonder,* 226–231, esp. 227; *Advertising Age,* March 2, 1959, 3; S. Jonathan Wiesen, "Miracles for Sale: Consumer Displays and Advertising in Postwar West Germany," *Consuming Germany in the Cold War,* ed. David Crew (Oxford: Berg, 2003), 151–178; Unternehmensarchiv Volkswagen AG, Wolfsburg (hereafter UVW),

1850 (Generaldirektion 1958), memo on promotional gifts, December 9, 1957; UVW, 263/394, comment on letter to Heinrich Nordhoff, "Werbung und Verkaufsförderung Inland," April 22, 1963.

38. Bernbach is cited in Nelson, *Small Wonder,* 232. See also his remarks in Dennis Higgins, *The Art of Advertising: Conversations with William Bernbach, Leo Burnett, George Gibson, David Ogilvy, Rosser Reeves* (Chicago: Advertising Publications, 1965), 11–25, esp. 14; *Advertising Age,* March 27, 1961, 87–96.

39. Dan R. Post, *Volkswagen: Nine Lives Later: the Lengthened Shadow of a Good Idea* (Arcadia: Motor Era Books, 1966), 193–197; Nelson, *Small Wonder,* 232–237; Rowsome, *Think Small: The Story of Those Volkswagen Ads,* 71–74; *New York Times,* February 19, 1950, 10; January 25, 1955, 19; January 31, 1958, 9; McCarthy, *Auto Mania,* 88–89.

40. *50 Jahre Volkswagen Werbung: Stern Spezial* (Hamburg: Stern, 2002), 2.

41. Nelson, *Small Wonder,* 234–235; Frank, *Conquest of Cool,* 63.

42. On cool, see Dick Pountain and David Robbins, *Cool Rules: Anatomy of an Attitude* (London: Reaktion, 2000).

43. For the figure, see Nelson, *Small Wonder,* 248.

44. David N. Lucsko, *The Business of Speed: The Hot Rod Industry in America, 1915–1990* (Baltimore: Johns Hopkins University Press, 2008); Robert C. Post, *High Performance: The Culture and Technology of Drag Racing, 1950–1990* (Baltimore: Johns Hopkins University Press, 1994); *"Road and Track" on Volkswagen,* 24–25, 84–85; "Shot Out at the Riverside Corral," *Hot Rod Magazine,* June 1967, 44–47. The car descriptions are from "The Beetle Bomb," *Time,* December 20, 1963, 64; *New York Times,* February 14, 1968, 38; "Bug Is Small, but Oh My!" *Sports Illustrated,* December 12, 1966, 22–23; *New York Times,* December 8, 1968, 6.

45. Nelson, *Small Wonder,* 321; Bob Waar, *Baja-Prepping VW Sedans and Dune Buggies* (Los Altos: H. P. Books, 1970); Gary Gladstone, *Dune Buggies* (Philadelphia: Lippincott, 1972).

46. Michael Mase, telephone interview with author, July 21, 2009. On the van, see Bernd Wiersch, *Der VW Bully: Die Transporter Legende für Leute und Lasten* (Bielefeld: Delius Klasing, 2009); Kirse Granat May, *Golden State, Golden Youth: The California Image in Popular Culture, 1955–1966* (Chapel Hill: University of North Carolina Press, 2002), esp. 74–113; Drew Kampion and Bruce Brown, *A History of Surf Culture* (Cologne: Taschen, 2003), esp. 69–108; Lawrence Culver, *The Frontier of Leisure: Southern California and the*

Shaping of America (Oxford: Oxford University Press, 2010), 170–197; *Small World,* Winter 1963/1964, 3–4; ibid., Fall 1964, 17.

47. Hale, *A Nation of Outsiders,* 84–131, 163–237. For Beetles at Woodstock and at happenings, see "Talk of the Town," *New Yorker,* August 30, 1969, 17–21; *New York Times,* August 13, 1967, 71.

48. Peter Abschwanden, "How I Got the Bug in My Eye," in *My Bug: For Everyone Who Owned, Loved, or Shared a VW Beetle,* ed. Michael J. Rosen (New York: Artisan, 1999), 19–22, here 19; Jean Rosenbaum, *Is Your Volkswagen a Sex Symbol?* (New York: Hawthorn, 1972), 19–20; *Small World,* Fall 1968, 12–13; ibid., Winter 1969, title page, 12–13; ibid., Winter 1970, 12–13.

49. John Muir, *How to Keep Your Volkswagen Alive: A Manual of Step-by-Step Procedures for the Compleat Idiot* (Santa Fe: John Muir, 1990 [1969]), 3.

50. The Beetle's relative technical simplicity worked to this effect, too. See *The Last Whole Earth Catalogue: Access to Tools* (New York: Random House, 1971), 248.

51. Susan Sackett, *The Hollywood Reporter Book of Box Office Hits, 1939 to the Present* (New York: Billboard Books, 1996), 202–204; British Film Institute, London, "The Love Bug," BFI microjacket collection, file *The Love Bug,* Walt Disney Studios, *The Love Bug* press book (Los Angeles, 1969); "The Love Bug," *Variety,* December 11, 1968, 10; BFI microjacket collection, file *The Love Bug,* untitled clipping, *Time,* April 4, 1969; *Hollywood Reporter,* December 9, 1968, 3, 8.

52. "Wunder der Wanze," *Der Spiegel,* May 26, 1965, 119–125, esp. 125; "Käfer-Strategie," *Auto Motor und Sport,* no. 25, 1962, 12–13, here 13; *Hessische Nachrichten,* July 24, 1957; *Schwäbische Landeszeitung,* April 6, 1957; "Die Dinosaurier," *Der Spiegel,* May 28, 1958, 54–55, here 54.

53. *Christ und Welt,* July 19, 1963; *Hannoversche Allgemeine Zeitung,* July 13, 1963. On "German quality work," see Sebastian Conrad, *Globalisierung und Nation im Deutschen Kaiserreich* (Munich: Beck, 2006); Joan Campbell, *Joy in Work, German Work: The National Debate, 1800–1945* (Princeton, NJ: Princeton University Press, 1989).

54. "Mit 30 immer noch ein flotter Käfer," *Quick,* January 22, 1967, 45; Johannes Paulmann, "Deutschland in der Welt: Auswärtige Repräsentation und reflexive Selbstwahrnehmung nach dem 2. Weltkrieg—eine Skizze," in *Koordinaten deutscher Geschichte in der Epoche des Ost-West Konflikts,* ed. Hans-Günther Hockerts (Munich: Oldenbourg, 2004), 63–78; Kay Schiller and Christopher Young, *The*

1972 Munich Olympics and the Making of Modern Germany (Berkeley: University of California Press, 2010), esp. 87–126.

55. Edelmann, *Heinz Nordhoff und Volkswagen,* 137; *Christ und Welt,* July 19, 1963; *Die Zeit,* June 11, 1965.

56. Eckberth von Witzleben, "Des Käfers Schritte: Die Volkswagen-Chronologie," in *Käfer: Der Erfolkswagen; Nutzen—Mythos—Alltag,* ed. Wilhelm Hornbostel und Nils Jockel (Munich: Prestel, 1997), 11–130, here 123; Detlef Siegfried, *Time Is on My Side: Konsum und Politik in der westdeutschen Jugendkultur der 60er Jahre* (Göttingen: Wallstein, 2006), 264–274. For hippie Beetles, see *Das deutsche Auto: Volkswagenwerbung und Volkskultur,* ed. Knuth Hickethier, Wolf Dieter Lützen, and Karin Reiss (Wiesmar: Anabas, 1974), 222–227.

57. "Harte Männer, weiche Muskeln," *Der Spiegel,* September 16, 1964, 108–109; *Werbung in Deutschland: Jahrbuch der deutschen Werbung '64,* ed. Eckard Neumann and Wolfgang Spraug (Düsseldorf: Econ, 1964).

58. "Luft und Luft," *Der Spiegel,* May 2, 1966, 103–104; *Die Welt,* July 8, 1966; *Münchner Merkur,* October 12, 1966.

6 "The Beetle Is Dead—Long Live the Beetle"

1. Unternehmensarchiv Volkswagen, Wolfsburg (hereafter UVW), 373/162,3, minutes, meeting of the executive board, December 12, 1971, 7; "Classic VW Beetle TV Ad: Der Weltmeister!" www.youtube.com/watch?v=Ym0pLJU9R2E (accessed August 17, 2011).

2. Charles Maier, "'Malaise': The Crisis of Capitalism in the 1970s," in *The Shock of the Global: The 1970s in Perspective,* ed. Niall Ferguson et al. (Cambridge, MA: Harvard University Press, 2010), 25–48; Charles Maier, "Two Sorts of Crisis? The 'Long' 1970s in the West and the East," in *Koordinaten deutscher Geschichte in der Epoche des Ost-West-Konflikts,* ed. Hans Günther Hockerts (Munich: Oldenbourg, 2004), 49–62; Martin Geyer, "Rahmenbedingungen: Unsicherheit als Normalität," in *Geschichte der Sozialpolitik in Deutschland seit 1945: Band 6, 1974–1982,* ed. Martin Geyer (Baden-Baden: Nomos, 2008), 1–110; Anselm Doering-Manteuffel and Lutz Raphael, *Nach dem Boom: Perspektiven auf die Zeitgeschichte seit 1970* (Göttingen: Vandenhoeck & Ruprecht, 2008), esp. 34–42; Gerold Ambrosius, "Sektoraler Wandel und internationale Verflechtung: Die bundesdeutsche Wirtschaft im Übergang zu einem neuen Strukturmuster," in *Auf dem Weg in eine andere Moderne? Die Bundesrepublik Deutschland in den siebziger und achtziger Jahren,* ed. Thomas Raitel, Andreas Rödder, and Andreas Wirsching (Munich: Oldenbourg, 2009), 17–30; Konrad H. Jarausch, "Verkannter Strukturwandel: Die siebziger Jahre als

Vorgeschichte der Probleme der Gegenwart," in *Das Ende der Zuversicht? Die siebziger Jahre als Geschichte,* ed. Konrad Jarausch (Göttingen: Vandenhoeck & Ruprecht, 2008), 9–26.

3. Morten Reitmayer and Ruth Rosenberger, "Unternehmen am Ende des 'goldenen Zeitalters': Die 1970er Jahre in unternehmens- und wirtschaftshistorischer Perspektive," in *Unternehmen am Ende des "goldenen Zeitalters": Die 1970er Jahre in unternehmens- und wirtschaftshistorischer Perspektive,* ed. Morten Reitmayer and Ruth Rosenberger (Essen: Klartext, 2008), 9–26; Kim Christian Priemel, "Industrieunternehmen, Strukturwandel und Rezession: Die Krise des Flick-Konzerns in den siebziger Jahren," *Vierteljahrshefte für Zeitgeschichte* 57 (2009): 1–31.

4. "Ist der VW veraltet? Teil I," *Stern,* no. 43, 1957, 52–61; "Ist der VW veraltet? Teil II," ibid., no. 44, 1957, 60–65; Christina von Hodenberg, *Konsens und Krise: Ein Geschichte der westdeutschen Medienöffentlichkeit, 1945–1973* (Göttingen: Wallstein, 2006), 183–186.

5. Letters to the editor, *Stern,* no. 46, 1957, 57; "Ist der VW veraltet?" *Der Spiegel,* September 30, 1959, 40–58; letters to the editor, ibid., October 7, 1959, 3–14.

6. Walter Henry Nelson, *Small Wonder: The Amazing Story of the Volkswagen Beetle* (Boston: Little, Brown, 1970), 349–354; Bernd Wiersch, *Volkswagen Typenkunde, 1945–1974* (Bielefeld: Delius Klasing, 2010), 50–79.

7. Manfred Grieger et al., *Volkswagen Chronik* (Wolfsburg: Volkswagen, 2004), 63, 73; Volkswagenwerk AG, *Bericht über das Geschäftsjahr 1963* (Wolfsburg: Volkswagenwerk, 1964), 10; Volkswagenwerk AG, *Bericht über das Geschäftsjahr 1966* (Wolfsburg: Volkswagenwerk, 1967), 8; Anne von Oswald, "Volkswagen, Wolfsburg und die italienischen 'Gastarbeiter,' 1962–1975," *Archiv für Sozialgeschichte* 42 (2002): 55–79; Roberto Sala, "Vom Fremdarbeiter zum 'Gastarbeiter': Die Anwerbung italienischer Arbeitskräfte für die deutsche Wirtschaft (1938–1973)," *Vierteljahrshefte für Zeitgeschichte* 55 (2007): 93–120; Ulrich Herbert and Karin Hunn, "Guest Workers and Policy on Guest Workers in the Federal Republic: From the Beginning of Recruitment in 1955 until Its Halt in 1973," in *The Miracle Years: A Cultural History of West Germany, 1949–1968,* ed. Hanna Schissler (Princeton, NJ: Princeton University Press, 2001), 187–218.

8. "Die große Jagd beginnt: Opel Kadett," *Auto Motor und Sport* (hereafter *AMS*), no. 18, 1962, 25–28, here 28. See also "Daten und Fahreigenschaften des Opel Kadett," *AMS,* no. 19, 1962, 14–17; "Ford Taunus 12M," *AMS,* no. 22, 1962, 20–25.

9. Steven Tolliday, “From ‘Beetle Monoculture’ to the ‘German Model’: The Transformation of Volkswagen, 1967–1991,” *Business and Economic History* 24 (1995): 111–132, 112; UVW, 69/530/2, Dokument B. 3358, Strukturverschiebungen auf dem Automobilmarkt.

10. Wiersch, *Volkswagen Typenkunde,* 56–62, 70–75; “VW 1500: Deutschlands Maßhalte-Auto,” *Der Spiegel,* October 31, 1962, 70–79; Kurt Lotz, *Lebenserfahrungen: Worüber man in Wirtschaft und Politik auch sprechen sollte* (Düsseldorf: Econ, 1978), 93.

11. UVW, file 69/530/2, Dokument B. 4216, Markenloyalitätsrate.

12. Lotz, *Lebenserfahrungen,* 94; *Kölnische Rundschau,* March 29, 1966.

13. Nelson, *Small Wonder,* 337, 355–356; Tom McCarthy, *Auto Mania: Cars, Consumers, and the Environment* (New Haven, CT: Yale University Press, 2007), 165–175; Jameson M. Wetmore, “Redefining Risks and Redistributing Responsibilities: Building Networks to Increase Automobile Safety,” *Science, Technology and Human Values* 29 (2004): 377–405, esp. 382–389; UVW, file 69/722/2, minutes, meeting of the executive board, June 7, 1967, 11.

14. *New York Times,* July 7, 1967, 1; January 12, 1968, 47; April 15, 1966, 20; John W. Garrett and Arthur Stern, *A Study of Volkswagen Accidents in the USA* (Buffalo, NY: Cornell Aeronautical Laboratory, 1968); Center for Auto Safety, *Small—on Safety: The Designed-in Dangers of the Volkswagen* (New York: Grossman, 1972).

15. Quoted in Tolliday, “From ‘Beetle Monoculture,’ ” 113.

16. Werner Conze, *Die Suche nach der Sicherheit: Eine Geschichte der Bundesrepublik Deutschland von 1949 bis zur Gegenwart* (Munich: Siedler, 2009), 362–363; Werner Abelshauser, *Deutsche Wirtschaftsgeschichte seit 1945* (Munich: Beck, 2004), 288–292; Detlef Siegfried, “Prosperität und Krisenangst: Die zögerliche Versöhnung der Bundesbürger mit dem neuen Wohlstand,” in *Mit dem Wandel leben: Neuorientierung und Tradition in der Bundesrepublik der 1950er und 60er Jahre,* ed. Friedrich Kießling und Bernhard Rieger (Cologne: Böhlau, 2011), 63–78, esp. 69–72.

17. Grieger et al., *Volkswagen Chronik,* 71, 75; UVW, 69/722/2, minutes, meeting of the executive board, April 6, 1967, 8–9; *Süddeutsche Zeitung,* May 20, 1967; Manfred Grieger, “Der neue Geist im Volkswagenwerk: Produktinnovation, Kapazitätsausbau und Mitbestimmungsmodernisierung, 1968–1976,” in Reitmayer and Rosenberger, *Unternehmen am Ende,* 31–66, here 34.

18. UVW, 69/722/2, minutes, meeting of the executive board, November 10, 1967, 8–9.

19. See Grieger, "Der neue Geist," 36–44; Lotz, *Lebenserfahrungen,* esp. 101–109; *Die Zeit,* April 16, 1971; June 25, 1971; *Süddeutsche Zeitung,* September 11, 1971.

20. Grieger et al., *Volkswagen Chronik,* 97.

21. Grieger, "Der neue Geist," 44–54; UVW, 69/730/1, minutes, meeting of the executive board, January 15, 1974, 2; ibid., minutes, meeting of the executive board, January 23, 1974, 3; ibid., minutes, meeting of the executive board, March 11, 1974, 5.

22. *Frankfurter Allgemeine Zeitung,* April 18, 1975; *Die Welt,* April 18, 1975; Deutscher Bundestag, Berlin, Pressedokumentation, file 102–5/10, *Heute* (7 pm edition), transcript, April 15, 1975; *Plusminus,* transcript, April 17, 1975; "Massenentlassungen, Millionenverluste, Managementkrise: Was wird aus VW?" *Der Spiegel,* April 14, 1975, 25–33; *Deutsche Zeitung,* April 18, 1975; *Die Zeit,* April 18, 1975; *Frankfurter Allgemeine Zeitung,* February 8, 1975.

23. "VW—war denn alles falsch?" *Stern,* no. 18, 1975, 170.

24. Jürgen Peter Schmied, *Sebastian Haffner: Eine Biographie* (Munich: Beck, 2010), 407–429.

25. Grieger, "Der neue Geist," 54–64; Tolliday, "From 'Beetle Monoculture,'" 121–123; *Frankfurter Allgemeine Zeitung,* July 8, 1976; *Stuttgarter Zeitung,* July 8, 1976.

26. *General-Anzeiger,* November 8, 1977; *Die Welt,* December 19, 1977.

27. *General Anzeiger,* November 8, 1977; *Frankfurter Rundschau,* December 31, 1977.

28. DDB Information Center, New York City, file corp-Volkswagen, *CBS Evening News with Morton Dean,* transcript, August 26, 1977; *New York Times,* August 21, 1977, 1–2.

29. *Süddeutsche Zeitung,* March 20, 1985; *Neue Ruhr Zeitung,* February 1, 1985; *Hannoversche Allgemeine,* March 26, 1985.

7 "I Have a *Vochito* in My Heart"

1. Unternehmensarchiv, Volkswagen AG, Wolfsburg (hereafter UVW), 174/641/2, telex, Helmut Barschkis to Dr. Prinz, November 26, 1971.

2. Alexander Gromow, *Eu amo fusca* (São Paulo: Ripress, 2003); Joel Wolfe, *Autos and Progress: The Brazilian Search for Modernity* (New York: Oxford University Press, 2010).

3. Ryszard Kapuściński, *The Emperor: Downfall of an Autocrat* (London: Penguin, 2006), 12–13, 162; *New York Times,* September 13, 1974, A1, A13.

4. Harm G. Schröter, "Außenwirtschaft im Boom: Direktinvestitionen bundesdeutscher Unternehmen im Ausland, 1950–1975," in *Der Boom 1948–1973: Gesellschaftliche und wirtschaftliche Folgen in der Bundesrepublik Deutschland und in Europa,* ed. Hartmut Kaelble (Opladen: Leske & Budrich, 1992), 82–106; Werner Abelshauser, *Deutsche Wirtschaftsgeschichte seit 1945* (Munich: Beck, 2004); Harold James, *Krupp: Deutsche Legende und globales Unternehmen* (Munich: Beck, 2011).

5. UVW, 174/435/3, note on conversation, January 13, 1965; *El Sol de Puebla,* December 2, 1967.

6. See Claudia Nieke, *Volkswagen am Kap: Internationalisierung und Netzwerke in Südafrika 1950 bis 1966* (Wolfsburg: Volkswagen AG, 2010), 5, 187–193; Ludger Pries, "Volkswagen: Accelerating from a Multinational to a Transnational Automobile Company," in *Globalization or Regionalization of the European Car Industry?* ed. Michel Freyssenet et al. (London: Palgrave Macmillan, 2003), 51–72, esp. 54–55; Frank Wellhöner, *"Wirtschaftswunder"—Weltmarkt—westdeutscher Fordismus: Der Fall Volkswagen* (Münster: Westfälisches Dampfboot, 1996), 259–304.

7. Jeffry A. Frieden, *Global Capitalism: Its Fall and Rise in the Twentieth Century* (New York: Norton, 2006), 303–306; Asociación Nacional de Distribudores de Automóviles, *Aspectos fundamentales de la fabricación y distribución de automóviles y camiones en México* (Mexico City: Arana, 1966), 32–33; Douglas C. Bennett and Kenneth E. Sharpe, *Transnational Corporations versus the State: The Political Economy of the Mexican Auto Industry* (Princeton, NJ: Princeton University Press, 1985), 117–154.

8. Enrique Cárdenas, *La política económica en México, 1950–1994* (Mexico City: El Colegio de México, 1996), 56–85; Elsa M. Gracida, *El desarrolismo: Historia económica de México,* vol. 5 (Mexico City: UNAM, 2004); INEGI, *Estadísticas historicas de México, cuarta edición,* vol. 1 (Aguascalientes: SNC, 2000), 334; Nora Lustig, *Mexico: The Remaking of an Economy* (Washington, DC: Brookings Institution, 1992), 17; Nacional financiera, *La economía Mexicana en cifras 1990: 11a edición* (Mexico City: INEGI, 1990), 47; *Handelsblatt,* September 15, 1969 (special issue, *Mexiko: Ein Wegweiser für den deutschen Geschäftsmann*); *Die Welt,* November 22, 1966.

9. UVW, 174/435/3, minutes of meeting Vorstands-Ausschuß für Tochtergesellschaften im Ausland und Montagewerke, February 12, 1964; UVW, 69/826/2, Situationsbericht der Volkswagen de Mexico, S.A. de C.V., January 1967; UVW, 69/826/2, memorandum by Helmut

Barschkis, Xalostoc, January 5, 1967; Politisches Archiv, Auswärtiges Amt, Berlin (hereafter PA), B65-IIIB4, vol. 181, report, German Embassy Mexico City, March 2, 1964; UVW, 174/640/2, letter, Gustavo Díaz Ordaz to Kurt Lotz, February 18, 1969; UVW, 174/640/2, letter, Helmut Barschkis to Kurt Lotz, July 27, 1970.

10. UVW, 174/435/3, minutes of meeting, Vorstands-Ausschuß für Tochtergesellschaften im Ausland und Montagewerke, August 28, 1963; UVW, 69/826/2, report, Volkswagen de México, February 2, 1968, 2.

11. Gerhard Schreiber, *Una historia sin fin: Volkswagen de México* (Puebla: Volkswagen de México, 1988), 395; UVW, 69/826/2, report, Betriebswirtschaftliche Abteilung, Volkswagen de México, February 2, 1968, 1; UVW, 174/435/3, minutes of meeting, Vorstands-Ausschuß für Tochtergesellschaften im Ausland und Montagewerke, June 22, 1967; UVW, 69/723/1, minutes, meeting of the executive board, November 10, 1967, 11; Archivo, Programa de Industria Automotriz, Benemérita Universidad Autónoma de Puebla (hereafter APIA), database *Sedán en la industria automotriz en México.*

12. Asociación Nacional, *Aspectos fundamentales,* 38, 48; Alonso Aguilar Monteverde and Fernando Carmona, *México: Riqueza y miseria* (Mexico City: Editorial Nuestro Tiempo, 1968); "La invasión de menores," *Automundo,* April 1971, 72–75, 74; UVW, 69/731/1, minutes, meeting of the executive board, August 28, 1973.

13. Schreiber, *Una historia sin fin,* 397; Asociación Mexicana de la Industria Automotriz (hereafter AMIA), *La industria automotriz de México en cifras, edición 1988* (Mexico City: AMIA, 1988), 81; AMIA, *Organo informativo,* vol. 349, December 1994, 4; AMIA, *Organo informativo,* vol. 373, December 1996, 5; Nacional financiera, *La economía Mexicana en cifras 1990,* 47, 100; Miguel Ángel Vite Pérez, *La nueva desigualdad social Mexicana* (Mexico City: Miguel Ángel Porrúa, 2007); Isabel Rueda Peiro, *La cresciente desigualdad en México* (Mexico City: UNAM, 2009).

14. Schreiber, *Una historia sin fin,* 80–86; PA, B65-IIIB4, vol. 181, report, German Embassy Mexico, April 26, 1967; Wil Pansters, *Politics and Power in Puebla: The Political History of a Mexican State, 1937–1987* (Amsterdam: CEDLA, 1990), 102–123; UVW, 69/826/2, telex, Helmut Barschkis to Verwaltungsrat Volkswagen AG, April 24, 1967.

15. UVW, 69/826/2, report, Volkswagen de México, February 22, 1968, 2; UVW, 69/731/1, minutes, meeting of the executive board, November 6, 1973.

16. PA, B65-IIIB4, vol. 181, report, German Embassy Mexico, February 19, 1968; car worker A.J., interview by author, Puebla, September

19, 2008; car worker S.L.A.C., interview by author, Puebla, September 27, 2008.

17. Schreiber, *Une historia sin fin,* 396; car worker F.G.L., interview by author, Puebla, September 23, 2008.

18. Car worker E.T.G., interview by author, Puebla, September 20, 2008; Yolanda Montiel, *Proceso de trabajo, acción sindical y nuevas tecnologías en Volkswagen de México* (Mexico City: Colección Miguel Othón, 1991), 97; car worker C.C.P.O., interview by author, Puebla, September 23, 2008; car worker F.G., interview by author, Puebla, September 20, 2008; A.J., interview.

19. *El Sol de Puebla,* December 20, 1967, 1–2; F.G., interview; A.J., interview.

20. A.J., interview; car worker J.D.D., interview by author, September 23, 2008; F.G., interview; J.D.D., interview.

21. Montiel, *Proceso de trabajo,* 48, 57, 60–61; J.D.D., interview; APIA, *Contracto colectivo de trabajo Volkswagen de México,* 1976, 29.

22. Huberto Juárez Núñez, "Global Production and Worker Response: The Struggle at Volkswagen," *Working USA* 9 (2006), 7–28; Steven J. Bachelor, "Toiling for the 'New Invaders': Autoworkers, Transnational Corporations, and Working-Class Culture in Mexico City, 1955–1968," in *Fragments of a Golden Age: The Politics of Culture in Mexico since 1940,* ed. Gilbert Joseph, Anne Rubinstein, and Eric Zolov (Durham, NC: Duke University Press, 2001), 273–326, esp. 287–291; Montiel, *Proceso de trabajo,* 249.

23. Kevin J. Middlebrook, "Democratization in the Mexican Car Industry: A Reappraisal," *Latin American Research Review* 24:2 (1989), 69–93; Elena Poniatowska, *La noche de Tlatelolco* (Mexico City: Biblioteca Era, 1998); Pansters, *Politics and Power in Puebla,* 125–128; Yolanda Montiel, *Breve historia del sindicato independiente de Volkswagen de México* (Mexico City: Fundación Friedrich Ebert, 2007), 19.

24. Montiel, *Proceso de trabajo,* 82, 170–178; Schreiber, *Una historia sin fin,* 126.

25. José Luis Ávila, *La era neoliberal: Historia económica de México,* vol. 6 (Mexico City: UNAM, 2006), 280; *El Sol de Puebla,* August 7, 1978, 1, 10; *La Jornada,* July 1, 1987, 11; *La Jornada,* July 3, 1987, 13; *tageszeitung,* August 12, 1987; *tageszeitung,* August 11, 1987.

26. *La Jornada,* August 12, 1987, 10, 32; *El Sol de Puebla,* August 12, 1987, 1, 10; *La Jornada,* August 6, 1987, 10, 32; August 12, 1987; *El Sol de Puebla,* August 6, 1987, 1, 10; August 12, 1987, 1, 3.

27. *La Jornada,* August 12, 1987, 11. See also *La Jornada,* August 27, 1987, 1.

28. Pries, "Volkswagen: Accelerating from a Multinational," 56–62; Juárez, "Global Production," 11–12; Hans-Ludger Pries, "Globalisierung und Wandel internationaler Unternehmen: Konzeptionelle Überlegungen am Beispiel der deutschen Automobilkonzerne," *Kölner Zeitschrift für Soziologie und Sozialpsychologie* 52 (2000): 670–695.

29. *La Jornada,* August 13, 1992, 23, 56; August 12, 1992, 1, 14; Huberto Juárez Núñez, "La impunidad empresarial y nuevas relaciones de trabajo en VW, 1992," *Trabajo y democracia hoy: Número 64, edición especial* (2001): 130; Rainer Dombois and Hans-Ludger Pries, *Neue Arbeitsregimes im Transformationsprozeß Lateinamerikas: Arbeitsbeziehungen zwischen Markt und Staat* (Münster: Westfälisches Dampfboot, 1999), 115–121.

30. On VW's position, see *La Jornada,* July 27, 1992, 44, 14; Schreiber, *Una historia sin fin,* 395; AMIA, *Organo informativo,* vol. 443, December 2001, 7; Juárez, "Global Production," 14.

31. *tageszeitung,* August 21, 1992; *Frankfurter Rundschau,* July 30, 1992; *Frankfurter Allgemeine Zeitung,* July 30, 1992; *Süddeutsche Zeitung,* August 3, 1992; *Tagesspiegel,* November 25, 1992; *Frankfurter Allgemeine Zeitung,* November 23, 1992; *Welt am Sonntag,* November 22, 1992.

32. Asociación Nacional, *Aspectos fundamentales,* 48; UVW, 174/435/3, minutes, conversation between Octaviano Campos Salas and Otto Höhne, n.d.

33. Dennis Gilbert, *Mexico's Middle Class in the Neoliberal Era* (Tucson: University of Arizona Press, 2007), 12, 29; "La invasión de menores," 74–75; Asociación Nacional, *Aspectos fundamentales,* 38.

34. Gilbert, *Mexico's Middle Class,* 38–40; Salvador de Lara Rangel, "El impacto económico de la crisis sobre la clase media," in *Las clases medias en la conyuntura actual,* ed. Soledad Loanza and Claudio Stern (Mexico City: El Colegio de México, 1987), 29–49; Huberto Juárez Núñez, "Ahora sí, auto para los pobres," *Crítica: Revista de la Universidad Autónoma de Puebla* 41/42 (1990): 67–69.

35. *Publicidad Mexicana: Su historia, sus instituciónes, sus hombres,* ed. José A. Villamil Diarte (Mexico City: Demoscopia, 1971), 237; *Automundo,* December 1970, 47; Jeffrey M. Pilcher, "Mexico's Pepsi Challenge: Traditional Cooking, Mass Consumption, and National Identity," in Joseph, Rubinstein, and Zolov, *Fragments of a Golden Age,* 71–90, esp. 81.

36. *El Sol de Puebla,* September 13, 1980, 1, 2, 8; "Somos exportadores," *Automundo,* June 1971, 14.

37. *El Sol de Puebla,* October 19, 1990, 1, 6.

38. *Automundo,* August 1977, 43; Jeffrey Pilcher, *¡Que vivan los tamales! Food and the Making of Mexican Identity* (Albuquerque: University of New Mexico Press, 1998), 130, 139–141.

39. J.D.D., interview; C.C.P.O., interview.

40. E.T.G., interview; F.G., interview; conversation of author with street food vendor, Calle Rio Lerma, Mexico City, September 13, 2008.

41. Schreiber, *Una historia sin fin,* 345–377. "¿Es mi mama un Volkswagen?" *Automundo,* May 1971, 46; A.J., interview; F.G., interview; E.T.G., interview.

42. Car worker C.G.L., interview by author, Puebla, September 23, 2008. The joke as told by Huberto Juárez Núñez, Puebla, September 22, 2008.

43. John Mason Hart, *Empire and Revolution: The Americans in Mexico since the Civil War* (Berkeley: University of California Press, 2002); Julio Moreno, *Yankee Don't Go Home! Mexican Nationalism, American Business Culture, and the Shaping of Modern Mexico, 1920–1950* (Chapel Hill: University of North Carolina Press, 2003); Stephen D. Morris, *Gringolandia: Mexican Identity and the Perception of the United States* (Lanham, MD: Rowman & Littlefield, 2005), 215–242.

44. Pierre Nora, *Les lieux de memoire,* 7 vols. (Paris: Gallimard, 1984–1992).

45. Distrust of the country's small workshops informs Gerardo Salgado Fonseca, *Cómo reparar su Volkswagen* (Mexico City: Editores Mexicanos Unidos, 1988), 7; conversation between F.G. and Huberto Juárez Núñez, Puebla, September 20, 2008; conversation between J.D.D. and C.C.P.O., Puebla, September 23, 2008.

46. Conversation between Huberto Juárez Núñez and A.J., Puebla, September 19, 2008; C.C.P.O., interview; A.J., interview.

47. S.L.A.C., interview.

48. Conversation with tour guide during factory tour, Puebla, October 1, 2008; E.T.G., interview; *Reforma,* September 20, 2008, sección automotriz, 18.

49. Francis Alys, *Rehearsal I* (*Ensayo I*) (1999), exhibited at *Francis Alys: A Story of Deception,* Tate Modern, London, June 15–September 5, 2010.

8 Of Beetles Old and New

1. *Mad Men,* "Ladies Man," first aired on AMC, July 26, 2007.

2. Frank Witzel, Klaus Walter, and Thomas Meinecke, *Die Bundesrepublik Deutschland* (Hamburg: Nautilus, 2009); "Episode 1," Disc 1, *Life on Mars,* directed by Bharat Nalluri (London: Kudos Film, 2006).

3. Steven M. Gelber, *Hobbies: Leisure and the Culture of Work in America* (New York: Columbia University Press, 1999), 139; Wolfgang Hardtwig, *Verlust der Geschichte—oder wie unterhaltsam ist die Vergangenheit?* (Berlin: Vergangenheitsverlag, 2010).

4. M. Santoro, "Das große Krabbeln," *Gute Fahrt,* no. 6, 2010, 73–76, here 76; Brett Hawksbee, *Bug Jam and All That* (n.p.: Sane VA Publications, 2007), 11; "Bugjam VW Festival," www.bugjam.co.uk (accessed August 29, 2011).

5. "Vintage Volkswagen Club of America," www.vvwca.com/aboutus (accessed August 29, 2011); Michael Mößlang, "Die Anfänge," *Der Käfer,* 3, 1985, 23–27.

6. Questionnaire 2, distributed at London Volksfest, North Weald Airfield, August 1, 2004.

7. On a proper motorized subculture, see Dick Hebdige, "Object as Image: The Italian Scooter Style," in *Hiding in the Light* (London: Routledge, 1988), 77–115.

8. Questionnaire 1, distributed at London Volksfest, North Weald Airfield, August 1, 2004; questionnaire 3, distributed at London Volksfest, North Weald Airfield, August 1, 2004; questionnaire 7, distributed at London Volksfest, North Weald Airfield, August 1, 2004.

9. Classifieds, *VolksWorld,* March 2006, 102; Frank Weigl, interview by author, Nuremberg, July 24, 2011; Gisela Feldner, interview by author, Nuremberg, July 24, 2011; A.J., interview by author, Puebla, September 19, 2008.

10. Questionnaire 3, distributed at Steintribünentreffen, Nuremberg, July 24, 2011; questionnaire 3, Nuremberg, 3; questionnaire 5, distributed at Steintribünentreffen, Nuremberg, July 24, 2011; Klaus Jahn, interview by author, Nuremberg, July 24, 2011.

11. Questionnaire 2, Nuremberg; questionnaire 1, Nuremberg; questionnaire 3, distributed at London Volksfest, North Weald Airfield, August 1, 2004; questionnaire 5, distributed at London Volksfest, North Weald Airfield, August 1, 2004.

12. Questionnaire 3, North Weald Airfield; Volker Petz, interview by author, Nuremberg, July 24, 2011; Joe Hughes, personal communication to author, Middletown, CT, March 27, 2008.

13. Interview with anonymous VW enthusiast, Nuremberg, July 24, 2011.

14. Keith Seume, *The Story of the California Look VW* (Beaworthy: Herridge and Sons, 2008), 23; David N. Lucsko, *The Business of Speed: The Hot Rod Industry in America, 1915–1990* (Baltimore: Johns Hopkins University Press, 2008), 85–102.

15. Paul Wager, *Beetlemania* (London: Grange Books, 1995), 44. See also Alessandro Pasi, *The Beetle: A History and a Celebration* (London: Aurum Press, 2000), 97–105; Susan Sontag, "Notes on Camp," in *Against Interpretation and Other Essays* (London: Penguin, 2009), 275–292, esp. 275, 281; Christopher Breward, "The Uses of 'Notes on Camp,'" in *Postmodernism: Style and Subversion, 1970–1990,* ed. Glenn Adamson and Jane Pavitt (London: V&A Publishing, 2011), 166–169.

16. Rich Kimball, "The Restoration of a Driver," *Dune Buggies and Hot VWs,* March 1986, 84–87, here 87; Steve Mierz, "The Real Deal," *VW Autoist,* January/February 1998, 6–9, here 7.

17. Santoro, "Das große Krabbeln," 74; questionnaire 7, distributed at London Volksfest, North Weald Airfield, August, 1, 2004.

18. Steven Biel, *Down with the Old Canoe: A Cultural History of the Titanic Disaster* (New York: Norton, 1996), 189–190; Gelber, *Hobbies,* esp. 23–56.

19. Weigl, interview; Bjarne Erik Roscher, interview by author, Nuremberg, July 24, 2011; Anonymous, interview by author, Nuremberg, July 24, 2011; Randy O. Frost and Gail Steketee, *Stuff: Compulsive Hoarding and the Meaning of Things* (Boston: Mariner Books, 2010), esp. 44–51; Daniel Miller, *The Comfort of Things* (London: Polity, 2008).

20. Questionnaire 9, North Weald; Weigl, interview; questionnaire 1, distributed at Steintribünentreffen, Nuremberg, July 24, 2011; questionnaire 3, Nuremberg; questionnaire 2, Nuremberg.

21. Sventlana Boym, *The Future of Nostalgia* (New York: Norton, 2001), 49–55.

22. Questionnaire 4, North Weald; questionnaire 5, North Weald; questionnaire 6, North Weald; questionnaire 3, North Weald; program, distributed at Steintribünentreffen, Nuremberg, July 24, 2011.

23. Roscher, interview; "Blitzkrieg Racing," http://blitzkriegracing.com (accessed September 3, 2011); "Welcome to the DKP Website," www.dkpcarclub.com/ (accessed September 3, 2011).

24. Questionnaire 2, Nuremberg; questionnaire 9, North Weald; Weigl, interview.

25. Wolfgang Klebe, "Im Sog der Beatles: Sigrid und ihr schwarzer VW," in *Mein erstes Auto: Erinnerungen und Geschichten,* ed. Franz-Josef Oller (Frankfurt: Fischer, 1999), 142–144, here 144; Christoph Stölzl, "Er läuft nicht mehr," *Die Zeit,* June 26, 2003, 47–48, here 47; Weigl, interview.

26. Cited in *Die Welt,* January 13, 1992.

27. David Kiley, *Getting the Bugs Out: The Rise, Fall, and Comeback of Volkswagen in America* (New York: Wiley, 2002), 114–149.

28. Ludger Pries, "Volkswagen: Accelerating from a Multinational to a Transnational Automobile Company," in *Globalization or Regionalization of the European Car Industry?* ed. Michel Freyssenet et al. (London: Palgrave Macmillan, 2003), 51–72; Ludger Pries, "Globalisierung und Wandel internationaler Unternehmen," *Kölner Zeitschrift für Soziologie und Sozialpsychologie* 52 (2000): 670–695.

29. James Mays, quoted in Matt de Lorenzo, *The New Beetle* (Osceola, WI: MBI Publishing, 1998), 23, 35; *Philadelphia Daily News,* March 30, 1998, local section, 5; Florian Illies, *Generation Golf: Eine Inspektion* (Berlin: Fischer, 2000).

30. Kiley, *Getting the Bugs Out,* 15–17.

31. "What a Concept," *VW-Autoist,* March/April 1994, 6–7, here 7; John Lamm, "Show Time," *Road and Track,* April 1994, 92; *Chicago Tribune,* February 13, 1994, transportation section, 1.

32. Pries, "Volkswagen," 63–64; Pries, "Globalisierung und Wandel internationaler Unternehmen," 673; *Wall Street Journal,* October 26, 1995, A4; A.J., interview.

33. See Tom Janiszewski, "Dawn of a New Era," *VW-Autoist,* March/April 1998, 6–11, here 8, 10; Bill Vlasic, "Still Groovy after All These Years," *Business Week,* January 12, 1998, 46; *New York Times,* January 11, 1998, C4.

34. Matt DeLorenzo, "The New Beetle," *Road and Track,* April 1998, 76–78, here 78; Matt Stone, "The Beetles: Yesterday and Today," *Motor Trend,* July 1998, 81–88, esp. 85; DeLorenzo, "New Beetle," 88–90.

35. DeLorenzo, "New Beetle," 77; Stone, "Beetles," 85; *New York Times,* March 22, 1998, B4; Kiley, *Getting the Bugs Out,* 239; *Ward's Automotive Yearbook 1999* (Southfield, MI: Ward's Communications, 1999), 276; *Ward's Automotive Yearbook 2001* (Southfield, MI: Ward's Communications, 2001), 271.

36. Dan Quelette, *The Volkswagen Bug Book: A Celebration of Beetle Culture* (Santa Monica, CA: Angel City Press, 1999), 19; *Philadelphia Daily News,* March 30, 1998, local section, 5.

37. Kiley, *Getting the Bugs Out,* 195–197. On retro-chic, see Raphael Samuel, *Theatres of Memory* (London: Verso, 1994), 83; Elizabeth E. Guffey, *Retro: The Culture of Revival* (London: Reaktion Press, 2006), esp. 17; "The Nostalgia Boom," *Business Week,* March 23, 1998, 58–64, esp. 60–62; "New Legs for a Bug," *Newsweek,* January 12, 1998, 46–48.

38. *New York Times,* January 11, 1998, C2; Quelette, *Volkswagen Bug Book,* 15.

39. "New Legs for a Bug"; *New York Times,* March 13, 1998, B6; *Advertising Age,* April 1, 1998, 26.

40. "Nostalgia Boom," 59–61; *New York Times,* January 11, 1998, C2; Jean-François Lyotard, *The Postmodern Condition: A Report on Knowledge* (Minneapolis: University of Minnesota Press, 1993), xxiv. The term "yuppie guilt" is from *New York Times,* January 11, 1998, C2.

41. *Automotive News,* January 19, 1998, 12; *Washington Post,* March 13, 1998, C1.

42. *New York Times,* June 13, 1998, A5; *New York Times,* September 1, 1998, A9; Constantin Goschler, *Schuld und Schulden: Die Politik der Wiedergutmachung für NS-Verfolgte seit 1945* (Göttingen: Wallstein, 2008), 413–450.

43. See *New York Times,* July 8, 1998, A1; Susanne-Sophia Spiliotis, *Verantwortung und Rechtsfrieden: Die Stiftungsinitiative der deutschen Wirtschaft* (Frankfurt: Fischer, 2003).

44. Gerald Posner, "VW Day," *New York Times Magazine,* October 4, 1998, 128; letter to the editor, *New York Times Magazine,* October 25, 1998, 6.

45. *Berliner Zeitung,* November 4, 1998, section Auto & Straße; *tageszeitung,* November 27, 1998, 8.

Epilogue

1. Horst Köhler, "Die Ordnung der Freiheit," speech delivered on March 15, 2005, www.bundespraesident.de/SharedDocs/Reden/DE/Horst-Koehler/Reden/2005/03/20050315_Rede.html (accessed January 2, 2012).

2. Harvey Molotch, *Where Stuff Comes From* (New York: Routledge, 2003), esp. 161–193; Mark Pendergrast, *For God, Country and Coca-Cola* (New York: Basic Books, 2000); James L. Watson, ed., *Golden Arches East: McDonald's in East Asia* (Palo Alto, CA: Stanford University Press, 1998); Rob Kroes, "American Empire and Cultural Imperialism: A View from the Receiving End," in *Rethinking American History in a Global Age,* ed. Thomas Bender (Berkeley: University of California Press, 2002), 295–313.

3. Priscilla Parkhurst Ferguson, *Accounting for Taste: The Triumph of French Cuisine* (Chicago: Chicago University Press, 2004); Vanessa Schwartz, *It's So French: Hollywood, Paris, and the Making of Cosmopolitan Film Culture* (Chicago: Chicago University Press, 2007); Emilio

Ambasz, ed., *Italy: The New Domestic Landscape; Achievements and Problems of Italian Design* (New York: New York Graphic Society, 1972); Laura E. Cooper and B. Lee Cooper, "The Pendulum of Cultural Imperialism: Popular Music Interchanges between the United States and Britain, 1943–1967," *Journal of Popular Culture* 27 (1993): 61–78.

Acknowledgments

"Entomology?" a colleague asked somewhat incredulously after I had answered his question about what I was working on. No, I assured him, I was not researching beetles, just one Beetle, albeit a very special species.

Many institutions and people have helped me follow the Volkswagen's origin and evolution. The historian gets nowhere without money, and I gratefully acknowledge the financial support provided by the British Academy and the Arts and Humanities Research Council, which funded archival trips and freed me from some of my teaching obligations. Henriette Bruns, Helen Matthews, and Nicola Miller offered important guidance when I applied for these grants.

Without the expertise of the staff working at the libraries and archives I visited, many traces the Beetle left across the world would have evaded my attention. I am deeply grateful to the professionals at the newspaper archives of the Federal Press Office as well as of the Bundestag in Berlin, the Politisches Archive of the Foreign Office in Berlin, the Stabis in Berlin and Munich, the Institut für Zeitgeschichte und Stadtpräsentation, Wolfsburg, the Library of Congress, the University of Maryland Libraries, the Free Library of Philadelphia, the Biblioteca Nacional de México, the Hemeroteca Puebla, and the National Archives in Kew. Helmuth Trischler repeatedly eased access at the Deutsches Museum in Munich. A special vote of thanks needs to go to Ulrike Gutzmann and Manfred Grieger at the Unternehmensarchiv Volkswagen in Wolfsburg. Beyond the world of the printed word, I am deeply appreciative of the people whom I interviewed and who filled out questionnaires, thereby opening up a wealth of information that would have been impossible to secure otherwise.

While developing my ideas I had the good fortune to discuss my findings at seminars and conferences in London, Cambridge, Newcastle, Brighton, Vienna, Munich, Tübingen, Erlangen, East Lansing, Middletown (Connecticut), and Mexico City. Particularly pleasing have been my conversations with Achim and Heidi Bade, Moritz Basler, Volker Berghahn, Michael Berkowitz, Kerstin Brückweh, Jane Caplan, Martin J. Daunton, Anselm Doering-Manteuffel, Geoff Eley, Seth Fein, Jerry Garcia, Martin Geyer, Christina von Hodenberg, Marisa Kern, Ethan Kleinberg, Frieder Kießling, Daniel Laqua, Peter Mandler, Eckard Michels, Stefan Moitra, Holger Nehring, Johannes Paulmann, Patrice Poutrus, Peter Schröder, Lewis Siegelbaum, Detlef Siegfried, Ben Smith, Martina Steber, Elliott Weiss, and Thomas Zeller. I am grateful for the constructive suggestions by the anonymous readers of my manuscript. At Harvard University Press, Brian Distelberg dealt with my inquiries with exemplary efficiency. Like many historians before me, I have nothing but praise for Joyce Seltzer's keen editorial eye. The hospitality I encountered in Mexico counts among the most inspiring experiences I had during this project. Thank you to Julio Castellanos Elías, Luis Antonio Ramírez, and most especially to Huberto Juárez Núñez, without whom the chapter on Mexico would be infinitely poorer.

Liz Buettner has been with this project since the design stage, and she will be pleased to see it in the rear mirror. Her loving encouragement, her intellectual company, and her talent for seeing the funny side have been my main propellants.

I dedicate this book to my mother and to the memory of my father, both Beetle drivers.

Index